W9-CMB-367

Environmental Change and Tropical Geomorphology

TITLES OF RELATED INTEREST

Physical processes of sedimentation
J. R. L. Allen

The formation of soil material
T. R. Paton

Petrology of the sedimentary rocks
J. T. Greensmith

Elements of geographical hydrology
B. J. Knapp

Geology for civil engineers
A. C. McLean & C. D. Gribble

Thresholds in geomorphology
D. R. Coates & J. D. Vitek (eds)

Soil processes
B. J. Knapp

Fluvial geomorphology
M. Morisawa (ed.)

Glacial geomorphology
D. R. Coates (ed.)

Theories of landform development
W. N. Melhorn & R. C. Flemal (eds)

Geomorphology and engineering
D. R. Coates (ed.)

Geomorphology in arid regions
D. O. Doehring (ed.)

Tectonic processes
D. Weyman

Geomorphological techniques
A. S. Goudie (ed.)

Terrain analysis and remote sensing
J. R. G. Townshend (ed.)

British rivers
J. Lewin (ed.)

Soils and landforms
A. J. Gerrard

Soil survey and land evaluation
D. Dent & A. Young

Adjustments of the fluvial system
D. D. Rhodes & G. P. Williams (eds)

Applied geomorphology
R. G. Craig & J. L. Craft (eds)

Space and time in geomorphology
C. E. Thorn (ed.)

Pedology
P. Duchaufour

Sedimentology: process and product
M. R. Leeder

Sedimentary structures
J. D. Collinson & D. B. Thompson

Geomorphological field manual
R. V. Dackombe & V. Gardiner

Earth structures engineering
R. J. Mitchell

Statistical methods in geology
R. F. Cheeney

Groundwater as a geomorphic agent
R. G. LaFleur (ed.)

The climatic scene
M. J. Tooley & G. S. Sheail (eds)

Quaternary paleoclimatology
R. S. Bradley

Geomorphology and soils
K. S. Richards, R. R. Arnett & S. Ellis (eds)

Models in geomorphology
M. Woldenberg (ed.)

Tectonic geomorphology
M. Morisawa & J. T. Hack (eds)

Quaternary environments
J. T. Andrews (ed.)

Introducing groundwater
M. Price

Geomorphology: pure and applied
M. G. Hart

Environmental chemistry
P. O'Neill

The dark side of the Earth
R. Muir Wood

Geology and mineral resources of West Africa
J. B. Wright, et al.

Principles of physical sedimentology
J. R. L. Allen

Planetary landscapes
R. Greeley

Environmental magnetism
F. Oldfield & R. Thompson

Environmental Change and Tropical Geomorphology

EDITORS
I. DOUGLAS & T. SPENCER

School of Geography, University of Manchester

A publication of the British Geomorphological Research Group

London
GEORGE ALLEN & UNWIN
Boston Sydney

George Allen & Unwin (Publishers) Ltd,
40 Museum Street, London WC1A 1LU, UK

George Allen & Unwin (Publishers) Ltd,
Park Lane, Hemel Hempstead, Herts HP2 4TE, UK

Allen & Unwin Inc.,
Fifty Cross Street, Winchester, Mass 01890, USA

George Allen & Unwin Australia Pty Ltd.
8 Napier Street, North Sydney, NSW 2060, Australia

First published in 1985

British Library Cataloguing in Publication Data

Environmental change and tropical geomorphology.
1. Landforms—Tropics
I. Douglas,Ian II. Spencer, Tom
III. British Geomorphological Research group
551.4′0913 GB446
ISBN 0-04-551074-1

Library of Congress Cataloging in Publication Data

Main entry under title:
Environmental change and tropical geomorphology.
"A publication of the British Geomorphological Research Group."
Papers originally presented at the Annual General Meeting of the British Geomorphological Research Group held at Holly Royde College, Manchester University in October 1981.
Bobliography: p.
Includes index.
1. Geomorphology—Topics. 2. Climatic changes—Tropics. I. Douglas, Ian. II. Spencer, Tom.
III. British Geomorphological Research Group.
IV. British Geomorphological Research Group. General Meeting (1981: Manchester University)
GB446.E58 1985 551.4′0913 84-24541
ISBN 0-04-551074-1 (alk. paper)

Set in 10 on 12 point Times by Mathematical Composition Setters Ltd.
Ivy Street, Salisbury, Wilts.
and printed in Great Britain by Mackays of Chatham

Dedication

To Gordon Warwick, founder member of the British Geomorphological Research Group and enthusiast for international collaboration in geomorphology, who died during the preparation of this volume, and to Amelia Nogueira Moreira and her companions who died tragically during a Radambrasil survey flight.

Preface

The growth of scientific investigations in the tropics since the 1960s by the scientists of the new and expanding institutions of low-latitude countries, by participants in international research projects such as those initiated by the International Biological Programme, by expeditions such as those of the Royal Geographical Society to Gunung Mulu in Sarawak, and by expatriate scientists from more poleward countries, has confirmed earlier suspicions that the tropics provide the key to understanding much biological and earth science. This is particularly true for geomorphology, which in higher latitudes suffers great seasonal contrasts in process intensity and type and which often in the past underwent the dramatic changes of glaciation and periglaciation.

Yet these same investigations in the tropics have shown that the legacy of past climatic changes is much more dramatic throughout the tropics than once believed. This book brings together the variety of evidence about such environmental changes, over a variety of timescales, and sets it against the current knowledge of the nature of geomorphic processes in the tropics. The ideas, observations and analyses presented here will stimulate students concerned with tropical and global environmental change, landform evolution and resource management to take a wider, deeper view of the tropics. The book will be of immense interest to all researchers in the earth and life sciences concerned with tropical continents and ecosystems and their adjacent seas.

The volume is organised into five sections. The papers in Part A are contextual, outlining the intellectual development of, and some of the research questions raised by, tropical geomorphology and exploring the linkages between tropical high diversity, environmental stability and disturbance regime. This debate leads on, in Part B, to explorations of the relationship between process operation rates and landform development at different timescales, from the contemporary to the Quaternary. The combination of the great expansion of radiometric dating, the discovery of new lines of evidence and the use of new airborne and satellite technologies has produced the explosion of evidence for Quaternary environmental change in the tropics summarised in Part C. This section has even wider implications in its demonstration of how it may be possible ultimately to model changes in global circulation over the last three million years from geomorphological and ecological evidence. On a still longer timescale, Part D investigates the role of tectonic style in long-term landform development. Some tropical landforms can be related to the breakup and subsequent drift of the Gondwanaland continent, yet continue to be deformed by present-day tectonics. Hopefully, the concluding remarks to the volume show how the book brings together material not previously synthesised, suggests tests of widely held theories of landform development and indicates avenues of future research.

The initial impetus for this volume came from the Annual General Meeting of the British Geomorphological Research Group (BGRG) held at Holly Royde College, Manchester University, in October 1981. The majority of the papers collected here were originally presented at that meeting and we are grateful to the members of the BGRG for providing a lively and critical forum for debate at that time. Several additional papers were invited to further develop a broad international coverage of topics and geographical areas.

We are grateful to all the authors for their forebearance during the lengthy editorial process and to the conscientious and skilful work of numerous referees around the world. We are indebted to our academic, secretarial and technical colleagues in Manchester for their willing and helpful assistance.

Ian Douglas
Tom Spencer
Manchester
February 1984

Acknowledgements

The editors and authors, individually and severally, gratefully acknowledge the help and support of the following with the work involved either in the preparation of individual chapters or in the book as a whole:

For editorial and translation assistance: Stephen Belbin; Bill Brice; and S. P. Harrison.

For laboratory and technical assistance: Achim Schnütgen; Karin Oltmann; Dr J. Freundich; Oxford University Computing Laboratory; Radcliffe Science Library, Oxford; Departments of Geography at the University of Bristol, Birkbeck College of the University of London, and Oxford University; Christ Church, Oxford; A. J. Clark of the Ancient Monuments Laboratory at the Department of the Environment in London; E. C. Flenley; and N. Watson.

For support of and assistance with field investigations: Keith Martell, Mine General Manager at Yengema; Basil Foster; Malcolm Bannister; Chris Clements; Owen King; Alieu Mhadi; John Rogers; Carnegie Trust for the Universities of Scotland; Natural Environmental Research Council; the Royal Society; University of Liverpool; ICI, Brunner Mond Division; Jack Colson; US Department of Energy, Contract No. DE-ACO_2-79EV10097 to T. Webb III at Brown University; Sarawak State Secretary; Director of the Sarawak Forestry Department; Cathay Pacific Airlines; all firms who sponsored the Mulu 80 expedition; Ghar Parau Foundation; A. Eavis; Ronnie King; Henry Pasau; Danny Lawi; H. A. Osmaston; A. C. Waltham; and M. K. Lyon.

For cartographic assistance: Jürgen Kubelke; Anne Lowcock; Graham Bowden; and particularly N. Scarle, who was responsible for the bulk of the diagrams drawn in Manchester.

For secretarial assistance: Beate Forthmann; Emma Dixon; and especially Jean Mellor, the editors' secretary in Manchester.

For permission to publish findings: National Diamond Mining Company of Sierra Leone; Sierra Leone Selection Trust (BP Minerals International); R. J. Morley; T. C. Whitehead; A. P. Kershaw (for Fig. 2.5); and Gebrüder Borntraeger (for Fig. 1.1).

Contents

Preface *page* viii
Acknowledgements x
List of tables xvi
List of contributors xviii

PART A INTRODUCTION

1 The history of geomorphology in low latitudes 3
I. Douglas & T. Spencer

2 The significance of environmental change: diversity, disturbance and tropical ecosystems 13
T. Spencer & I. Douglas
2.1 Introduction 13
2.2 Rainforest structure and change 16
2.3 Coral reef response to environmental stress 21
2.4 Variability in extreme events 25
2.5 The temporal and spatial distribution of hurricanes and tropical cyclones in the Caribbean basin 27

PART B ENVIRONMENT AND PROCESS

Process and time 36
Editorial comment

3 Present-day processes as a key to the effects of environmental change 39
I. Douglas & T. Spencer
3.1 Introduction 39
3.2 Nutrient cycling in tropical forests 45
3.3 Movement of water 51
3.4 Sediment and solute transport 55
3.5 Specific environments: New Guinea and Malaysia 66
3.6 Response of geomorphic processes to environmental change 69

4 Aspects of present-day processes in the seasonally wet tropics of West Africa 75
J-M. Avenard & P. Michel
4.1 Introduction 75
4.2 Morphoclimatic conditions 76
4.3 Processes and their effects on slopes 82
4.4 Fluvial processes 87

5 The influence of climate, lithology and time on drainage density and relief development on the tropical volcanic terrain of the Windward Islands 93
R. P. D. Walsh
5.1 Introduction 94
5.2 Measurement of drainage density and relief development 97
5.3 Analysis and interpretation of drainage density 106
5.4 Factors affecting drainage density and relief evolution 116
5.5 Conclusions 121

6 Surface and underground fluvial activity in the Gunung Mulu National Park, Sarawak 123
P. L. Smart, P. A. Bull, J. Rose, M. Laverty, H. Friederich & M. Noel
6.1 Introduction 123
6.2 The study area 124
6.3 The development of the regional and limestone drainage 127
6.4 The clastic cave sediments 134
6.5 Fluvial terraces 138
6.6 Discussion 142
6.7 Conclusions 147

PART C ENVIRONMENTAL CHANGE

Evidence for environmental change 150
Editorial comment

7 Relevance of Quaternary palynology to geomorphology in the tropics and subtropics 153
J. R. Flenley

8 Geomorphic implications of late Quaternary hydrological and climatic changes in the Northern Hemisphere tropics 165
F. A. Street-Perrott, N. Roberts & S. Metcalfe

8.1 Introduction 165
8.2 The water balance of closed lakes 165
8.3 Methods used in study 166
8.4 Fluctuations in lake level in Africa and Arabia 168
8.5 Variations in moisture conditions in Central America 177
8.6 Geomorphic implications 180

9 Evidence from lake sediments for recent erosion rates in the highlands of Papua New Guinea 185
F. Oldfield, A. T. Worsley & P. G. Appleby
9.1 Introduction 185
9.2 Aims and requirements 186
9.3 Previous results 189
9.4 Erosion rates 191
9.5 Sediment sources 192
9.6 Conclusions and prospects 196

10 Evidence of Upper Pleistocene dry climates in northern South America 197
J. Tricart
10.1 Introduction 197
10.2 Evidence of climatic change in central Amazonia 197
10.3 Climatic changes on and around the margins of Amazonia 204
10.4 Conclusion 216

11 Pleistocene aridity in tropical Africa, Australia and Asia 219
M. A. J. Williams
11.1 The humid tropics: stable or dynamic 219
11.2 The duration and amplitude of the Quaternary climatic fluctuations 221
11.3 Intertropical ice-age aridity 222
11.4 The evidence of Quaternary climatic change 224
11.5 The late Quaternary climates 227
11.6 Future palaeoclimatic research in the tropics 233

PART D LANDFORM EVOLUTION

Tectonic style and tropical landforms 236
Editorial comment

12 Environmental change and episodic etchplanation in the humid tropics of Sierra Leone: the Koidu etchplain 239
M. F. Thomas & M. B. Thorp

12.1 Introduction 239
12.2 Geological and geomorphological background 242
12.3 Landform elements and deposits 245
12.4 Episodic etchplanation and environmental change 261
12.5 Continuous versus episodic etchplanation 264
12.6 Some questions of geomorphic theory 266

13 A provisional world map of duricrust 269
M. Petit
13.1 Introduction 269
13.2 Critical review of previous work 269
13.3 Mapping of the duricrusted areas at the 1:20 000 000 scale 272
13.4 Conclusion 279

14 Tectonic background to long-term landform development in tropical Africa 281
M. A. Summerfield
14.1 Introduction 281
14.2 Domal uplifts and rift systems 285
14.3 Mechanisms of uplift and rifting 286
14.4 Morphotectonics of continental margin development 291
14.5 Some more general implications 292
14.6 Conclusions 293

15 Soil and slope development in the wet zone of Sri Lanka 295
H. Bremer
15.1 Introduction 295
15.2 Steep slopes and caves 295
15.3 Moderate slopes and regolith 297
15.4 Shallow slopes 298
15.5 Slope development 299
15.6 Conclusion 300

16 Relief generation and soil in the dry zone of Sri Lanka 303
H. Späth
16.1 Climatic characteristics 303
16.2 Geological overview 303
16.3 The oldest relief generations 305
16.4 Landform development in the Upper Tertiary 306
16.5 The youngest planation surface 306
16.6 Fossil sediments and palaesols and their implications for landform evolution 311
16.7 Discussion 314

PART E CONCLUSION

17 Findings, answers to questions and implications for the future 319
I. Douglas & T. Spencer
17.1 Typical tropical landforms 319
17.2 The effects of climatic contrasts on geomorphic processes 320
17.3 The role of plate tectonics and other tectonic events 321
17.4 Parallel sequences of landform evolution in different continents 322
17.5 The effect of sea-level changes 322
17.6 Ecosystem dynamics and landform development 323
17.7 Geomorphic events and scales of change 326

References and bibliography 329

Index 373

List of tables

2.1 Rainforest disturbance regimes *page* 20
B.1 Approximate correlation of timescales and climate-related events 35
3.1 Seasonality of sediment discharge 44
3.2 Annual return of nutrients by total small litterfall in selected natural tropical forests 46
3.3 Litter production in tropical forests 48
3.4 Nutrients in rain, stemflow and throughfall in tropical forests 49
3.5 Rates of denudation and sediment transport by surface wash in tropical areas 56
3.6 Dissolved and solid loads of major tropical rivers 58
3.7 Sediment yields in tropical environments 60
3.8 Environmental change and fluvial processes in north Queensland 70
4.1 Variability of annual and mean monthly rainfall at Pô, Upper Volta 79
5.1 Rainfall data for some Windward Island locations 96
5.2 Lithology and drainage density in late Upper Pleistocene areas of Dominica 106
5.3 Ruggedness index values for volcanic centres of different ages in Dominica 117
5.4 Recent climatic change in the Windward Islands 118
5.5 Drainage density data from tropical areas 120
6.1 Idealised sedimentary sequence in the Clearwater Cave complex, with related cave environmental conditions 134
6.2 Relative abundance of lithologies in pebbles from the Clearwater Cave complex 136
7.1 Altitudinal zonation of vegetation in Sumatra 159
8.1 Basins in the Northern Hemisphere tropics included in the Oxford databank on late Quaternary lake-level fluctuations 177
9.1 Characteristics of the lakes and catchments in Papua New Guinea 190
9.2 Summary of sedimentation rates for selected periods in cores from each lake 190
9.3 Some criteria for selecting lakes suitable for calculating whole-lake/whole-catchment sediment budgets 191

11.1 Sources of evidence for palaeoclimatic reconstruction 225
11.2 Climates of tropical Africa between 25 000 and 17 000 BP 229
11.3 Climates of tropical Africa between 12 000 and 7000 BP 230
D.1 Comparison of some pedological and geomorphological features of erosional landscapes in northern Australia, Sri Lanka, peninsular Malaysia, Borneo, New Guinea and Hawaii 237

List of contributors

Dr P. G. Appleby, Department of Applied Mathematics, University of Liverpool, Liverpool, L69 3BX, England

Professor J-M. Avenard, Centre de Geographie Appliquée, Université Louis Pasteur, 43 rue Goethe, 67000 Strasbourg, France

Professor Dr H. Bremer, Geographisches Institut der Universität zu Köln, D-5000 Köln 41, West Germany

Dr P. A. Bull, Christ Church, University of Oxford, Oxford, OX1 1DP, England

Professor Ian Douglas, School of Geography, University of Manchester, Manchester, M13 9PL, England

Dr J. R. Flenley, Department of Geography, University of Hull, Hull, HU6 7RX, England

Dr H. Friederich, Department of Geography, University of Bristol, Bristol, BS8 1SS, England

Dr M. Laverty, School of Geography, University of Oxford, Oxford, OX1 3TB, England

Miss S. Metcalfe, School of Geography, University of Oxford, Oxford, OX1 3TB, England

Professor P. Michel, Centre de Géographie Appliquée, Université Louis Pasteur, 43 rue Goethe, 67000 Strasbourg, France

Dr M. Noel, Institute of Oceanographic Sciences, Godalming, Surrey, GU8 5UB, England

Professor F. Oldfield, Department of Geography, University of Liverpool, Liverpool, L69 3BX, England

Professor M. Petit, Institut de Géographie, Universite Paris Val-de-Marne, Avenue du Général de Gaulle, 94000 Creteil, France

Dr N. Roberts, Department of Geography, Loughborough University, Loughborough, Leicestershire, LE11 3TU, England

Mr J. Rose, Department of Geography, Birkbeck College, University of London, London, W1P 1PA, England

Dr P. L. Smart, Department of Geography, University of Bristol, Bristol, BS8 1SS, England

Dr H. Späth, Geographisches Institut der Universität zu Köln, D-5000 Köln 41, West Germany

Dr T. Spencer, School of Geography, University of Manchester, Manchester, M13 9PL, England

Dr F. A. Street-Perrott, School of Geography, University of Oxford, Oxford, OX1 3TB, England

Dr M. A. Summerfield, Department of Geography, University of Edinburgh, High School Yards, Edinburgh, Scotland

Professor M. F. Thomas, Department of Environmental Sciences, University of Stirling, Stirling, FK9 4LA, Scotland

Dr M. B. Thorp, Department of Geography, University College Dublin, Dublin 4, Eire

Professor J. Tricart, Centre de Géographie Appliquée, Université Louis Pasteur, 43 rue Goethe, 67000 Strasbourg, France

Dr R. P. D. Walsh, Department of Geography, University College Swansea, Singleton Park, Swansea, SA2 8PP, Wales

Associate Professor M. A. J. Williams, Department of Geography, Monash University, Clayton, Victoria, Australia

Mr A. T. Worsley, Department of Geography, University of Liverpool, Liverpool, L69 3BX, England

Part A
INTRODUCTION

1
The history of geomorphology in low latitudes

I. Douglas & T. Spencer

The landforms of tropical areas aroused comments by many European travellers from the beginning of the sea voyages of the late 15th century. However, reports that contributed to the later development of geomorphological ideas did not begin to appear until the growth of the science of geology from the late 18th century; thus Charles Darwin was greatly influenced by Lyell's *Principles of Geology*, taking the first volume with him on 'The Beagle' and receiving the second and third while in South America. The great naturalists and travellers of the 19th century set the scene for recording the earliest impressions of the character of tropical landform evolution. Exploration of the interior of Africa led to encounters with strange features such as those found by Thomson (1882) in the Rovuma valley, which now forms the Mozambique/Tanzania border:

> The chief feature ... is the number of extraordinary isolated hills which give variety to the otherwise monotonous landscape. They appear in every possible shape as peaks, domes, cones, needles, etc. They all rise abruptly from the surrounding countryside, and present a scene not easily forgotten. (Thomson 1882, p. 69)

Gosse (1874) expressed similar surprise when he first saw Ayers Rock:

> One immense rock rising abruptly from the plain ... appears more wonderful every time I look at at it.

Such landforms were later described as inselbergs by the German explorer Bornhardt (1900) and have continued to fascinate geomorphologists ever since (Bremer & Jennings 1978).

Other travellers were impressed by the great depths of rotted rock found in tropical regions:

> Near Rio, every mineral except the quartz has been completely softened, in some places to a depth little less than one hundred feet. The minerals retain

> their positions in folia ranging in the usual direction; and fractured quartz veins may be traced from the solid rock, running for some distance into the softened, mottled, highly coloured, argillaceous mass. (Darwin 1890, p. 417)

Chorley *et al.* (1964) give James D. Dana credit for being the first to publicise to Europeans the influence of tropical climate on landforms. From observations on the United States Exploratory Expedition, under the leadership of Lt Charles Wilkes, Dana (1850) illustrated the fundamental roles of lithology and aspect in the landforms of Oahu and Hawaii, showing how the flutings of the windward side of the Koolau Range of Oahu were deep, winding, narrow valleys cut by running water during heavy rainstorms. Quoting W. Hopkins (1844) on the 'sixth power law of traction', Dana comments:

> There is every thing favourable for degradation which can exist in a land of perpetual summer; and there is a full balance against the frosts of colder regions in the exuberance of vegetable life, since it occasions rapid decomposition of the surface, covering even the face of a precipice with a thick layer of altered rock, and with spots of soil wherever there is a chink of shelf for its lodgement.

The seasonal contrast between the relative calm of baseflow conditions and the rapid rise of streams following intense rainfall was shown to be responsible for changing of 'a lofty volcanic dome ... to a skeleton island like Tahiti'.

While these explorations were proceeding, the colonial administrators of the Indian subcontinent were coping with the practical problems of managing irrigation schemes and using local building materials. Buchanan's (1807) writings on laterite and early studies of fluvial dynamics (Tremenheere 1867) were among the products of this colonial period. Expansion of the British Empire further east led to comments on granite weathering in Hong Kong (Johnston 1844) and the need for legislation to prevent soil erosion being caused by excessive forest clearance on steep land in the Malay Peninsula (Daly 1882). The web of information and research linkages between home and abroad is well illustrated by the distribution of botanic gardens corresponding with Kew, London in 1889: seven in the West Indies, and five each on the Indian subcontinent, in Africa and in Australia. Interests in economic botany spread as far as Singapore, Fiji and New Zealand (Brockway 1979). In Indochina and Indonesia, French and Dutch colonists collected similar information, but perhaps it was the surge of activity by the newly united German nation that led to the first general statements about tropical landforms.

Explorers in the tradition of Alexander von Humboldt and natural scientists from Germany, Europe's leading scientific nation at the end of the 19th century, set out to the new colonies in the Pacific and Africa. In New Guinea,

Sapper worked primarily in New Britain, New Ireland and Bougainville, producing not only regional accounts (Sapper 1909, 1910a,b,c) but also papers on erosion processes (Sapper 1914) and volcanoes (Sapper 1917a,b, 1921). Behrmann (1917) took part in a major expedition to the Sepik area and later produced a regional study (Behrmann 1924) and the first general accounts of tropical geomorphology in New Guinea (Behrmann 1921, 1928). In East Africa, Passarge (1904, 1923), Obst (1915) and Bornhardt (1900) described the level surfaces and isolated inselbergs, while in the wetter, forested Cameroons, Guillemain (1908, 1914), Passarge (1910) and Thorbecke (1911, 1914) examined the interrelationships of recent volcanics, older intrusive rocks and inselberg landscapes associated with a short but marked dry season. Soil studies by Vageler (1911, 1912) led to early ideas about soil–landform interdependence in East Africa (Opp 1983). Passarge (1903) also travelled extensively in South America, but the major German contribution to tropical geomorphology from that continent was made later by Freise (1930, 1932, 1934, 1936, 1938) in a series of investigations of processes of denudation and climatic conditions under rainforest. With this strong scientific background, it was not surprising that the first general discussion on tropical geomorphology occurred in Germany at a meeting chaired by Thorbecke (1927) in Düsseldorf. At this conference, the contrast between equatorial continuously humid climates and tropical climates with a dry season was emphasised and the contributors were asked to examine the extent to which present-day landforms were dependent on present-day climates. Little evidence for climatic change was produced at the Düsseldorf meeting and general opinion held that the rainforest areas at least had been unchanged throughout the Quaternary. These arguments were paralleled by Grund's (1914) comments, based on Danes' descriptions of tropical karst in Java, Jamaica and Australia and a strong Darwinian method, that the 'Cockpitslandschaft' was one of old age. Nevertheless, Gregory (1894a,b) had already found evidence of past glaciations on Mount Kenya in areas now forested, indicating at least a considerable lowering of the treeline in the Quaternary.

Even though the early German tropical geomorphology was impressive, perhaps the detailed observations by equally perceptive American travellers to South America (Branner 1896, Hayes 1899) provided some of the most fundamental truths about geomorphological processes in low latitudes. Branner's (1896) evidence that, in the tropics, rain has four times as much nitric acid as in the temperate zone has been questioned. But his comments that the process of exfoliation in Brazil is not essentially different from that common in other parts of the world, and that a given quantity of rainfall evenly distributed throughout the year would do less work than if it were concentrated into a few months, have been substantiated by more recent observations, such as those of Ollier (1965) on granite weathering and Fournier (1960) and Douglas (1967) on denudation rates. Hayes (1899) showed the importance of climatic contrasts over relatively small distances in the tropics. In Nicaragua he found that the contrast between the humid east and the seasonally dry west of the country gave rise directly to

> very striking differences in vegetation, and either directly or indirectly, to difference in the appearance and structure of the soils, in the topographic forms of the land surface and in the effectiveness of various physiographic processes

Hayes went on to argue that:

> After a careful study of the region it was considered that the absence of frost more than counterbalances the enormous rainfall, and that degradation of the surface is, on the whole, slower than in temperate regions, where the rainfall is less than a quarter of that in Nicaragua, but where the surface soil is thoroughly loosened by the action of frost.

The beginnings of tropical geomorphology thus raised some fundamental questions that have persisted, inadequately answered, throughout subsequent studies. As summarised by Stoddart (1969a), they are as follows:

(a) Are there unique tropical landforms, such as inselbergs?
(b) Are there tropical climatic regions corresponding to morphological regions, as suggested in Nicaragua?
(c) Have changes in climate produced recognisable sequences of tropical landforms, as the evidence of glaciation on Mount Kenya might imply?
(d) How does tropical geomorphology differ from other sorts of geomorphology, in terms of the intensity of processes causing deep weathering, or the absence of frost or otherwise?

A further important contrast was made by the Germans whose experience in New Guinea and Africa revealed the great difference between the selva landscapes of continually wet (*Immerfeuchten*) tropics with a characteristic feral relief (Cotton 1958) with V-shaped valleys (*Kerbtäler*) and the broad gentle valleys (*Flachmuldentäler*) of the seasonally wet (*Wechselfeuchten*) savanna climates (Louis 1961). This recognition, although not explicitly spelt out, of the role of tectonics and structure for tropical landforms, and thus the operation of geomorphic processes, later proved to be vital for understanding low-latitude landforms. However, the period after the 1920s saw few general appraisals of tropical geomorphology, apart from Sapper's (1935) important synthesis and Krynine's (1936) review. Instead, the development of the colonies led to a growth of detailed regional and local work in geomorphology and related sciences.

Most geomorphological work in British colonies was a byproduct of geological (Scrivenor 1931, Dixey 1922a,b) or soil (Harrison 1933, Milne 1935, Hardy 1939) surveys, but the character and evolution of inselbergs and laterite were re-examined, and attempts were made to explain river systems and erosion surfaces in terms of the Davisian cycle-of-erosion concept then dominating geomorphological thought in Britain, for example in Uganda (Wayland 1921), peninsular India (Wadia 1966) and peninsular Malaysia (Richardson 1947). Erosion surfaces in Africa (Dixey 1942, Wayland 1934), Sri Lanka (Adams 1929, Wadia 1941, 1945) and India (Heron 1938) were found to have much in common. The 1928–9 Great Barrier Reef Expedition was a pioneering interdisciplinary field study which provides a benchmark in the ecological and geomorphological investigations of coral reefs. J. A. Steers subsequently developed the geomorphological work, both on the Reef and in Jamaica (Steers 1929, 1940).

In Indonesia, Dutch scientists pursued detailed studies of Java and Sumatra (Rutten 1928), some of which benefited from excellent topographic maps (Pannekoek 1941). The pressure for commercial development in the Belgian Congo saw a steady growth of pedological and geomorphological studies (Cornet 1896, Bayens 1938, Robert 1927), while French pedological studies (Aufrère 1932, 1936, Blondel 1929a, de Lapperent 1939) led eventually to the integration of pedogenesis and landscape evolution in Erhart's (1956, 1961, 1966) concept of alternating periods of stability and landscape evolution. Important French contributions to karst geomorphology came from Indochina (Cusinier 1929) where Blondel (1929b) found that the limestone areas had no specifically tropical attributes, but could be explained by the general laws of karst geomorphology.

American work continued apace, with many important regional syntheses, especially as a result of work in Puerto Rico (Hubbard 1923, Lobeck 1922, Meyerhoff 1927, 1933, 1937), Hawaii (Wentworth 1927, 1928, Palmer 1927, Jones 1938) and the Philippines (Pendleton 1940). Important general comments were made at this time on such low-latitude landforms as coral reefs (Daly 1910, 1934), inselbergs and rift valleys (Willis 1936) and rock weathering (Blackwelder 1925).

In Australia, local geological surveys were revealing geomorphological features such as duricrusts (Woolnough 1927, Jutson 1934) and distributing seasonal flood-channel systems (Whitehouse 1940) not previously well described in the literature. Gradually a picture of the role of climate in landform evolution that went beyond the simple glacial, arid and so-called 'normal' climates described by W. M. Davis (1909) was emerging. Significantly, Davis never described a 'tropical cycle'. By 1940, de Martonne (1951) was able to incorporate a discussion of the relationship between climate and geomorphological processes into his text on physical geography. De Martonne (1940) also wrote on the landforms around Rio de Janeiro and on geography as zonal geography (de Martonne 1946), producing an important report on slope development in the humid tropics (de Martonne & Birot 1944). These observations, allied to those of such writers as Sapper and Behrmann, led to arguments that special tropical conditions gave rise to unique landform assemblages whose occurrence in extratropical zones must indicate the former existence of tropical climates. Pendleton (1940) expresses the characteristic views of the time:

> The larger amount of precipitation in the humid regions, and the abundant opportunity for it to take up carbon dioxide from the rapidly decaying organic matter on the forest floor, together with the continuously high temperatures, are important factors in the very deep weathering of the rocks of the humid tropics, weathering which has resulted in the complete alteration of the entire rock mass to sometimes tens of metres below the zone of physical or mechanical comminution or fracture of the rock. For these reasons it is often impossible in many regions to find unweathered rock at or near the surface, and these are causes also which at times make it possible for erosion, once it gets through the surface protection, to do great damage.

These views were incorporated, for example, into the explanation of karst landforms. H. Lehmann (1954) argued that the contrast between the 'hollow-forms' of the temperate regions and the 'full-forms' of the tropics was due not to age but to the level of solutional activity, this being a function of greater runoff from higher precipitation and increased soil carbon dioxide concentrations as a result of higher temperatures (Lehmann 1964, 1970). Such views were reinforced by Lehmann's pupils, e.g. Gerstenhauer (1960) in Tabasco, Mexico, and Sunartadirdja and Lehmann (1960) in Sulawesi. Simultaneously, and paradoxically, Jean Corbel was claiming low rates of limestone erosion, based upon the inverse relationship between carbon dioxide solubility and temperature (Corbel 1957, 1959, 1971, Corbel & Muxart 1970). These generalisations were usually made outside any formal research design and with few measurements of process operation rates. In fact, the major controls on water quality appear to be aclimatic and it is difficult to distinguish between tropical and temperate soils in terms of carbon dioxide content (Smith & Atkinson 1976). Similarly, steep slopes, effective wash of debris, parallel retreat of slopes and the developmental deep weathering and relatively weak fluvial erosion are the characteristics of humid tropical landforms described in the textbooks of the 1950s and 1960s (Thornbury 1954, Birot 1966). Steep slopes, effective surface work of debris, parallel retreat of slopes and the development of large plains, often with duricrusts, were said to be the dominant characteristics of savanna regions. Particularly important was the development of the theory of double planation processes, by the inward and downward migration of the weathering front (B in Fig. 1.1) which accompanies the surface stripping of the weathered mantle by overland flow during storms (across surface A in Fig. 1.1) (Büdel 1957). Büdel's claim that tropical seasonal erosion surface development, or etchplanation, is one of the fundamental landforming processes, together with valley development under periglacial conditions, remains a major hypothesis to be tested by studies involving the whole range of geomorphological techniques.

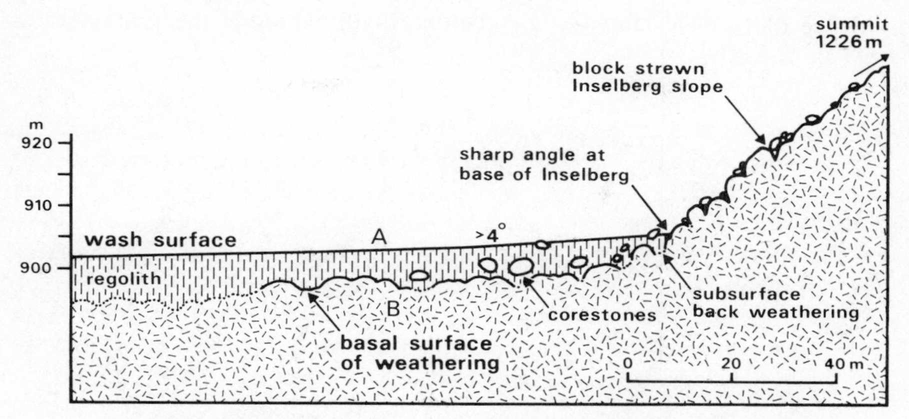

Figure 1.1 Double planation surface development (after Büdel 1957, 1982).

However, the landforms of low latitudes are far from as uniform as the simple division into constantly humid and seasonally wet tropical morphogenetic zones might suggest. Much of the evidence about landforms described in the preceding paragraphs came from specific investigations in small areas: inselbergs and double planation surface theories from the vast plateaux of Africa and the Deccan, steep slope maintenance by landsliding from Hawaii, central America and New Guinea. Differences in tectonics and structure produce striking variations in regional landforms and in the dominance of processes. These tectonic and structural variations are well illustrated by the way in which Machatschek (1955) divided his study of tropical landforms into six major relief zones:

(a) The south and south-east Asian mountain chain zone.
(b) The Old World Gondwanaland area.
(c) Oceania.
(d) Central America (essentially Tertiary fold mountains).
(e) Andean South America (essentially Tertiary fold mountains).
(f) South America other than the Andes (essentially Gondwanaland).

Machatschek shows how tectonic contrasts have produced geomorphological contrast. The pattern of plate tectonics (Fig. 1.2) is thus another fundamental factor in tropical geomorphology whose dynamics interact with those of climate to create the landscape diversity of low latitudes. Only recently has the theme suggested by Machatschek been given a closer examination (Ollier 1979, 1981, Summerfield 1981, Ch. 14 this volume). Just as tectonics and structure provide variable bases for landform development, so a simple division into constantly humid and seasonally wet environments is too simplistic. The tropics incorporate such diverse climatic regimes as the Doldrums, the monsoons and the Trades and suffer from incursions of polar air and incidence of hurricanes, cyclones and typhoons.

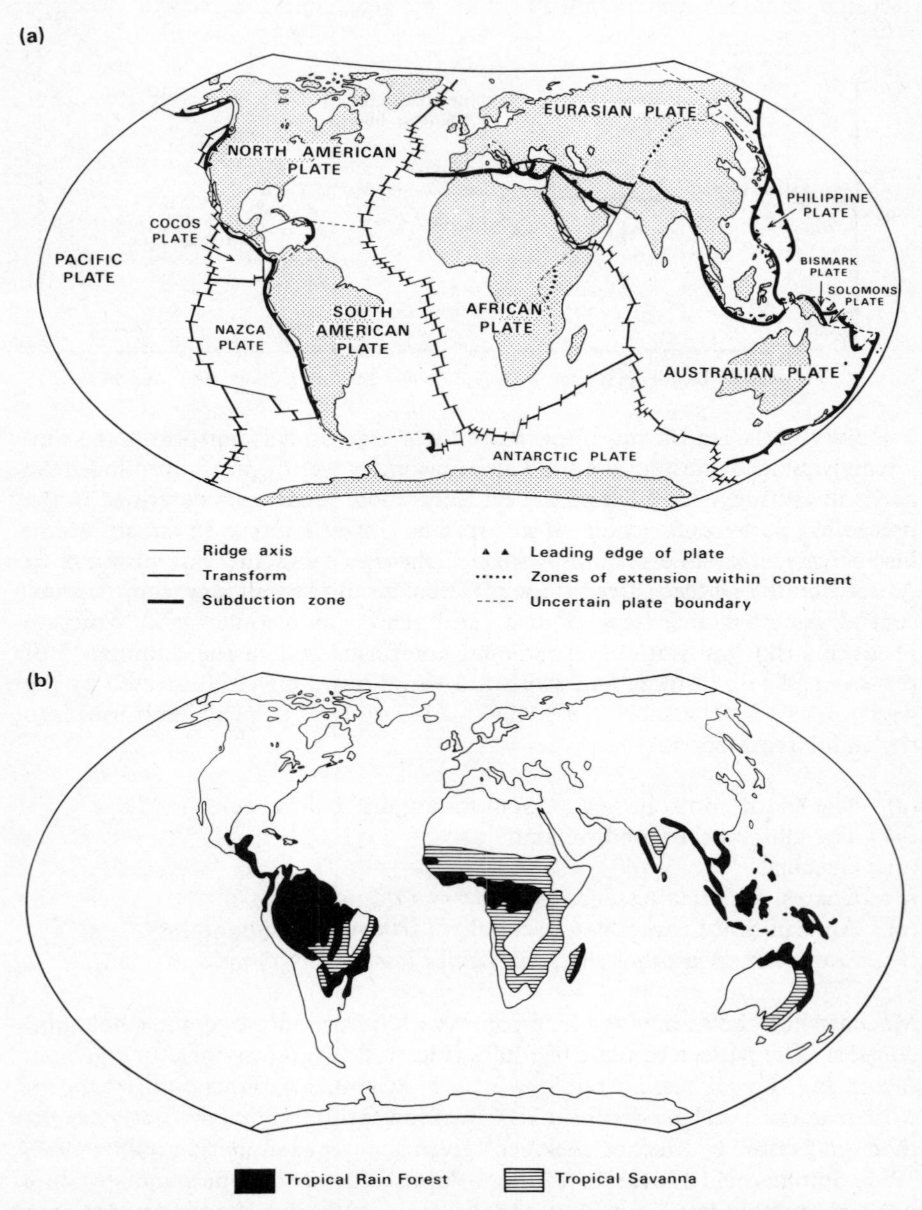

Figure 1.2 (a) Plate tectonic structure of the continents and oceans (after Strahler & Strahler 1977), and (b) areas of rainforest and savanna vegetation (based on maps in the Oxford Atlas).

However, by the 1950s and 1960s, the need for accurate data on processes became apparent and detailed studies of geomorphic processes (Rougerie 1960) and nutrient cycling (Nye & Greenland 1960) began, culminating in large interdisciplinary projects such as those at the El Verde forest in Puerto Rico (Odum 1970), on the Amazon (Sioli 1961) and at the Pasoh Reserve in Malaysia (Peh 1978, Leigh 1982). Such work has greatly improved understanding of the dynamics of ecosystems and the nature of geomorphic processes in rainforest and savanna environments.

At the same time, greater possibilities for travel and improvements in remote sensing have made it possible for many more geomorphologists to visit tropical areas and to study aerial photography and satellite imagery. Sideways-looking airborne radar (SLAR) provides a means of examining landforms beneath a dense forest cover, thereby assisting Tricart (1974a,b, 1975) to recognise fossil sand dunes in what is now a rainforest area of north-east Brazil. This recognition of dry phases during the Quaternary in equatorial rainforest areas has led to a reappraisal of both the biology and the geomorphology of rainforest areas. Climatic changes and associated variations in sea level have been strong influences on all tropical landform evolution. Debate continues as to whether processes operate rapidly enough to obliterate legacies from past drier periods in the wettest parts of the present-day humid tropics, but generally the task of unravelling the legacy of the past is as challenging in the tropics as elsewhere.

Among the key issues that emerge from this brief review of the evolution of the study of tropical landforms are the following:

(a) Are there typical tropical landforms?
(b) How do climatic contrasts, especially in the amount and seasonal distribution of rainfall, affect geomorphic processes in low latitudes?
(c) What is the role of plate tectonics and other tectonic events, both associated with plate boundaries and otherwise, in tropical landform evolution?
(d) Is it possible to distinguish parallel sequences of landform evolution in the tropical areas of different continents?
(e) What has been the effect of sea-level changes?
(f) How closely are the dynamics and evolution of escosystems related to landform development?

The questions hark back to the fundamental issues raised earlier in this chapter. This book is an attempt to illustrate how much is now known about these issues and to what extent answers can be provided to these questions.

2

The significance of environmental change: diversity, disturbance and tropical ecosystems

T. Spencer & I. Douglas

2.1 Introduction

It is somewhat ironic that tropical geomorphology often stresses the distinctiveness not of tropical landforms but of tropical climates and, in particular, tropical vegetation. Obviously there are dangers in arguing that tropical landforms should be peculiar because tropical climates are distinctive (Stoddart 1969a). Nevertheless, the closeness of the relationship between vegetation and landforms does have potential analytical value in that it is useful to view the 'humid tropical denudation system' as a functioning ecosystem (Douglas 1969). This leads to an awareness of the complexity of tropical ecosystems and to a recognition of the high diversity of tropical plants and animals.

Tropical diversity, here defined as taxonomic richness, may be explained by historical or biotic/environmental factors. 'Why have the tropics generated or accumulated more species than the extratropical regions?' 'Why/how do so many species live next to each other in the tropics?' It is not the intention of this chapter to cover the range of possible explanations for high tropical diversity (for reviews see Baker 1970, Stott 1978, Rosen 1981), but it is important to consider the suggested causal links between diversity, stability and time. A long-standing argument has been that organic complexity confers ecological stability (e.g. Odum 1963), with equilibrium being more readily maintained where there are more interacting species (e.g. MacArthur 1955). This line of reasoning can be extended by incorporating the theory of succession and arguing that, as diversity increases over time, so more stable ecosystems should outlast less stable ones (MacArthur 1955, Hutchinson 1959). Thus the labelling of the tropical rainforest as a climax community leads to assumptions about ecological stability and also, presumably, has implications for the distinctiveness of geomorphic processes in the tropics and the development of particularly 'tropical' landscapes.

However, such arguments now need complete reassessment with the realisation of the scale of environmental change in the tropics during the Pleistocene.

We can now safely say that large areas of the tropics were subjected to a much drier, somewhat cooler and probably windier climate between 2×10^4 and 12.5×10^4 years BP (Goudie 1983a). A quite stunning picture, well established from palaeoenvironmental reconstructions using palynological and geomorphological evidence and from the modern biogeography of plants and animals, has emerged of the retreat of the tropical rainforest at this time into remnant enclaves, or 'refugia', in South America and Africa. The position in Malesia is less clear. Although the details of refugia size and location have yet to be worked out (for criticisms of 'refuge theory' see Endler 1982, Benson 1982), there is a reasonable degree of agreement between different lines of evidence (e.g. Fig. 8 of Street 1981, Hamilton 1982, Fig. 3.13 of Goudie 1983b). Clearly any argument which is based upon the idea that the rainforest is diverse because of long-continued environmental stability is no longer adequate. If, alternatively, diversity is a product of youth and environmental stress, then the research questions become ones of the rate of speciation, the speed and style of forest reconstitution in interglacial episodes and the responses communities to environmental shocks.

Ideally, geomorphological research might provide some insights into this last question, based on observations on a shorter time-scale of 10^1 to 10^3 years. Unfortunately, for most of the tropics, climatic records are not yet 10^2 years long. Observations on ecosystems tend to be similarly brief and then the relationship between climatic events and ecosystem development have only been explored in a few instances. The following section of this chapter examines our knowledge of the influences of extreme natural events on tropical rainforests: this in turn is succeeded by a more explicit analysis of the influence of particular storms on coral reef structures.

The apparently different rainforest and coral reef ecosystems have several characteristics in common. The influence of the physical environment on their growth is often reflected in the morphology of specific plant and animal communities. Thus, for example, encrusting and digitate corals dominate in high-energy environments, branching forms in moderate-energy environments and massive forms where wave action is low. An even more fundamental control in this ecosystem is the rapid reduction in illumination and radiant energy with depth (Wells 1957), as this leads to a decline in coral calcification and, therefore, adaptations in skeletal geometry. *Montastrea annularis*, for example, changes from a rounded, massive form to a flattened plate as water depth increases (Barnes 1973, Dustan 1975). However, at the same time, growth in the forest and on the reef alters the local environment, modifying the physical processes operating within it. This ability to adapt through growth is important to trees and corals as they are static apart from a very brief period at the start of their life history. The fundamental concern is one of competition for light energy; this becomes a limiting resource to those components of the ecosystem denied access to it. Access to light becomes competition for space. This space filling is not a straightforward, successional process, but a dynamic, ever-changing situation brought about by the interaction of environmental events with community structure. This combination of process and pattern repeatedly redefines the space in the ecosystem, the growth pattern of the community and, overall, the distinctive nature of the rainforest and the reef.

Changing community response to small-scale, relatively frequent environmental changes would suggest that the two ecosystem types are quasi-stable responses to a moderate level of climatic variability. The alternative view is that disturbances are required to maintain diversity. In particular, Connell (1978), using field data from Australian rainforests and reefs, has outlined a hypothesis suggesting that it is disturbances at 'intermediate' levels of frequency, severity and spacing which promote and preserve taxonomically rich

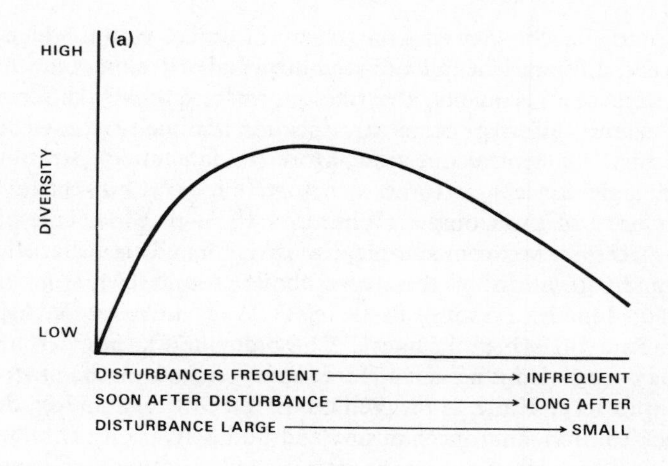

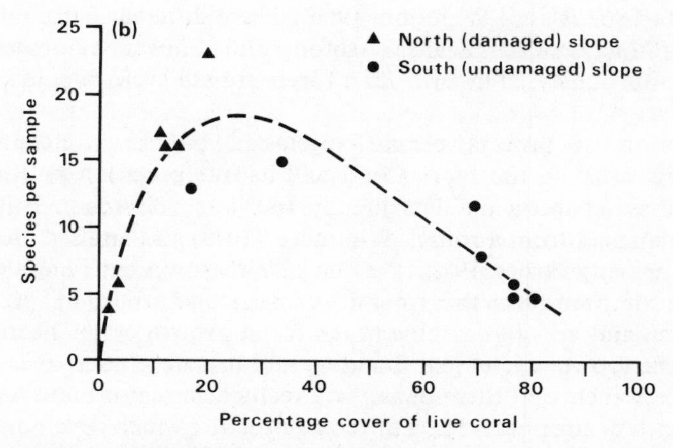

Figure 2.1 The 'intermediate disturbance' hypothesis: (a) distubance type and species diversity; (b) species diversity of corals in subtidal outer slopes at Heron Island, Queensland. 20 m line transects 3–4 months after 1972 hurricane (after Connell 1978).

communities (Fig. 2.1a). To geomorphologists, this plot is a reminder of the 'work done' curve from Wolman and Miller's (1960) classic paper on magnitude and frequency, not least in that the scale of analysis is only implicit in the diagram (in this case field plots rather than drainage basins) and the axes are not formally scaled with any common unit of measurement. Whether Connell's (1978) relationships are of more general application and whether they can be more closely defined using geomorphological evidence will be considered later below.

2.2 Rainforest structure and change

The rainforest may be viewed as a structural object within which such processes as regeneration, leaf fall and seed dispersal are taking place. In such an approach, the fixed elements, the trees, form a vertically differentiated, or stratified, canopy. Whether or not stratification is a reality (e.g. Richards 1969) is not, in fact, the central question. More fundamentally, stratification is a 'static typologic concept of forest structure that gives no recognition to the dynamic nature of the canopy' (Whitmore 1978, p. 640). Tree architecture changes as trees mature; for example, leaf division and area characteristics can be correlated with height of the crown above ground level (e.g. Brazil: Cain *et al.* 1956; Jamaica: Asprey & Loveless 1958, Loveless & Asprey 1957; Trinidad: Beard 1945–6b; Malaysia: Ashton 1978). Furthermore, crown morphology may also change as a tree matures and as the environment of growth varies (e.g. Hallé 1978, Veillon 1978, Oldeman 1978); this may be traced back to branching mechanisms and bud activation (Tomlinson 1978). Deep, conical tree crowns are characteristic of young trees and can be distinguished from the flatter crowns of the mature forest (e.g. Brunei: Ashton 1964; Santa Cruz Island: Whitmore 1966). These different morphologies may well have a functional significance (Ashton 1978). Thus the rainforests may be more profitably analysed in terms of a forest growth cycle than in a stratification framework.

The notion of process behind vegetation pattern stretches back to Aubréville's work on the Ivory Coast and its interpretation by Richards, to Jones' studies in Nigeria and, less directly, to Watts' research on British woods and heathlands. More recently, Whitmore (1978) has utilised the work of foresters, or 'empirical ecologists' as he calls them, in marrying descriptions of floristic composition to the dynamics of particular groups of species. Gaps, regeneration and growth combine in the forest growth cycle. There are three stages to the growth cycle; gap, building and mature phases. It is important to stress, however, that these phases are really abstractions and not separate entities. Each of the phases has a different structure which develops from, and into, the other phases. Thus as trees grow up and die, the forest canopy is continually changing and at any one point in time is a mosaic of different structural phases. The gap phase is succeeded by the pole forest of the building phase, which is in turn replaced by the spreading crowns of the mature forest. The building phase can commence with existing seedlings or saplings producing upward growth or with the establishment of new trees from seeds germinating within the gap. At its simplest, established seedlings and saplings most often grow to maturity in small gaps but are unable to survive the drastic changes in microclimate that follow the conversion of shaded conditions into large gaps. These big gaps are colonised by those species that are absent in the dense forest but are equipped to flourish on open sites. Thus different species are successful in gaps of different sizes. Gap size, therefore, has an important influence on species composition and spatial arrangements within the rainforest as a whole. The larger the gap, the more the microclimate within it differs from that under a closed forest canopy. A comparison of microclimatic conditions in the forest understorey and within large gaps (Denslow 1980) shows that differences are pronounced, particularly near ground level. Thus, for example, the change in humidity between 10 cm and 1.5 m is as great as the change between 1.5 m and conditions just above the canopy (Schulz 1960). In the largest gaps a class of pioneer species can be identified. These are light-demanding species which can only establish in the open and cannot regenerate under shade, including their own shade. Schulz (1960) has suggested that large-gap strategists require in excess of 10–20% full sunlight for establishment. The pioneers produce large quantities of small, dispersable seeds. The seeds may be lying dormant in a 'seed bank' or be transported to the gap as 'seed rain' (e.g. Cheke *et al.* 1979, Hartshorn 1978). Gemination may be triggered by the high soil temperatures resulting from prolonged exposure to strong sunlight in large gaps (Bazzaz & Pickett 1980, Vazquez-Yanes 1974). The small roots of the seedlings are able to penetrate baked and compacted ground and, after establishment, height growth is rapid, with large leaf areas and high leaf production rates (Coombe 1960, Coombe & Hadfield 1962). At the other extreme are the shade-tolerant species. They are noted for their big seeds with substantial food reserves. Big seeds may also restrict frugivory to those few animals able to handle large seeds (e.g. Janzen 1978). Seedlings are large, produce deep roots and persist in deep undergrowth, growing slowly or not at all, waiting for a gap to appear in the canopy above. Then growth is rapid (e.g. Schulz 1960, Liew & Wong 1973, Nicholson 1965). Obviously the light-demanding and shade-demanding 'syndromes' are not exclusive categories and in some localities further classes have been identified. 'Late secondary species' (Whitmore 1982) can neither colonise bare sites nor regenerate *in situ* but may become dominant at an intermediate stage of gap transformation. In addition, factors such as the timing of gap occurrence and the proximity of gaps to seed source must introduce stochastic controls on forest dynamics and vegetation pattern.

The 'coarseness' or 'openness' of the vegetation mosaic depends upon the initial gap size. Whitmore (1982) has reviewed the evidence for the critical gap size at which light-demanding rather than shade-tolerant species successfully establish. At Mount Gede, Java, artificial clearings of 1000 m^2 were recolonised by primary forest trees but clearings of 2000 and 3000 m^2 were seeded by invading pioneers. Other studies in Surinam, Costa Rica and on the Ivory Coast have confirmed the success of shade-bearing species in gaps of less than 1000 m^2.

What, therefore, controls gap size? The smallest gaps are formed by the slow, *in situ* death of large individual trees. The progressive loss of portions of the crown and bole results in a gap 15–18 m in diameter. Gap sizes resulting from tree fall in the La Selva rainforest vary in area from 54 ± 43 m^2 on slopes on old alluvium to 120 ± 115 m^2 at forest swamp sites (Table 26.1 of Hartshorn 1978). The rate of tree fall is approximately 1 tree ha^{-1} yr^{-1} (Hartshorn 1980). Turnover rates, or the estimation of the number of years required to cover a plot with gaps, average 118 ± 27 years (Hartshorn 1978); this is within the range of turnover rates for Malayan forests of 35 to 180 years (Poore 1968). At a slightly larger scale, many trees are blown over by the wind before they die of old age. This process may be exacerbated by the presence of large epiphyte loads (Strong 1977). The falling bole cuts a swathe through the forest which is enlarged by the disintegrating crown into an elongated gap. Windfall gaps on a 2 ha plot of lowland evergreen forest at Sungai Menyala, Malaysia, have a mean area of 258 m^2 (estimated from Whitmore 1975). Existing wind gaps are also sometimes extended by further wind-throw at their margins. The largest wind gaps have been recorded after strong squall events in the peat forests of Sarawak (Whitmore 1978). These reach a size of 80 ha in area. Lightning strikes also create gaps; these are found, for example, in the peat swamp forests of Malaysia, where they reach 0.6 ha in size, and in the mangroves of Papua New Guinea. The most extensive gaps in tropical forest are those produced by tropical cyclones and hurricanes. Cyclones have been shown to be important in determining forest structure in the Solomon Islands (Whitmore 1974), in the tropical lowland rainforest of north Queensland, Australia (Webb 1958) and on Caribbean islands, notably Puerto Rico (Crow 1980) and the 'hurricane forest' (Beard 1945–6a) of the Lesser Antilles chain. On Savari Island, Samoa, cyclones in 1961, 1966 and 1968 destroyed over half the potentially exploitable forest (Whitmore 1978). At the El Verde forest, Puerto Rico, three hurricanes during the late 1920s and early 1930s caused defoliation, windfall and stem breakage. This natural thinning process meant that 'growing space' was still available in the 1940s, this being represented by the continuing influx of secondary species and the rapid rate of biomass and basal area accumulation (Crow 1980). In such environments, the forest canopy may be adjusted to the cyclone environment and maintained, through non-perpetuating tree species, in a kind of 'hurricane climax' (Webb 1958). In Puerto Rico after hurricane Betsy, in 1956, severely damaged areas supported a dense cover of herbaceous vines and climber towers developed on trees that had lost their crowns (Crow 1980). Similarly, light-demanding woody climbers are characteristic of the 'cyclone vegetation' on the storm-damaged northern and western islands of the Solomons archipelago (Whitmore 1974). In north Queensland, Australia, the recurrence interval for severe or general cyclone damage is between 3 and 40 years. The area's 'cyclone scrubs' are characterised by 'low, uneven canopies and dense vine tangles' and 'taller rain-forests with dense vine understories or with scattered sclerophyll emergents' (Webb 1958). If fire is added as a third variable to disturbance and time, then the gap creation and restoration process becomes a more complex sequence, especially at sites where seasonal drought is probable, as on the hillsides near Cairns, Queensland. Here three prominent vegetation types occur: tall grassland, open eucalypt forest and rainforest. Stocker considers that the first two are derived from the third and postulates the following developmental sequence:

(1) A cyclone severely disturbs an area of rainforest, thereby allowing a fire in the following dry season to penetrate some of the disturbed area, and in the process killing many trees and enabling some grass invasion.
(2) Although many of the rainforest trees regenerate by coppice, they are gradually eliminated by repeated fires in subsequent years and grasses form an almost pure stand.
(3) With repeated burning, the fertility of the surface soil declines and the grasses lose much of their earlier vigour. Consequent lower fire intensities allow fire-resistant species such as *Eucalyptus tereticornis* Sm. and *E. intermedia* R. T. Baker to invade the grassy slopes.
(4) Competition from established fire-resistant tree species further reduces grass growth, again lowering fire intensities and thus allowing a gradual reinvasion of rainforest species and the eventual return of rainforest of typical structure and floristics (Stocker & Mott 1981).

As well as having direct effects on rainforest structure, cyclones may also disrupt vegetation more indirectly. For many localities, for example, the maximum precipitation may be associated with a slow-moving or decaying hurricane cell. High porewater pressures in deep and unstable regoliths, surface wash and gullying may all lead to the local collapse of the forest canopy. However, landsliding as a product of normal weathering processes does not appear to be a major disturbance factor in rainforest communities, although it is difficult to find areas apparently unaffected by man's acceleration of slope failure. Calculations for areal disruption on Hawaiian slopes suggest that shallow landslides, although frequent, are small and only affect 0.6–0.7% of the study areas per century (based on Wentworth 1943, Scott & Street 1976). Similarly, the Bewani Mountains in New Guinea have denudation rates of 3% of forest area every 100 years (Simonett 1967). In some areas, however, landsliding may also result from aclimatic, seismic triggers. Aerial photography has been used in two cases to estimate earthquake-initiated slope failure. In 1976, two shallow offshore earthquakes triggered landsliding along the Pacific coast of Panama. The distribution of slides was patchy but affected in total 54 km^2 of a study area of 450 km^2 (or 12% of total area). The Panama–Colombia region experiences an earthquake of force 6.0 or greater at an average recurrence of one event every 250 years per 10^3 km^2; thus if the

1976 event was typical, earthquakes indirectly disturb 2% of this Central American rainforest per 100 years (Garwood *et al.* 1979). This suggests that in this region earthquake-initiated landslides are not important influences on rainforest structure. By contrast, in the Torricelli Mountains of northern Papua New Guinea, the frequency of earth tremors has been calculated to be one event every 56–111 years over an area of 10^3 km^2. Calculations based on known earthquake-triggered slope failures suggest an average area of damage of 100 km^2 per event. Multiplying disturbance frequency by this area proposes that 8–16% of the Torricelli forests are subjected to landsliding and canopy disruption stemming from seismic activity. Furthermore, high-magnitude seismic shocks can be highly destructive; after a force 7.0 tremor in November 1970 near Madang, Aldelbert Range, Papua New Guinea, 60 km^2 of a 240 km^2 study area was cleared, with the coalescence of slides locally producing 60% vegetation clearance (Pain & Bowler 1973).

Table 2.1 attempts to outline the differing disturbance regimes of some rainforests in terms of the varying importance of tree falls, earthquake-triggered landslides and weathering-limited landslides. Estimates vary as a result of the rates of disturbance and recovery times employed, but clearly earthquakes are an important source of disturbance in New Guinea, whereas they are not in

Table 2.1 Rainforest disturbance regimes.

Type of disturbance	(i) Rate of disturbance (% area per 100 yr)[a]	(ii) Recovery time (yr)	(iii) Total area disturbed (%) [(iii) = (i) × (ii)]	Source
tree falls	62–125	20	12–24	Garwood *et al.* (1979)
		30	19–38	
		50	31–62	
weathering landslides				
Hawaiian Is.	0.6–0.7			Wentworth (1943), Scott and Street (1976)
New Guinea	3	200–300	6–15	Simonett (1967), Garwood *et al.* (1979)
earthquake landslides				
Panama (magnitude 6.0)	2	200–500	4–10	Garwood *et al.* (1979)
New Guinea (magnitude 6.0)	8–16	200	16–33	Simonett (1967), Garwood *et al.* (1979)
		300	24–49	
		500	41–89	
New Guinea (magnitude 7.0)	12.5 locally 30.0			Pain and Bowler (1973)

[a]Rate of disturbance = frequency × mean size of disturbance, expressed as per cent of forest area per 100 years.

Panama. It may also be significant that 38% of the Malesian rainforest and 14% of the American rainforest lie within zones of high seismic activity, but that less than 1% of the African rainforest can be so classified (Garwood *et al.* 1979). Neither New Guinea nor Panama are usually struck by cyclones; it would be interesting to compare the structure of these rainforests with the vegetation cover resulting from storm damage. One would expect different disturbance mechanisms to supply gaps of different size and distribution. Landslide locations, for example, may be partially determined by lithology; in both Papua New Guinea and Malaysia, slides are more frequent on granite than on sedimentary rocks. This review suggests that 'cyclone vegetation' is an open, or rather coarse, mosaic. By comparison, the forest canopy is less disturbed in the great forests of Sumatra, Malaya, Borneo and the southern Philippines. Here canopy disruption is confined to the gaps produced by the death of an individual or small group of trees and replacement is by shade-tolerant species. Extensive storm damage is rare yet preserved in the forest structure; thus a cyclone in 1880 in Kelantan, Malaysia, is still recognisable in the canopy (Wyatt-Smith 1954, Whitmore 1975).

The arguments above show that environmental variability is reflected in vegetation structure but that it is difficult to relate floristic diversity to disturbance regime clearly on the basis of the fragmentary observations currently available. Proper investigations will require experimental designs in, and regional comparisons between, the world's rainforests. Fortunately, however, studies of environmental fluctuations on coral reefs are more useful in that they often relate to specific castastrophic events.

2.3 Coral reef response to environmental stress

The hermatypic, or reef-building, corals are characterised by the presence of symbiotic, unicellular dinoflagellates, or zooanthellae, in their endodermic layer. As well as being important in photosynthesis and nutrient cycling, the algae play a key role in promoting calcification and skeletal construction. However, corals are not simply autotrophic but also derive energy from the ingestion of reef zooplankton. Many corals are successful carnivorous suspension feeders with specialised food capture mechanisms. On any reef, therefore, coral morphology reflects the relative importance to each species of zooplankton and photosynthetic products; species that rely upon light tend to have a branching, often multilayered morphology and small polyps, whereas the active carnivores are massive and have large polyps to capture their mobile prey (Porter 1974, 1976, Johannes & Tepley 1974). The radial expansion of massive, hemispherical colonies averages 4–20 $mm\,yr^{-1}$. This growth is dwarfed by the rapid growth tip extension of the branching species which regularly reaches 100 $mm\,yr^{-1}$ and can exceed 200 $mm\,yr^{-1}$ (Buddemeier & Kinzie 1976). Thus the arborescent growth of some corals enables them to overtop slower-growing massive species. Overgrowth ability can be quantitatively assessed and shading hierarchies headed by dominant species can be identified. However, on most reefs rapid- and slow-growing colonies coexist and mechanisms must operate to interrupt the overtopping process and so prevent competitive exclusions. These mechanisms include the imposition of environmental gradients and the presence of interspecific aggression between corals (e.g. Lang 1973, Sheppard 1979, Wellington 1980). The combination of 'normal' environmental impact coupled with biological interaction gives a stable but dynamic environment. Repeated resurveys, over a five year period, of photographic transects on Curaçao have shown that physical and biological processes lead to a high degree of constancy in cover but with considerable spatial change in the reef components and the routing of sediment flows, as a result of new larval settlement, colony growth, coral head dislodgement and coral death (Bak & Luckhurst 1980). More radical restructuring of the reef community results from the imposition of more extreme environmental conditions.

Coral reefs may experience a variety of irregular and abnormal stresses. Coral morphologies and distributions are usually adjusted to expected levels of predation and bioerosion but there may be occasions when this balance is upset by population explosions of particular predators. The most well known of these in recent years has been the crown-of-thorns starfish, *Acanthaster planci*, and its infestation of Indo-Pacific reefs (e.g. Endean 1973). Similarly, Caribbean reefs appear to be suffering from a population explosion of the sea urchin, *Diadema antillarum* (e.g. Sammarco 1980). However, this chapter is more concerned with abnormal events in the physical environment. These can be broadly divided into tidal effects, radical water temperature changes and the occurrence of tropical storms.

The infrequent and unpredictable emergence of reefs, when low spring tides coincide with anomalous falls in regional sea level, can result in the extensive mortality of corals and reef organisms (e.g. Fremantle: Hodgkin 1959; Guam: Yamaguchi 1975). Thus low tides on the Eilat reef flat, Red Sea, in 1970 caused 80–90% mortality of hermatypic corals (Loya 1976). The danger of emergence is also often compounded by high midday air temperatures (e.g. Puerto Rico: Glynn 1968); corals become stressed above 28°C and water temperatures above 36°C are lethal. At the other extreme, low temperatures result in corals losing their ability to capture food and then their ability to reproduce. At the margins of the reef seas in the northern hemisphere, winter outbreaks of polar air lead to low air temperatures and the rapid chilling of extensive areas of shallow bay and carbonate bank waters below the critical threshold of 16°C for up to several days. Such events, and their effects, have been described from the Florida–Bahamas reef tract (Roberts *et al.* 1982, Walker *et al.* 1982, Davis 1982) and the Persian Gulf (Shinn 1976). In the Dry Tortugas in January 1977, one such event led to 90% mortality in *Acropora cervicornis* to a depth of 15 m. Such extreme events clearly determine patterns of reef growth at high latitudes; this is shown by the repeated presence of temperature-induced stress bands in the sections through individual coral heads (e.g. Hudson 1981).

However, neither predation nor coral death by emergence or temperature stress alters the physical arrangement of the reef. It is important, therefore, to consider the mechanical destruction of coral communities by storm activity. Besides these direct effects, tropical hurricanes and cyclones also affect coral health by increasing turbidity and reef sedimentation (Goreau 1964) and, through cyclone-associated heavy rainfall, lowering reef salinities (Goodbody 1961, Goreau 1964).

The effects of tropical cyclones and hurricanes on Indo-Pacific reefs have been reviewed by Stoddart (1971). His summary can be supplemented by more recent observations on Guam (Randall & Eldredge 1977), Fiji (Cooper 1966), Hawaii (Dollar 1982) and Funafuti Atoll, Tuvalu (Maragos *et al.* 1973). In the Atlantic reef province, hurricane damage has been described from Florida (Ball *et al.* 1967, Perkins & Enos 1968), the Bahamas (Perkins & Enos 1968) and Puerto Rico (Glynn *et al.* 1964). Of particular value have been those studies which have been able to compare post-storm reef structure with earlier baseline suveys. The most comprehensive study, including the estimation of recovery rates, has been that of Stoddart (1963, 1965, 1969b, 1971, 1974) on the reefs and cays of the Belize barrier reef in the aftermath of the severe hurricane Hattie in 1961. Recently, studies on the impact of hurricane Allen (1980) on reefs on the north coast of Jamaica (Woodley *et al.* 1981, Knowlton *et al.* 1981, Porter *et al.* 1981) and of hurricanes David and Frederic (1979) on the shallow reefs of St Croix, US Virgin Islands (Rogers *et al.* 1982) have specifically considered the effects of storm events on reef diversity and structure where a considerable inventory of ecological data had already been assembled.

Several general conclusions can be made from this range of studies. First, minor storms appear to have only short-term effects on reefs, whereas major storms need several decades for subsequent reef recovery (Stoddart 1977). Secondly, the extent and pattern of storm damage may be controlled by regional bathymetry and the interaction between this and the direction of storm approach. The simplest pattern was shown by hurricane Hattie. As this severe storm approached almost normal to the barrier reef, coral destruction was banded into zones of damage away from the storm track (Stoddart 1971). Normally, the pattern is more complex. This can be shown by comparing the effects of two hurricanes of similar intensity in the Florida Keys–Bahamas region, hurricane Donna (1960) and hurricane Betsy (1965). Hurricane Donna, coming over Florida Bay and reinforcing a high spring tide, produced a 4 m storm surge and considerable, if short-lived, supratidal deposition. By comparison, hurricane Betsy, coming from the east and lacking any high-tide synchroneity, generated a smaller surge and caused little supratidal deposition (Perkins & Enos 1968). Thirdly, one can identify areas more susceptible to damage than others *within* reefs. This can be illustrated by the impact of hurricane Allen on the reefs of Discovery Bay, north coast of Jamaica. This particular storm generated winds of 285 $km\,h^{-1}$ at its centre and approximately 110 $km\,h^{-1}$ at Discovery Bay: 12 m waves, producing bottom velocities of 5.4 $m\,s^{-1}$ in *c.* 15 m of water, levelled 1–3 m high colonies of *Acropora palmata* in 0–5 m of water and led to physical disturbance down to 50 m. At depths of less than 10 m, coral cover was reduced from 51–54% to 10–12% (Woodley *et al.* 1981). Spatial patterns of damage were broadly controlled by aspect, with east- and north-east-facing reefs faring worse that west-facing reefs, and by slope, with sloping or level surfaces suffering more than vertical faces. Locally, damage was worse in sand channels, and near sand patches, than on reef spurs. At an even smaller scale, large individual coral heads either produced local shelter effects or, if detached, trails of damage. Shallow fore-reef areas were also more severely damaged than deep reefs. The general depth of hurricane attack can be deduced from the composition of hurricane deposits on reef cays and islands. These often take the form of coral rubble ridges and shingle ramparts (e.g. Funafuti Atoll: Baines *et al.* 1974, Baines & McLean 1976). Hurricane forces are theoretically capable of entraining and transporting cobble-sized fragments (Hernandez-Avila *et al.* 1977) and the relatively unabraded nature of the material suggests a rapid transit time from reef to shore. The species composition of coral rubble on Grand Cayman Island, Caribbean, suggests that reefs are particularly pruned and thinned at depths of 7–10 m (Rigby & Roberts 1976).

What does this damage mean in terms of the composition of the reef? This is a valid geomorphological question as coral types determine reef structure and reef structure influences physical processes. It has already been suggested in this chapter that, over time, slow-growing massive corals will be competitively excluded by the overgrowth strategies of the faster-growing branching corals. However, the periodic removal of the ramose species by storm activity may prevent this outcome (Wells 1957, Lang 1973) and high diversity may be maintained. This argument has been put forward for the Heron Island reefs, Great Barrier Reef (Connell 1973, 1978 & Fig 2.1b) and initially derived some support from Grigg and Maragos' (1974) study of coral colonisation of dated, submerged lava flows off Hawaii. In both cases exposed sites are characterised by low coral cover and relatively high diversity, whereas sheltered locations have high cover but are taxonomically poor, presumably because succession and competitive exclusion are less likely to have been interrupted (Fig. 2.1b). More recently, Dollar (1982) has attempted, more formally, to show that 'intermediate disturbances' do maintain species diversity on Hawaiian reefs; unfortunately the coral community is a rather simple one, with three species accounting for 97% of all coral cover.

In terms of skeletal architecture, the arborescent species are more susceptible to breakage; the slender branches, rapidly growing and diverging branch tips and relatively high-density aragonitic skeleton of *Acropora cervicornis*, for example, promote its breakdown into coral 'sticks' (Folk & Robles 1964, Gilmore & Hall 1976, Tunnicliffe 1981). At Discovery Bay in 1980 mortality was greatest amongst the acroporid corals with a 99% reduction in living planar area (Woodley *et al.* 1981). Thus the usual competitive order was reversed. 'Before the storm, shallow water mortality was strongly influenced by biological controls of competition through overgrowth; during the storm, mortality was predominantly controlled by physical factors, with heaviest losses amongst the commonest, competitively superior (but with respect to the storm, morphologically inferior) species' (Porter *et al.* 1981 p. 250). So in the long term, this rebalancing should preserve reef diversity. However, and alternatively, it has also been argued that storms may reduce diversity by favouring the clonal expansion of common shallow-water species. If storm-dispersed clonal fragments are both large enough and coherent enough to project above the substratum, then these fragments may regenerate, initially at up to 0.05 cm day^{-1} (Rogers *et al.* 1982), and form the base of new communities, neatly circumventing the problem of larval settlement on shifting reef sediments. Thus fragmentation may confer evolutionary advantage, allowing rapid species dispersion and decreasing the likelihood of clonal extinction by disease or predation. Furthermore, competitors may be smothered by stacks of fractured acroporid branches (Highsmith 1982). However, such comments need to be qualified in two respects. First, the initial survival of fragments may be misleading. Over 98% of the *Acropora cervicornis* fragments which initially survived hurricane Allen at Discovery Bay were lost over the next five months, probably as a result of further abrasion, predation and disease (Knowlton *et al.* 1981). Reported survival rates for *Acropora palmata* appear much better, with 64% mortality after hurricane Greta (1978) and only 50% fragment loss seven months after hurricanes David and Frederic (1979) on the reefs of St Croix (Rogers *et al.* 1982). This is probably partly due to the ability of *A. palmata* branch tips to reorientate towards the light on toppled colonies but the major factor in improved survivorship is larger fragment size. Colony breakup is only advantageous where fracture produces a coral 'stick' above a critical size. Thus, for example, survival of the few, large fragments of *Acropora palmata* is much higher than that in the more finely fragmented *Pocillopora damicornis* (Highsmith *et al.* 1980).

In conclusion, storms may either 'open up' the reef structure, thereby maintaining reef diversity, or promote the spread of already dominant species and thus decrease community diversity in the long term. Furthermore, if diversity is a function of the frequency, timing and magnitude of catastrophic events (Fig. 2.1), then as these variables interact there will be no simple relationship between disturbance and taxonomic richness and the identification of high diversity-generating 'intermediate disturbances' (Connell 1978 p. 303) will not be easy. Indeed, it would be strange if Connell's parabolic relationship was generally recognisable in community structure; in no sense does it represent a workable schema for identifying disturbance regimes.

2.4 Variability in extreme events

Despite the reservations expressed above, biologists have clearly recognised the importance of extreme natural events and disturbance regimes and have shown some understanding of the interactions between disturbance magnitude, the interval between events and community structure. For these reasons, their research findings should be more widely studied by geomorphologists. However, there has been a tendency in the biological literature to regard catastrophic events as entirely unpredictable; thus the impact of high-magnitude episodes has been of interest whereas their spatial and temporal distribution has not.

In physical geography, studies on landform development have concentrated on changes *between* climates rather than changes *within* climates. Thus there has been a great deal of effort expended on characterising glacial and interglacial climates, particularly in reconstructing environmental conditions at the last glacial maximum and contrasting them with postglacial climates. By comparison, the number of studies on 'within climate' or 'historical' timescales has been small. Nevertheless it is clear from a variety of environments that, where geomorphic processes are indicators, perhaps only indirectly, of general atmospheric circulation, then there are often discernible temporal patterns of landform response (e.g. lake levels: Goudie 1983b; glacier fluctuations and associated mass movements: Grove 1966, 1972). In some cases, landform change may be dependent on the work achieved by large-magnitude events; they too can show distinctive fluctuations in frequency. It is, therefore, important for geomorphologists to view the environmental base as a fluctuating one and not to regard climate as a fixed backdrop subject only to major shifts of 'climatic change'.

The continuing debate over the duration of the Sahelian drought shows that it is not always easy to delimit particularly large environmental fluctuations and to decide whether they represent abnormal episodes or whether they indicate more secular shifts in climate.

One repetitive and recognisable event, the El Niño phenomenon, concerns the appearance of higher-than-normal sea surface temperatures (SSTs) in the central and eastern Pacific. The development of such a warm episode, which occurs at intervals of 2–10 years, coincides with a major change in pressure patterns known as the Southern Oscillation (Quinn 1974) and has been correlated with droughts over India, severe winter weather over North America (Philander 1983a) and coastal erosion in Australia (Grant 1981, Bryant 1983). Normally, the SE Trades drive the waters of the Pacific westwards and induce the upwelling cold, nutrient-rich water near the coast. However, this circulation is interrupted when the warm El Niño current flows polewards along the coasts of Ecuador and Peru from January to March (Wyrtki *et al.* 1976). In certain years – and there have been nine episodes since 1950 – temperatures are exceptionally high during the warm season and persist into the normal cold upwelling season. At this time, temperature anomalies expand westward to affect the entire tropical Pacific Ocean. The 1982–3 episode, however, was anomalous on account of its initial appearance in the western Pacific Ocean and its subsequent eastward expansion, its exceptionally large amplitude and its long persistence (Philander 1983b). The areas of upwelling are associated with zones of aridity and the low islands of the equatorial Pacific are usually dry and treeless. In August 1982, however, the usual westward equatorial surface flow reversed (Halpern *et al.* 1983) and an enormous flux of heat from the warm western Pacific to the cool eastern Pacific raised sea surface temperatures by 5°C above the 20-year monthly mean, deepened the thermocline and raised sea level by 40 cm. Furthermore, in the equatorial Pacific, the usual eastward undercurrent weakened and temporarily disappeared (Firing *et al.* 1983). The 1982–3 Niño event, therefore, produced massive departures in the precipitation regimes of Kiribati and the Line Islands and, further east, 3225 mm of rainfall fell on Santa Cruz Island between December 1982 and October 1983 against a long-term mean of 374 mm (1965–81) (Laurie 1983). The tropical storm genesis region of the south Pacific shifted to the east of its normal location and French Polynesia was subjected to the unusual occurrence of five cyclones between December 1982 and April 1983. In contrast to this heavy rainfall, serious drought conditions developed in a zone from the Philippines to the Hawaiian Islands, and Australia and Indonesia experienced severe, and in places record, aridity. El Niño episodes clearly have important geomorphological and ecological consequences. In particular, many of the arid equatorial islands of the Pacific and those near the Peruvian coast are sites of guano deposition and phosphate rock. Cold, nutrient-rich, upwelling water supports high fish stocks and large seabird populations; island aridity suppresses vegetation growth and allows for the nesting of ground-dwelling boobies, terns and frigates; and lack of rainfall also prevents the post-depositional removal of avian guano (Stoddart & Scoffin 1983). If this sequence is cut off at source by the arrival of warm ocean water then the results can be ecologically catastrophic: thus during the 1982–3 El Niño event the bird population of Christmas Island, estimated at 17 million, disappeared (Rasmussen & Hall 1983). Furthermore, if the distribution of ocean currents is ultimately expressed in the presence or absence of guano accumulations, then the phosphate rock stratigraphy of the islands of the equatorial Pacific might be used to reconstruct Pleistocene climates (Stoddart 1976).

The ocean–atmosphere interactions of the El Niño–Southern Oscillation phenomenon occur on an almost global scale. However, as Walsh (1977) has pointed out, this scale leads to problems of historical reconstruction, particularly as large areas of ocean are involved. More amenable to analysis are smaller regions where the environmental record is more complete and can be used to study changing spatial and temporal incidence of landscape-forming events. The final section of this chapter reviews the variation in hurricane and cyclone activity in the North Atlantic cyclone basin, providing an extension to the arguments on diversity and disturbance already discussed above.

2.5 The temporal and spatial distribution of hurricanes and tropical cyclones in the Caribbean basin

It is necessary, initially, to outline the present spatial and temporal distribution of storm activity in the Caribbean. Although hurricanes and tropical cyclones have developed in the North Atlantic in all months except January and April, over 90% of storms occur in the period from June to October, with the more severe storms being clustered at the end of the season (Gray 1968, Cry 1967). There is also a change in the locus of storm development as the season progresses. The classic storm track can be traced to a source near the Cape Verde Islands in the eastern Atlantic, but this is only characteristic of the period of high cyclone frequency from mid-August to mid-September. Prior to this period, hurricanes form in the western Caribbean and the Gulf of Mexico; after mid-September, activity also contracts back to this area.

Superimposed on these usual seasonal fluctuations, one finds marked variations in the temporal frequency and spatial incidence over the longer timescale of the documented record. A plot of a running 10-year total (Figure 2.2) shows that there was a decline in storm frequency in the early 20th century, culminating in the 1920s, and then a sudden upsurge in activity in the late 1920s/early 1930s before a further decline in the 1960s/early 1970s. It seems highly unlikely that advances in storm detection can account for these substantial fluctuations. Changes in the North Atlantic have lagged behind fluctuations in cyclone frequency in the Bay of Bengal but have preceded variations in the cyclone basins of the southern hemisphere (Fig. 2.3) (Milton 1974).

As cyclones are relatively localised in position, any change in track position will have important geomorphological consequences. Figure 2.4 shows the frequency of tropical hurricanes and cyclones for the 110-year period from 1971 to 1980, mapped as decadal totals for $5^\circ \times 5^\circ$ squares between 5 and 30°N and 45 and 100°W. Ideally, storm intensity should also be taken into account, but this is difficult to discover and define. While the long-wavelength variations in the temporal frequency of cyclones are reflected in Figure 2.4, it also shows changes in the positioning of the storm tracks. Thus, for instance, the area of maximum storm incidence for 1901–10 was south of 20°N, whereas between 1951 and 1960 it was north of 25°N. Walsh (1977 & Ch. 5), on the basis of

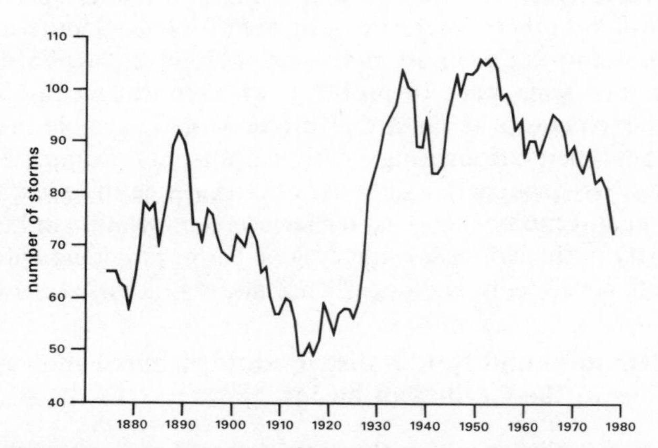

Figure 2.2 Decadal running total for the frequency of Atlantic hurricanes and tropical cyclones, 1875–1978 (Source: Neumann *et al.* 1978; USDC North Atlantic Hurricane Tracking Chart 1978—.)

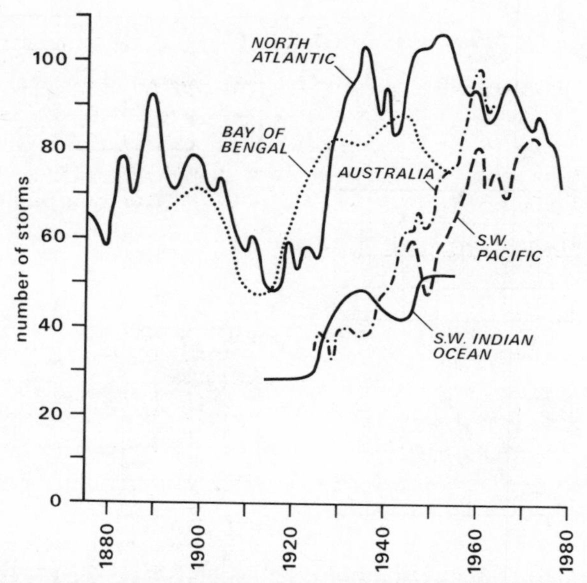

Figure 2.3 Cyclone frequencies, Bay of Bengal, North Atlantic, SW Pacific, Australian and SW Indian Ocean regions, 1880–1980 (Partly after Milton 1974).

a comprehensive analysis of cyclone tracks since 1650 in the Lesser Antilles chain, has suggested that declining cyclone frequency is associated with the equatorward displacement of storm tracks. Thus at times of high frequency, storm routes lie over the Leeward Islands, at medium frequency over the Windward Islands and Barbados at low frequency over Trinidad and Tobago. Figure 2.4, covering the whole Caribbean basin, extends this picture by showing that the change from high to low incidence takes place first through contraction to the west and north-west and then subsequently by a southerly migration of the peak of activity.

There are two major factors responsible for the variation in cyclone incidence. The first of these is water temperature. Cyclones only develop where sea surface temperatures are in excess of 26.5°C and it is well established that cyclones are less frequent at times of lower sea surface temperature. Several authors have detected sea surface temperature minima between 1900 and 1925, and particularly between 1910 and 1920, and a sharp rise in temperature in the early 1930s (Fieux & Stommel 1975, Pattridge & Woodruff 1981). These changes would neatly explain the observed changes in storm frequency. Similarly the low storm frequencies of the 1950s and 1960s corresponded to an average decline in sea surface temperature of nearly 1°C in the North Atlantic (Perry 1974, Wahl & Bryson 1975, Colebrook 1976). The second factor controlling cyclone incidence is the position of the equatorial trough and the subtropical high-pressure belt. This is reflected in the spatial patterning of cyclone tracks. As the majority of the cyclones develop just to the north of the equatorial trough, track position is thus a measure of the southerly

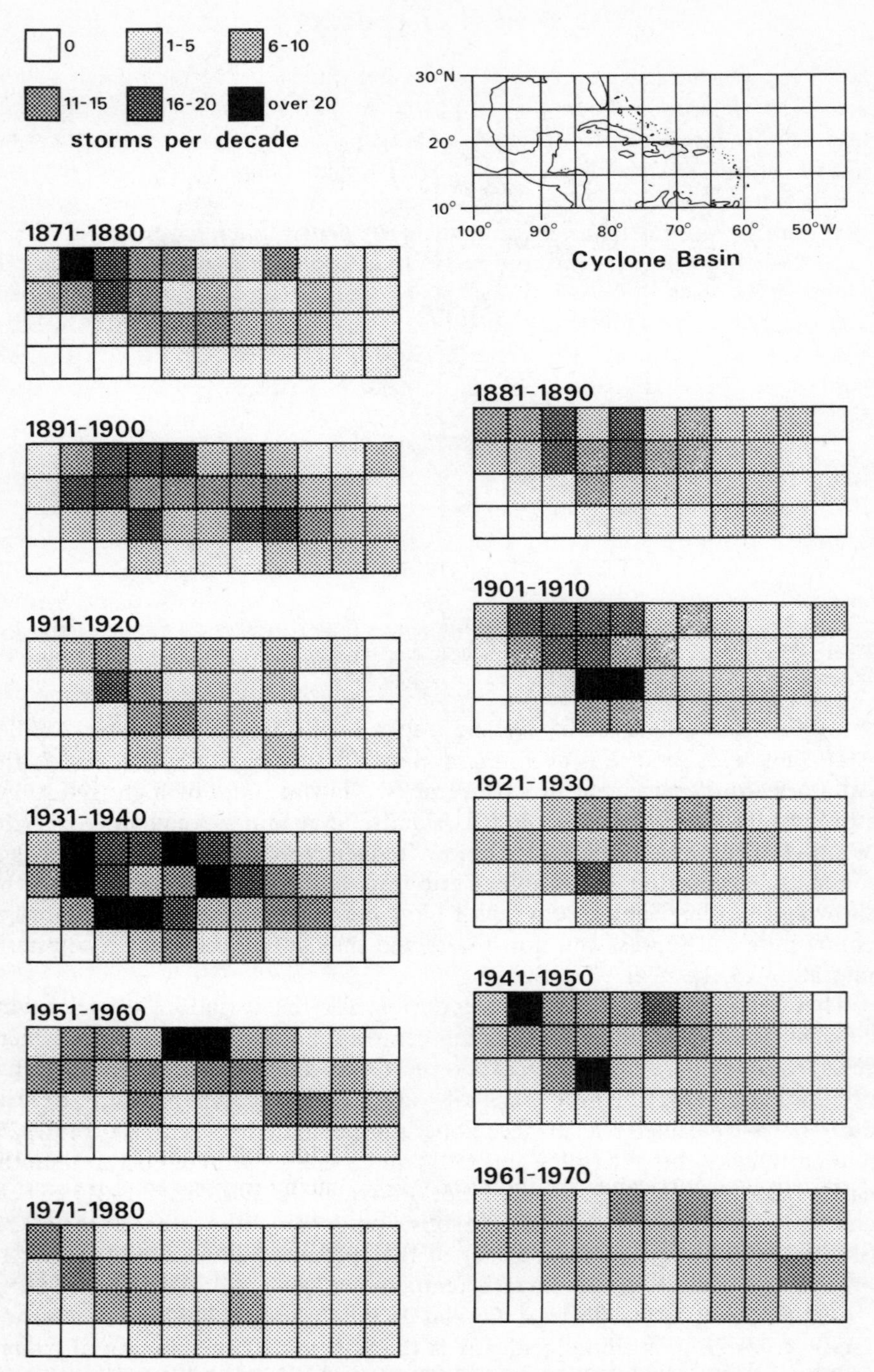

Figure 2.4 Changes in the spatial distribution of hurricanes and tropical cyclones in the North Atlantic cyclone basin by decade, 1871–1980 (After Neumann *et al.* 1978; USDC North Atlantic Hurricane Tracking Chart 1978—.)

shift, or alternatively the lack of northward extension, of the trough (Walsh 1977). As a Figure 2.4 effectively plots storm paths rather than simply storm numbers, the westward contraction of storm incidence, prior to the southerly shift in storm activity, suggests that more storms begin to 'recurve' north or north-east at this time. Thus low storm frequencies may also be associated with an increase in the meridionality of the general circulation.

As cyclone frequency appears to be so closely correlated with sea surface temperature, then it should be possible to estimate palaeostorm activity from historical data and from palaeoclimatic reconstructions. It is more difficult to establish past patterns of spatial incidence; one can only assume that the contemporary patterns for high and low storm frequency also applied in the past. Wendland (1977) has estimated that the annual mean frequency of cyclones in the late 18th/early 19th century was about five per year compared to the recent (1962–71) frquency of nine per year. Walsh's (1977) data set from the Lesser Antilles show that cyclone frequencies were even lower in the period from 1650 to 1764, which coincided with the deterioration in climate in Europe known as the 'Little Ice Age' (Lamb & Johnson 1961, Lamb, 1969). Sanchez and Kutzbach (1974) have sought to draw a direct parallel between the climate of the American subtropics since 1960 and the 'Little Ice Age', and the low cyclone frequencies of the 1960s and 1970s support their argument.

Over a longer timescale it would seem plausible that cyclone frequencies were similar to or greater than at present during the warmer phases of the Holocene (e.g. Neo-Atlantic: 1000–760 years BP; Sub-Atlantic: 2890–1690 BP; Atlantic 4680–2690 BP; Central Great Plains Chronology of Hoffman & Jones 1970) and the interglacials of the Pleistocene. Frequencies of 'glacial' cyclones can be estimated from the deep-sea core record. Oxygen isotope analysis on core material from the equatorial Atlantic (Imbrie & Broecker 1970), the Caribbean Sea and Gulf of Mexico (Dansgaard & Tauber 1969, CLIMAP Project Members 1976, Prell & Hays 1976, Brunner 1982) and the Tongue of the Ocean, Bahamas (Lynts *et al.* 1973) has shown that glacial sea surface temperatures were 2–6°C cooler that at present. Wendland (1977) has calculated that a cooling of 6°C would reduce tropical cyclone frequencies to zero in even the warmest months of the year and that the minimum cooling of 2°C would result in a drop in mean cyclone activity from 2.9 to 0.7 events in the warmest month.

Autumnal peaks in rainfall at Key West, Florida (1906–58) have been used to argue that hurricanes are important in balancing the seasonal distribution of water vapour in the northern hemisphere (Landsberg 1960). As a corollary to this, Moran (1975), from rainfall records, and Coleman (1980), from longer climatic records and river discharge data, have suggested that the 1960s/early 1970s drought of peninsular Florida was due to the decreased frequency of cyclone passage, accentuated by the presence of limestone bedrock and cover sands. If these recent drought conditions are a guide to the effects of decreased storm activity, then the Caribbean basin should have been a more arid region during glacial periods. This notion is supported by core mineralogy and sedimentology: increased influx of terrigenous material during the last glaciation (Prell 1978), the absence of feldspars in glacial sediments (Damuth & Fairbridge, 1970), and low kaolinite-to-quartz ratios (Bonatti & Gartner, 1973) and the composition of aeolian detritus in glacial sediments (Parmenter & Folger 1974). Palaeosalinities in the Gulf of Mexico during glacial summers were also raised (Brunner 1982). Increased dryness under glacial conditions is indicated by lake deposits in Venezuela (Bradbury *et al*. 1981, Sarmiento & Kirby 1962) and by pollen analysis in central Florida (Watts 1975) and the Carolinas (Watts 1980).

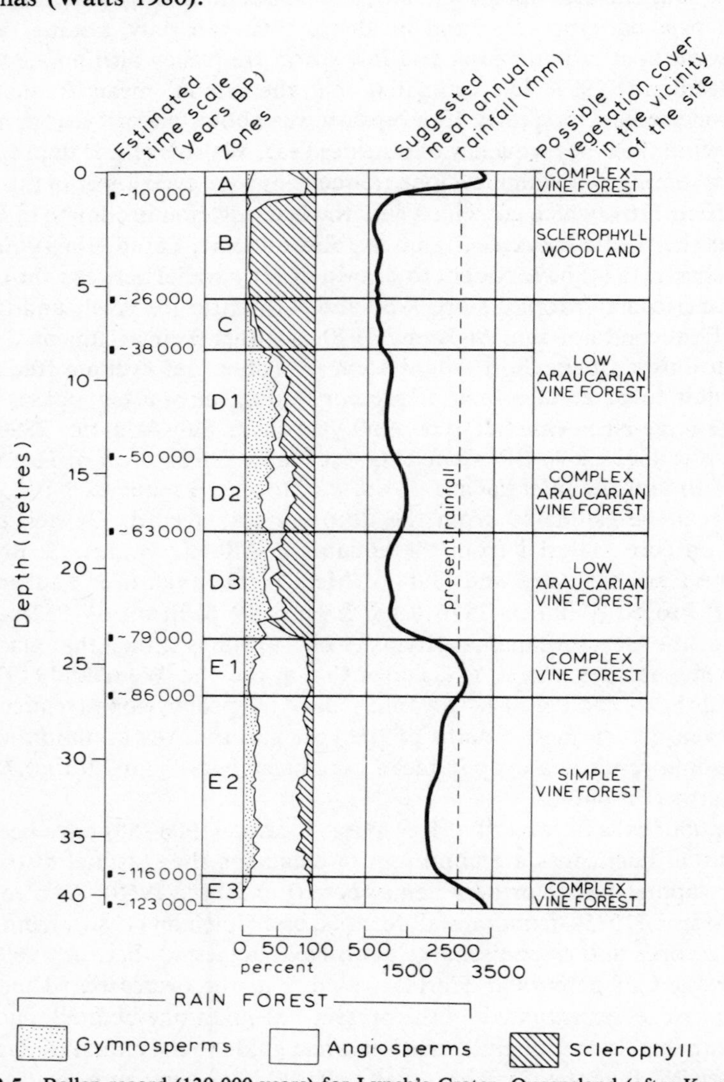

Figure 2.5 Pollen record (130 000 years) for Lynch's Crater, Queensland (after Kershaw 1976, 1980).

Reconstructions of this kind for Australasia have stressed the additional factor of the lowering of sea level and the exposure of shallow shelf areas (e.g. Webster & Streten 1978). Cyclones in this region are important contributors of precipitation and may be responsible for increased coastal erosion and river flooding in recent years (Grant 1981). Storms affecting Queensland either originate in the Coral Sea and move south into the Tasman Sea or form in the Gulf of Carpentaria and move across Cape York to re-form in the Coral Sea; in both these areas decadal frequencies are of the order of 15 to 20 events (Lourenz 1977). These shelf areas would have been dry plains during glacial episodes (e.g. Coventry & Hopley 1980). Allied with lower sea surface temperatures (-4 to -6°C: Bowler *et al.* 1976), rainfall totals must have been considerably reduced during full glacial conditions.

The vegetational response can be determined from Kershaw's (1976) remarkable pollen record from the Atherton Tableland (Fig. 2.5). Lynch's Crater shows the replacement of a species-rich vine forest by, first, a low forest dominated by *Araucaria* spp. and then a sclerophyll woodland characterised by *Eucalyptus* and *Casuarina*. There is no doubt that vegetation change and climatic change can be linked at this site, although human influences also need to be dissected out from the sequence. Northern Australia is rather marginal for rainforest, and it would be wrong to suggest that similar changes affected the equatorial rainforests further north. Nevertheless, extensions of the disturbance hypothesis to Pleistocene histories offer a challenging exercise which would seem worth pursuing.

Part B
ENVIRONMENT AND PROCESS

Process and time

Editorial comment

A fundamental notion about environmental change is the unequal distribution of change through time. Geomorphic processes work at varying intensity. Some, such as those associated with chemical reaction and transportation process, function virtually continuously, while others, such as the movement of coarse debris, occur rarely, being associated with irregular, high-energy events. Much of what is known about geomorphic processes under tropical climates is based on short periods of observation, often of a few years' duration or less. Such short-term event sequences may provide a misleading impression of the long-term variability of processes, particularly where thresholds are important (Church 1980). Indeed, most of the process information collected in the tropics so far is for periods too short to indicate the pattern of landform evolution. However, enough data have been collected from diverse tropical environments to indicate that processes under rainforest are far from uniform and that the inferences about what might have occurred under changed ecological conditions in the past must take account of both the tectonic situation and the character of the tropical climate involved, especially of rain-producing mechanisms.

The diversity of process is illustrated by the information on rainforest dynamics and erosion rates presented by Douglas and Spencer in Chapter 3, which shows sediment yields three orders of magnitude greater in rivers in geologically young, tectonically active areas than in those of old land surfaces in Africa and Australia. Human activity is a further major influence on variations in erosion rates as Avenard and Michel note with reference to the Sahel drought in the Senegal valley in Chapter 4. All these processes operate on pre-existing landforms, either perpetuating or truncating the evolution of such features. As Avenard and Michel show, in West Africa processes act in a framework of renewed erosion from the edges of granitic domes, inselbergs and relict duricrust-protected surfaces. Processes thus vary in intensity in both time and space, a point elaborated recently by Bruijnzeel (1983) with reference to the applicability of the variable-source-area concept to a small forested Javanese catchment. Direct runoff is only contributed by 1 to 8% of the catchment area, but the actual areas and water pathways involved vary with storm magnitude and antecedent conditions. If climate affects streamflow in this way, it will also affect stream length and drainage density, a question explored by Walsh in Chapter 5. The range of ages of volcanic surfaces in the Windward Islands of the West Indies, from a mid-Pleistocene end to volcanic activity in Grenada to major eruptions as recently as 1978 on St Vincent, provides an excellent opportunity to see how drainage density varies with land surface age and climate. However, other factors intervene; Walsh finds that although in the older volcanic areas drainage densities increase from 3 to 4 km km^{-2} at rainfalls of 1000 mm yr^{-1} to 10 km km^{-2} at a rainfall of 7000 mm yr^{-1}, in the younger volcanic areas drainage density and the rate of drainage network development appear to vary strikingly with lithology. Thus variations both in the energy and quantity of erosive agents and in the resistance of earth surface materials influence the rate and nature of tropical landform evolution. No particular single timescale of process operation can be suggested for more than a single lithological unit.

Some combinations of lithologies offer excellent opportunities for the study of changes in geormorphic processes through time. Such an opportunity, provided by limestones and shale of the Mulu area of Sarawak, is seized by Smart and his colleagues in Chapter 6. The levels of caves in the limestones are compared with terraces along adjacent rivers draining from the nearby shales and

Table B.1 Approximate correlation of timescales and climate-related events discussed in this book.

Years BP	West Africa (Avenard & Michel Ch. 4, Thomas & Thorp Ch. 12, Smith 1982, Giresse 1978)	Lake-level stages (Street-Perrott *et al.* Ch. 8)	Terminology used by Tricart (Ch. 10)	Terminology used by Williams (Ch. 11)	Events in north Queensland (Douglas & Spencer Ch. 3, Kershaw 1976, 1978)	African mountain glaciations (Hurni 1982)	Isotope stages and high sea levels (Shackleton & Opdyke 1976, Bloom *et al.* 1974)
0		A; more arid		drier in India		more arid	
5000	Nouakchottian	humid; B2			more humid	climatic optimum	
		arid; B1; humid		interglacial-type climate;			
10 000	active erosion; Holocene transgression	C; transition	Flandrian transgression	more humid in India		erosion phase in Simen Mts.	
							isotope stage 1
15 000	Ogolian; dry period	D; arid	Pre-Flandrian regression		dry	last cold period in Simen Mts. Ethiopia	
20 000		E		drier and windier in Africa, Australia and India; equivalent of full glacial climate (former supposed Gamblian pluvial)	low sea level 20 000 BP		
25 000	Post-Inchirian regression		Grimaldian regression				
30 000							-40 m high sea level
							isotope stage 2
35 000	Inchirian maximum						

sandstones. Periodic changes in fluvial activity both in the caves and on adjacent alluvial plains indicate shifts in rainfall regimes as well as the influence of changing sea level. Below the gorge where the Melinau River cuts through the limestone, a huge alluvial fan is still being constructed by the river. Fan development is thought to be an interglacial process, while in more arid periods the fan becomes dissected, partially in response to lower sea levels increasing river gradients and partially due to climatic change.

Such changes in fluvial processes are akin to those in the Senegal in West Africa (Ch. 4), in the Gulf of Carpentaria area of Australia (Ch. 3), in the Amazon (Ch. 10) and in Sierra Leone (Ch. 12). However, despite their similarity, such changes have not necessarily been synchronous. Establishment of the relative and absolute timescales of process change involves the evaluation of the varied lines of evidence discussed in Part C of this book. Local relative scales are often used in the book and correlation between them is not easy. Table B.1 attempts a very approximate comparison. Despite the variations in the quantity and quality of the evidence, major changes in geomorphic process regimes have occurred virtually everywhere in the tropics; however, the challenge of disentangling the oscillations of process intensity of differing magnitude remains. Much of the evidence of the nature of processes still relies on inference from sediments and the need for more prolonged observations of the dynamics of tropical environments, of the type inspired by the International Biological Programme and the International Hydrological Decade, cannot be overemphasised. More detailed analysis and supplementation of the data now being collected by national agencies in the tropics would be one way of making relatively rapid progress.

3

Present-day processes as a key to the effects of environmental change

I. Douglas & T. Spencer

3.1 Introduction

Ultimately, the landforms of any area can be explained as the outcome of the interplay of tectonic events and the influence of changing climates. In the tropics, climate is often viewed as the distinguishing characteristic that makes landforms differ from those outside the tropics. However, although the nature of tropical climates is vital for the nature of geomorphic processes in the tropics, the results of those processes can only be explained by a consideration of the tectonic style of the areas in which they operate. The landforms of the subduction zones at the Pacific Ocean margins where Pacific, Cocos, Nazca, Bismark and Solomons Plates are passing under adjacent continental plates (Fig. 1.2) provide a tectonically active, rugged, feral relief (Cotton 1958) quite distinct from that of the old, virtually level surfaces cutting across the fragments of Gondwanaland in South America, Africa, southern India and Australia. As elsewhere, landforms in the tropics have the diversity that stems from unequal dominance and persistence of legacies from the past and rates of present-day endogenetic and exogenetic processes.

While the map of plate tectonic structure provides an indication of likely areas of tectonic activity, the geomorphic significance of climate is more difficult to differentiate. In terms of energy to detach particles, measures of rainfall erosivity have been used, but results from different tropical areas (Fig. 3.1) may not be strictly comparable. Even though the generalisation inherent in Fournier's (1960) earlier adoption of p^2/P (the square of the maximum monthly mean rainfall divided by the mean annual rainfall) is confirmed, the areas with seasonal monsoon rainfalls tend to have the highest erosion potential. However, as the previous chapter has shown, the magnitude and frequency of geomorphic processes are variable depending on rainfall regimes and the role of major rare events, such as tropical cyclones. These crucial events have to be related to tectonic style.

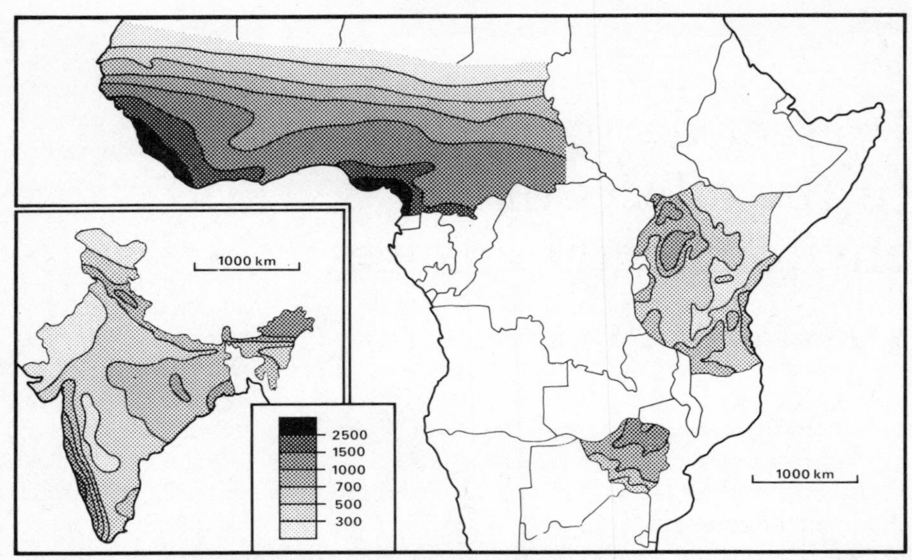

Figure 3.1 Rainfall erosivity for Africa and south Asia (based in part on Jansson, M.B. 1982, UNGI Rapport 57).

One of the problems that has arisen in the evaluation of tropical landforms is the assumption, inherent among geographers from the days of Herbertson's (1915) major natural regions, that climate and vegetation are so closely related that one is a surrogate for the other. Vegetation, however, has a history. Even though climate may change, vegetation will adapt more slowly to changed conditions. In many areas of the tropics the vegetation may well, where not disturbed by human activity, reflect a legacy from a past set of conditions. Vegetation also varies with substrate, as well as locally through the occurrence of random disturbance as described in the previous chapter. Climate and vegetation thus have to be distinguished.

The view of climate in the tropics must be in terms of water availability, especially the seasonal rhythms and extreme events produced by rainfall. The sort of classification suggested by Jackson (1977) is probably more meaningful than the wider classifications of Herbertson (1915), Büdel (1977) or Troll and Paffen (1964) (Fig. 3.2). However, all of these tend to neglect relief effects and are unable to cope with climatic variations from year to year. The variation of precipitation in the humid tropics is well known. Chia (1968) quotes a weighted average rainfall for the state of Selangor, Malaysia, for the period 1956–65 of 2345 mm, with a standard deviation of 180.59 mm and a coefficient of variation of 5%. Analysis of rainfall records for north-east Queensland using Maurer's (1928) variability index shows values of over 300 north-west of Cairns compared with values of less than 180 just over 70 km to the south (Douglas 1973). This difference is due in part to orographic effects and also to the greater importance of cyclonic rains in the northern area and the dominant role of southeasterlies further south.

Detailed analysis of the variation of rain from one minute to the next is possible when suitable recording rain gauges are available. Such analyses show that the maximum rainfall intensity for a 1 min duration in Bangkok would be 160 mm h^{-1} for a once in two-year return period and 280 mm h^{-1} for a once in 50-year period. The corresponding maximum 5 min duration intensities would be 130 and 255 mm h^{-1}, respectively (Pinkayan & Ketratanaborvorn 1975).

When a reasonably dense recording rain gauge network is available, great spatial variations are apparent as, for example, in the Sungai Kelang catchment, Malaysia, where two recording gauges less than 15 km apart show quite different patterns of rainfall on 30 August 1967. At the University of Malaya, fairly steady rain persisted for 3.5 h, with slightly higher-intensity rain at the beginning of the storm, while at the Jabatan Parit dan Taliayer (Drainage and Irrigation Department) Research Station to the north-east, an initial period of high-intensity rain gave way abruptly to drizzle after 30 min (Lim 1969). This contrast arises from the passing of a discrete storm cell close to the second station. These cells of intense activity in tropical storms are responsible for much of the local spatial and temporal variability of rainfall in low latitudes. Such within-a-storm variability leads to irregular wetting of the vegetation canopy and ground surface and to runoff from discrete areas of river catchments.

The localised storms are in great contrast to frontal systems and cyclones, which produce prolonged heavy rain over large areas. The systems that brought the December 1969 floods to Singapore (Chia & Chang 1971) and the January 1971 floods to most of Malaysia involved particular cyclonic vortices within general stationary frontal systems. Falls of over 300 mm in 48 h occurred on both these occasions. However, even in such large-scale systems, rainfall may be irregulary distributed, as the pattern of rainfall from the 24 November 1978 storm in Sri Lanka shows (Fig. 3.3).

Tropical cyclones may produce high rainfalls at coastal sites, but once they cross the coast they become widespread rain depressions, often producing the dominant rainfalls in savanna and seasonally wet tropical areas such as northern Australia. Such rains produce the runoff events that cause ephemeral channels obscuring the arid margins of tropical areas such as Coopers Creek in Australia to flow to their inland distributary and depositional zones.

River flow is more varied than rainfall, as at times there is abundant rainfall providing excess water for discharge to rivers, while during dry periods the water available to streamflow is limited by the evapotranspirative demands of the forest. The Layawan River in Mindanao, Philippines, an area affected by tropical cyclones, has an annual average runoff of 3054 mm with a standard deviation of 1099 mm and a coefficient of variation of 36%, annual runoff depths ranging from 2176 to 6180 mm. Maximum monthly discharge 1950–61 was 746.1 mm, minimum 108.6 mm, with coefficient of variation ranging from 54% in May to 25% in November. The more equatorial Sungai Gombak at Kuala Lumpur, Malaysia, shows a similar variation with an annual average

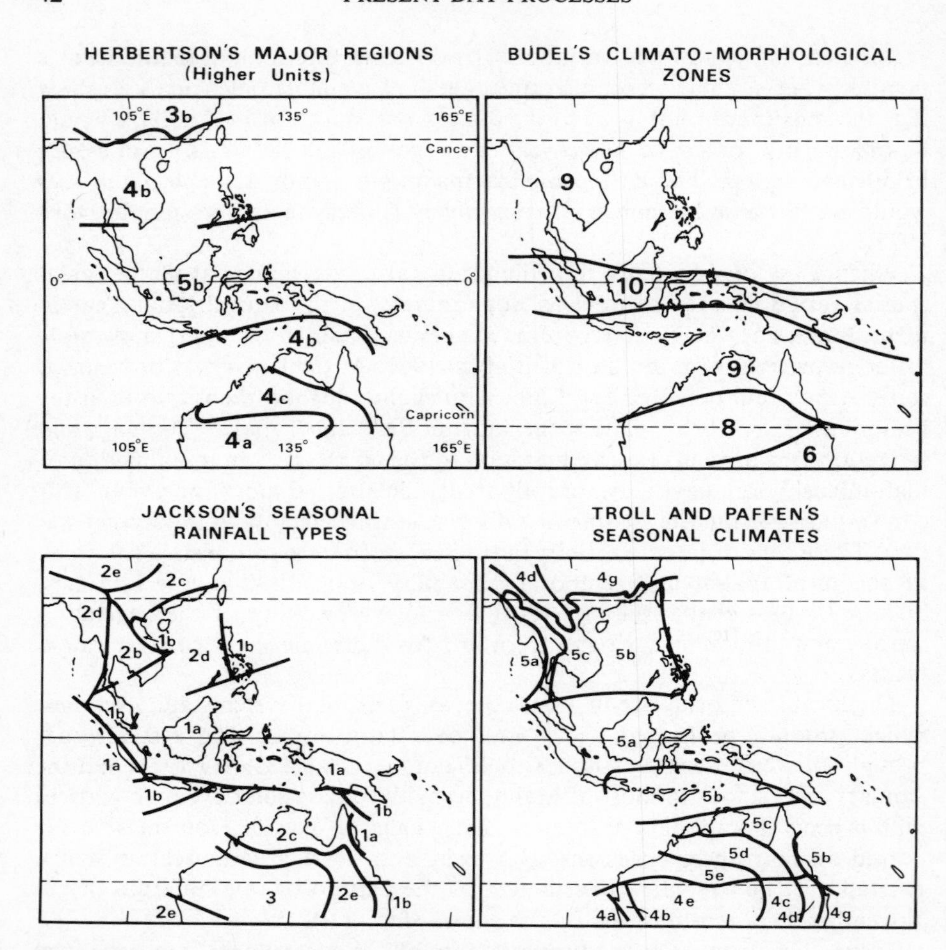

Figure 3.2 Four suggested classifications of south-east Asian climates.

Herbertson's (1915) regions:

3b Eastern margin or China type
4a Western desert or Sahara type
4b Monsoon summer rain type of eastern or southern margins
4c Summer rain type of interior or Sudan type
5b Wet equatorial mountainous islands or Malay type

Büdel's (1977) zones:

6 Subtropical zone of mixed relief development, monsoonal regime
8 Warm arid zone of surface preservation and traditionally continued development, largely through fluvio-aeolian sandplains
9 Peritropical zone of excessive planation
10 Inner tropical zone of partial planation

Jackson's (1977) seasonal rainfall types:

1 *Humid tropics*
1a While a seasonal variation may occur, in general terms there is no dry season. Annual totals usually > 2000 mm and all months at least 100 mm

1b No pronounced dry season, annual totals < 2000 mm but a drier period with a few months having < 100 mm. This is a transitional category between 'humid' and 'wet and dry' areas. It occurs on the poleward margins of 1a, but also on the windward (east) coasts of continents and islands where it may extend over a considerable width of latitude and is associated with the 'disturbed' Trade winds. Because advection of moisture onshore is involved, orographic effects often create marked local variations

2 *Wet and dry tropics.* Any division is fairly arbitrary, each type in most cases showing a gradual transition to an adjoining type

2a Annual totals 1000–2000 mm. Two rainy seasons with short dry seasons or months with lower rainfall, usually a few months < 50 mm

2b Annual totals 650–1500 mm. Two short rainy seasons separated by a pronounced dry season (a few months < 25 mm) and a short drier season

2c Annual totals 650–1500 mm. One fairly long rainy season (normally 3–5 months each with > 75 mm) and one long dry season

2d Annual totals > 1500 mm. One season of exceptionally heavy rain and one long dry season. Typically illustrated by some south-east Asian monsoon situations

2e Annual totals 250–650 mm. One short rainy season (3–4 months each with > 50 mm) and one long dry season

3 *Dry climates.* Annual totals < 250 mm, little rain at any time, but can be concentrated in a very short 'wet' season perhaps of only a few weeks.

Troll and Paffen's (1964) seasonal climatic types:

4a Dry-summer Mediterranean climates with humid winters
4b Dry-summer steppe climates with humid winters
4c Steppe climates with short summer humidity
4d Dry-winter climates with long summer humidity
4e Semi-desert and desert climates
4g Permanently humid climates with hot summers
5a Tropical rainy climates
5b Tropical humid-summer climates
5c Wet and dry tropical climates
5d Tropical dry climates
5e Tropical semi-desert and desert climates

---

runoff of 846.4 mm, standard deviation of 122.7 mm and coefficient of variation of 14% (Douglas 1971).

When the contrast between wet and dry months is examined, variation, even in an equatorial rainforest area like Selangor, Malaysia, is great. The monthly runoff in the Langat in 1966–8, for example, ranged from 34 mm in March 1968 to 355 mm in December 1966. Such a variation may indicate the extent of the depletion of the water stored in the forest ecosystem during dry periods. The forest is never uniformly wet. In Selangor the forest is sometimes dripping with water and droplets fall from every leaf. At other times, particularly in February and March, the forest may be dry, with dead leaves crackling underfoot and the streams shrinking to a mere portion of their normal width, exposing stretches of sand and boulders, and rising further down valley than in wet periods. When it is recalled that at flood times the streams overflow their channels, completely covering all riverine sands and inundating some of the adjacent vegetation, the reduction of river flow in dry periods is readily apparent.

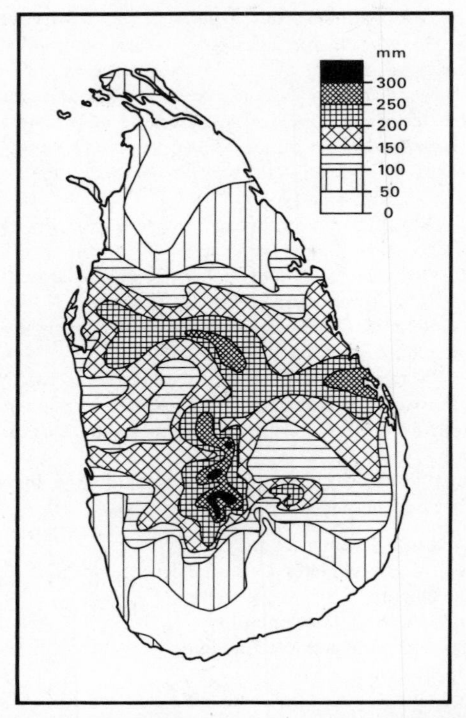

Figure 3.3 Pattern of 24 h rainfall over Sri Lanka, 23–24 November 1978 (after Van Lengkerke 1981).

Table 3.1 Seasonality of sediment discharge. Number of days per year required to transport 50% of annual suspended sediment load.

Site	Catchment area (km^2)		Source
experimental basin near Tiassale, Ivory Coast	0.02	0.5	Mathieu (1971)
Waikele at Waipahu, Oahu	118.4	2	Chinn *et al.* (1983)
Kamooalii Stream near Kaneohe, Oahu	9.87	2	Chinn *et al.* (1983)
experimental basin, Kountkozout, Niger	0.04	4	Vuillaume (1969)
Wild, north-east Queensland	585	5	Douglas (1976)
Amitioro, near Tiassale, Ivory Coast	170	~5	Mathieu (1971)
Wailuku at Hilo, Hawaii	663	5	Chinn *et al.* (1983)
Freshwater Creek, north-east Queensland	44	6	Douglas (1976)
Barron, north-east Queensland	225	11	Douglas (1976)
Millstream, north-east Queensland	92	15	Douglas (1976)
Pattani, Ban To, southern Thailand	888	18	Natl. En. Auth. (1965)
Gombak, Kuala Lumpur, Malaysia	140	24	Douglas (1969)

When the amounts of sediment carried by tropical rivers are examined, the importance of variations from day to day and year to year are readily apparent. In the Sungai Gombak, half the annual sediment load is carried in 24 days (Table 3.1), but further away from the equator, in monsoonal areas (e.g. Pattani River) or cyclone-affected areas (e.g. Queensland and Hawaii), half the load is carried in many fewer days. Analyses of the geomorphic work done in storm events would seem to be the key to understanding present-day geomorphic processes in the tropics. However, the heavy rains act through the vegetation on materials that have been altered by physical, chemical and biological processes which operate, with varying intensity, more continuously than the fall of precipitation. An understanding of the dynamics of tropical ecosystems is therefore important for the understanding of geomorphic processes.

3.2 Nutrient cycling in tropical forests

> In the rain forest the vegetation itself sets up processes tending to counteract soil impoverishment and under undisturbed conditions there is a closed cycle of plant nutrients. The soil beneath its natural cover thus reaches a state of equilibrium in which its impoverishment, if not actually arrested, proceeds extremely slowly. (Richards 1952)

Fresh plant nutrients are continually being set free by the decomposition of the parent rock. Provided the horizon in which they are set free is not too deep for the tree roots to reach, part of these nutrients is taken up by the vegetation in dilute solution. Some of these substances are fixed in the skeletal material of the plant – the cell walls – others remain dissolved in the cell sap. Eventually all of them are returned to the soil by the death and subsequent decomposition of the plant or its parts. The top layers of the soil are thus being continually enriched in plant nutrients derived ultimately from the deeper layers. The majority of the roots, including nearly all the 'feeding' roots, are in the upper layers of the soil. Most of the nutrients set free from the humus can be taken up again by the vegetation almost immediately and used for further growth. This nutrient loss tends to be slight.

Work in Ghana by Nye and Greenland (1960) confirmed this generalisation, showing that the litter of tropical rainforest is considerably richer in nitrogen and other elements than that of temperate forests and that these elements are made available to plants by rapid decomposition. In the last 20 years much more work has been done on mineral cycling in tropical forests, especially as part of the International Biological Programme. Data from these studies (Tables 3.2 and 3.3) show wide variations in the annual return of nutrients and in litter production. While a critical assessment of experimental design and analytical techniques would be necessary to check the comparability of observations, the two tables suggest that annual return of nutrients by small

Table 3.2 Annual return of nutrients (kg ha^{-1}) by total small litterfall in selected natural tropical forests.

Locality and forest type	N	P	K	Na	Ca	Mg	Source
(a) lower montane rainforests							
Central Java	–	3	27.5	2.5	128	22.5	Bruijnzeel (1982)
Papua New Guinea	82	4.7	25	–	80	17	Edwards (1982)
Puerto Rico	–	1	10.5	2	42	9	Odum (1970), Jordan (1970)
(b) lowland rainforests							
Zaire							
mixed forest	220	7	48	–	110	53	Laudelot and Meyer (1954)
Brachystegia forest	220	9	62	–	91	44	
Macrolobium forest	150	9	87	–	84	49	
Musanga cecropioides forest (15-year-old secondary)	140	4	100	–	120	43	
Ghana							
60-year-old secondary semideciduous forest	200	7.2	68	–	210	45	Nye (1961)
Guatemala							
rainforest	170	5.8	20	–	88	64	Ewel (1976)
Ivory Coast							
Banco plateau: evergreen forest	170	8	28	–	61	51	Bernhard (1970)
Banco valley: evergreen forest	160	14	81	–	85	36	
China, Yunnan							
tropical rainforest	169	12	29	–	110	51	Rodin and Bazilevich (1967)
Brazil							
Mocambo: terra firme forest	160	4.1	17	7.0	33	27	Klinge (1977)
Manaus: terra firme forest	110	2.1	13	5.5	18	13	
Belize							
seasonally dry tropical hardwood forest	160	9.2	59	4.9	370	32	Lambert *et al* (1980)
Colombia							
BPa: primary forest	140	4.2	17	–	90	21	Fölster and de las Salas (1976)
BPb: primary forest	100	3.4	29	–	120	12	Fölster *et al* (1976)
BS16: 16-year-old secondary forest	110	2.4	19	–	53	18	
Sarawak							
forest over limestone	140	4.5	16	1.5	370	33	Proctor *et al.* (1983)
alluvial forest	110	4.1	26	0.73	290	20	
dipterocarp forest	81	1.2	33	0.75	13	8.9	
heath forest	55	1.6	18	0.55	83	12	
Australia							
site 1: rainforest	130	12	66	2.4	230	28	Brasell *et al.* (1980)
site 2: rainforest	120	10	51	5.0	160	34	
Ivory Coast							
Yapo: evergreen forest	110	4	26	–	105	23	Bernhard-Reversat *et al.* (1978)

Table 3.2 *Continued*

Locality and forest type	N	P	K	Na	Ca	Mg	Source
Malaysia							
Pasoh: rainforest	100	2.8	31	–	70	18	Lim (1978)
India, Western Ghats							
Agumbe: rainforest	64	13	11	–	41	19	Rai (1981)
Bannadpare: rainforest	46	1.1	7.6	–	46	35	
Kagneri: rainforest	30	1.7	16	–	46	36	
South Bhadra: rainforest	36	8.3	5.2	–	26	11	
Panama							
tropical moist forest	–	8.6	130	–	240	22	Golley *et al.* (1975)
Tanzania							
rainforest	–	8	35	–	104	23	Lundgren (1978)
Trinidad							
Mora forest, N. Range (evergreen seasonal forest)	60	3	11	–	65	15	Cornforth (1970)
Venezuela							
San Carlos: Caatinga forest	42	2.6	27	–	31	8.8	Herrera (1979)
(c) freshwater swamp forests							
Malaysia	85	1.9	31	12	130	37	Furtado *et al.* (1980)
Brazil							
Cata: Ingapó forest	96	3.4	24	5.6	62	28	Klinge (1977)
Awa: Várzea forest	110	3.4	23	5.7	76	26	

litterfall may be less in lower montane than in lowland rainforests, while litter production in true humid tropical rainforests ranges from about 6 to 16 $kg\,ha^{-1}\,yr^{-1}$. Parent material strongly influences some of the ions found in litter, while in some areas loss of nutrients by throughfall and stemflow may exceed that elsewhere (Table 3.4). The wide variations of ions in precipitation, 21.8 $kg\,ha^{-1}\,yr^{-1}$ of calcium ions at El Verde compared with 4.2 $kg\,ha^{-1}\,yr^{-1}$ at Pasoh, for example, cause some of the diversity of the nutrient content of throughfall. Nevertheless, they do not explain why 169.4 $kg\,ha^{-1}\,yr^{-1}$ of potassium appears to be washed from the foliage of the Banco valley site, while only 23.8 $kg\,ha^{-1}\,yr^{-1}$ of potassium is so derived at Pasoh.

One of the outstanding studies has been that of the El Verde forest in Puerto Rico led by Odum (1970). Here nutrient cycling is seen as part of a total dynamic system involving all the processes of plant growth and decay. One of the important questions asked in this study was how far transpiration pumps minerals that might not otherwise cycle adequately.

Earlier studies had identified the role of evaporation as the means for drawing water up the long vascular tracts of lianes and trunks of tropical forest components including mangroves. Evapotranspiration had been regarded as a principal factor affecting tree height, and a principal reason for the effect of

Table 3.3 Litter production in tropical forests.

Forest	Production ($t\ ha^{-1}\ yr^{-1}$)	Source
Macrolobium forest, Zaire	15.3	Laudelot and Meyer (1954)
Musanga cecropioides forest, Zaire (15-year-old secondary)	14.9	Laudelot and Meyer (1954)
seasonally dry tropical hardwood forest, Belize	12.6	Lambert *et al.* (1980)
mixed forest, Zaire	12.4	Laudelot and Meyer (1954)
Brachystegia forest, Zaire	12.3	Laudelot and Meyer (1954)
forest on limestone, Sarawak	12.0	Proctor *et al.* (1982)
primary forest, BPa, Colombia	12.0	Fölster and de las Salas (1976)
evergreen forest, Banco plateau, Ivory Coast	11.9	Bernhard-Reversat *et al.* (1978)
tropical rainforest, Yunnan, China	11.6	Rodin and Bazilevich (1967)
alluvial forest, Sarawak	11.5	Proctor *et al.* (1982)
60-year-old secondary semideciduous forest, Ghana	10.5	Nye (1961)
terra firme forest, Mocambo, Brazil	9.9	Klinge (1977)
rainforest, site 2, Australia	9.9	Brasell *et al.* (1980)
tropical moist forest, Panama	9.8	Golley *et al.* (1975)
evergreen forest, Yapo, Ivory Coast	9.6	Bernhard-Reversat *et al.* (1978)
16-year-old secondary forest, BS16, Colombia	9.5	Fölster and de las Salas (1976)
muhulu, dry evergreen forest, Shaba, Zaire	9.3	Freson *et al.* (1974)
freshwater swamp forest, Malaysia	9.2	Furtado *et al.* (1980)
heath forest, Sarawak	9.2	Proctor *et al.* (1982)
evergreen forest, Banco valley, Ivory Coast	9.2	Bernhard-Reversat *et al.* (1978)
rainforest, site 1, Australia	9.1	Brasell *et al.* (1980)
rainforest, Guatemala	9.0	Ewel (1976)
igapó forest, Cata, Brazil	9.0	Klinge (1977)
lowland rainforest, Pasoh, Malaysia	8.9	Lim (1978)
dipterocarp forest, Sarawak	8.8	Proctor *et al.* (1982)
primary forest, BPb, Colombia	8.7	Fölster and de las Salas (1976)
várzea forest, Awa, Brazil	7.8	Klinge (1977)
terra firme forest, Manaus, Brazil	7.3	Klinge (1977)
lower montane rainforest, Indonesia	6.8	Bruijnzeel (1982)
inundation forest, Brazil	6.8	Adis *et al.* (1979)
deciduous forests near Udaipur, India	4.5	Vyas *et al.* (1976)
dry, mixed deciduous forest, Chandraprobha region, Varanasi, India	4.3	Rai and Srivastava (1982)
evergreen rainforest, Agumbe, Western Ghats, India	4.2	Rai (1981)
evergreen rainforest, Bannadpare, Western Ghats, India	4.1	Rai (1981)
evergreen rainforest, Kagneri, Western Ghats, India	4.0	Rai (1981)
evergreen rainforest, South Bhadra, Western Ghats, India	3.4	Rai (1981)

Table 3.3 *Continued.*

Forest	Production (t ha^{-1} yr^{-1})	Source
miombo, fire-and-hatchet induced woodland, Shaba, Zaire	2.7	Freson *et al.* (1974)
lowly stocked *Shorea–Buchnania* and *Terminalia–Shorea* stands, India	1.2–1.6	Gaur and Pandey (1978)

temperature on forest structure in altitudinal sequences. In the El Verde area, transpiration turned out to be low at 2.1 mm day^{-1}, compared with a range of 2 to 6.5 mm day^{-1} for a variety of rainforests. At El Verde, the limit of transpiration seems to be the competition of too much water from leaf interception and clouds. Some investigators doubt the link between transpiration and the movement of minerals. El Verde has substantial adaptation for both transpiration and root absorption, so each seem important in this forest. Low transpiration does not necessarily mean that it is not important and the fact that it is not high enough may be the reason for compensatory root development. Although diurnal variations in the water content of leaves in dry areas may be 10% or more, at El Verde it is hardly measurable. Whereas normal patterns for transpiration and stomatal response to photosynthesis result in closed stomata at night, at El Verde considerable night transpiration resulted, which may be an adaptation to a low saturation-deficit climate where transpiration might otherwise be inadequate to move nutrients from stems to crown. The decrease in water content from shade to sun leaves is consistent with the gradient in water potential.

Table 3.4 Nutrients in rain, stemflow and throughfall in tropical forests (kg ha^{-1} yr^{-1}).

					Banco[e]		
K	El Verde[a]	Sg Gombak[b]	Pasoh[c]	Kade[d]	plateau	valley	Amazon[f]
potassium							
rain	18.2	12.5	6.4	13.5	5.8	5.1	
stemflow	71.8		0.4				
throughfall	80.1	52.5	24.2	178.5	65	174.5	
calcium							
rain	21.8	14.0	4.2	9.3	21.3	21.6	0.26
stemflow	7.3		0.03				5.79
throughfall	23.9	28.9	3.8	31.3	36	28.4	10.50
magnesium							
rain	4.7	3.3	0.7	0.9	3.6	3.6	0.18
stemflow	1.6		0.2				2.05
throughfall	6.8	5.3	1.3	2.2	36.4	50.4	6.78

Sources: [a]Odum (1970); [b]Kenworthy (1971); [c]Manokaran (1980); [d]Nye and Greenland (1960); [e]Bernhard-Reversat (1975); [f]Brinkmann (1983).

Generally, as Zimmerman (1978) suggests, not enough is known about the mechanisms of water movement in tropical trees. Transpiration costs to trees go up as leaves are exposed at the canopy. Specific trees adjust to this; the leaves of the *Shorea* spp. emergents in Malaysia, for example, are oriented edge on to the radiant energy flow and have adapted to survive as emergents during limiting conditions imposed by lack of water. This is achieved by stomatal control of rates of transpiration.

All the evidence thus suggests adaptation to local edaphic conditions. In the Ivory Coast, two sites in the Banco forest were studied. One was on the plateau where soils are well drained and poor in nutrients, the other in the valley bottom where the phosphorus, potassium and calcium contents of the vegetation and soil are higher. More P, K, Ca and Mg is washed off the leaves of the valley forest than on the plateau. Much more phosphorus and potassium is returned to the soil in the valley site than in the plateau site.

Detailed observations elsewhere tend to confirm this edaphic adaptation. A study of the structure and growth of primary rainforests on a variety of sites in Sarawak (Ashton 1978) shows that net production of wood by volume on a shallow and very freely draining leached soil and a deep fertile loam are almost the same, even though the standing volume on the latter is almost double that on the former. There is no significant difference in mean girth increment in any size class between the two forests, although sampling error among the relatively few larger trees may obscure the lower girth increments that might be expected on the drier site. These results can be explained by the fact that the wetter site carries a dense stand of large emergent trees that accounts for most of the wood production, the understorey being diffuse, while the drier site carries such a dense stand of small trees that their wood production offsets the absence of forest giant trees.

There may be a need for some reconsideration of the nature of the rapid, efficient, closed rainforest nutrient cycle. Proctor (1983) has made the following points:

(a) Most nutrients are concentrated in the trunks and large branches of trees and wide variations occur between species.
(b) Some nutrients, especially calcium, are preferentially concentrated in the bark and large variations occur between species.
(c) There is a lack of significant correlation between mineral-nutrient concentrations in tree leaves and those of the trunk and bark and thus interpretations of the nutrient cycle based on leaves (foliar analysis) only should be treated with caution.
(d) Assessment of soil nutrients is vital, but analytical techniques may be unreliable (e.g. for N and P).
(e) Forests are long-lived and there is at present no good way of deciding nutrient availability over a period of years.
(f) There are great problems in assessing nutrient inputs to, and outputs

from, rainforest ecosystems. In stream-water analyses we seldom know the nutrient losses in sediment and in floating organic matter.

(g) Rates of decomposition in lowland forests at Gunung Mulu (rainfall of 5000 mm yr^{-1}) are much slower than elsewhere. Generalisations about rapid decomposition rates in tropical forests must be made with caution.

(h) Little is known about the storage of nutrients in trunks and branches and their recycling.

(i) Little evidence is available to support the view that biomass and forest stature are related to soil nutrient status.

Some evidence to support Proctor's point about the relatively low significance of soil nutrient status comes from recent work by Klinge and his colleagues in Amazonia (Klinge *et al.* 1983), who compare foliar nutrient levels in Várzea and Igapó forests. Várzea forests are exposed to flooding by rivers carrying turbid water (e.g. Solimões/Amazonas, Madeira). Such water has relatively high solute concentrations, with Ca^{2+} being the dominant cation. Igapó forests are exposed to flooding by rivers carrying clear water (e.g. R. Tapajós) or black water (e.g. Rio Negro). The Rio Negro shows low solute concentrations; the dominant cation is H^+ followed by Na^+.

The Várzea tree foliage is richer in chemical elements than the Igapó foliage. In comparison with other tropical forests, Várzea foliar concentrations are relatively high, Igapó relatively low. Although much lower than the Várzea foliage, the Igapó foliage is relatively rich in nitrogen when compared to the mean of tropical forest in general. Thus nitrogen is an abundant element in the generally nutrient-poor Igapó ecosystem.

Local variations in soils, drainage and landform thus have their influence on the forest nutrient status and turnover. In turn that biogeochemical cycle influences the weathering, solution, detachment and transport of mineral particles. Examination of the processes of water movement from the ground surface to rivers will demonstrate further contrasts.

3.3 Movement of water in tropical rainforest

Movement from the slopes to the rivers

The canopy of the rainforest intercepts considerable portions of rainfall, but the proportion of rain that never reaches the ground, because it settles on the foliage and eventually evaporates, decreases with the intensity and duration of the storm. It has also been suggested that variations in the amount of throughfall over short distances in tropical forests, often persisting for several years, arise from the spatial distribution of shoots in the canopy and are independent of the characteristics of individual storms (Hopkins 1965). Such a situation is particularly likely to occur where rainforests are disturbed irregularly by tropical cyclones, as is the case in the 'cyclone' forests of north-east Queensland (Webb 1958). Here throughfall accounts for over 70% of total rainfall (Gilmour *et al.* 1980). However, in general, the proportion of rain intercepted is a physical process governed by the size and intensity of the storm (Pereira 1973).

Stemflow may locally be important at the bases of trees through the washing away of fine material on the downslope side of tree trunks, but few quantitative measures of such processes are available. Splash of the soil surface by canopy drip or throughfall is an important process for soil particle detachment. Loosened soil particles are trapped by litter and may be detained on vegetal fragments until the organic matter is decomposed or is itself moved by surface wash.

Surface wash can be measured by traps placed flush with the soil surface to collect material washed down from further upslope. In terms of denudation, surface wash comprises four sequential but distinct processes: the detachment of soil particles by raindrops and leaf drips; the displacement of such detached particles downslope by gravity or splash; the detachment of soil particles by water flowing over the ground surface; and the downslope transport of all detached particles by water flowing over the surface in an unconcentrated form (Young 1972).

Trap experiments to measure these processes set up in several tropical rainforest locations in Malaysia (Leigh 1978), in north Queensland (Gilmour *et al.* 1980), in Amazonia (Nortcliff & Thornes 1978, 1981) and in Dominica (Walsh 1980a) yield conflicting results (Fig. 3.4).

Surface wash in the Pasoh forest reserve in Malaysia is an important contributor to river flow, water moving over the surface at the base of a 17–22° slope being about half the quantity moving at depth, while on a gentler slope, surface flow is about equal to the subsurface flow.

In the Babinda area of north Queensland, the surface 10 cm of soil under rainforest has a high permeability and offers no impedance to the highest rainfall intensities encountered. However, permeability decreases rapidly with depth, such that moderate rainfall intensities will not infiltrate readily below 20 cm and high-intensity rainfalls will not infiltrate below 10 cm. The surface soil thus becomes saturated rapidly and so produces saturated overland flow. A temporary perched water table over the whole catchment is the important factor initiating overland flow. Overland flow can occur over widespread areas of the catchment once the top 20 cm store has been filled.

At the Amazonian site, water falling on the slopes moves rapidly downwards through the soil. Saturated conditions are only reached in the floodplain areas, from which an area, or zone, of saturation appears to move upwards. While baseflow is the main contribution from the hillslope to the streamflow, overland flow on the floodplain is the dominant cause of the very rapid response of the stream to rain. The water 'rises out' of the floodplain and discharges immediately into the channel, or, in the higher reaches, flows down the floodplain parallel to the stream.

Both Gilmour *et al.* in north Queensland and Nortcliff and Thornes in Amazonia note contrasts between wet-season and dry-season hydrological

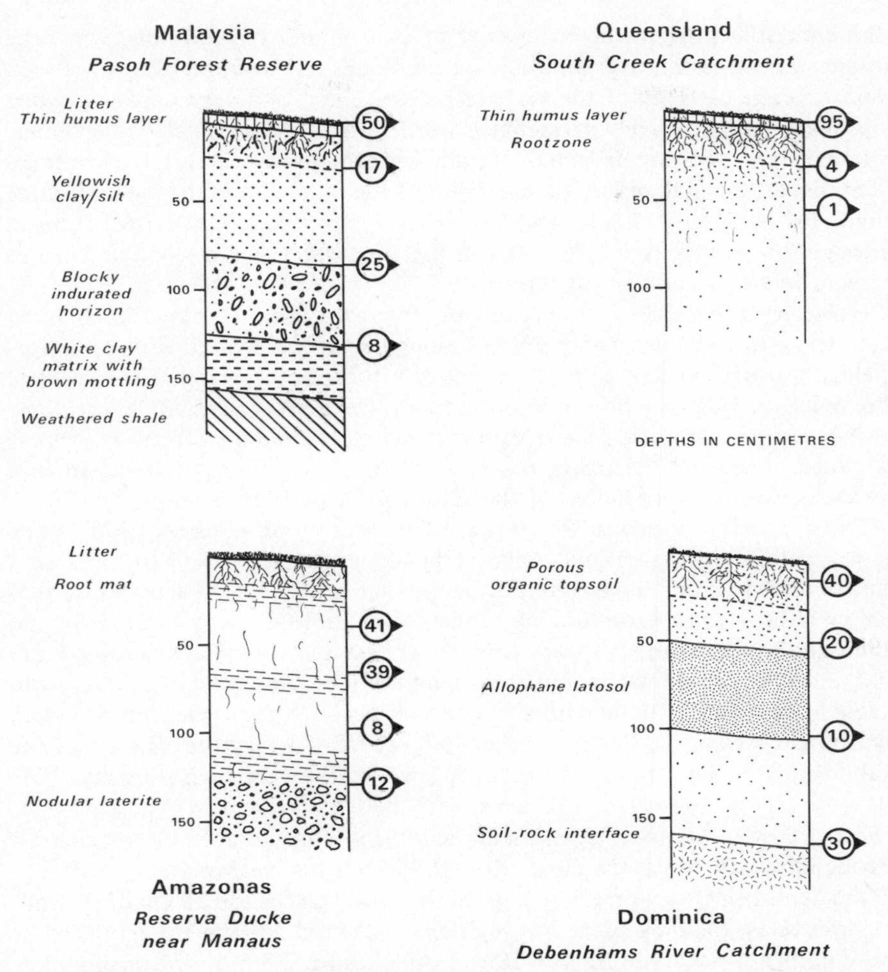

Figure 3.4 Proportion of runoff draining different surface and soil horizons in four tropical rainforest sites.

behaviour. In Queensland in the dry season, a greater proportion of the rain percolates into the deeper layers of the soil, while in the dry season in Amazonia there is sometimes an increase in the downslope movement, rather than vertical percolation of the water. Here only under very dry conditions is stream runoff dominated by subsurface flow from the hillslopes. Clearly, the hydrology of these two rainforest areas is quite different in detail.

In Dominica, a rugged volcanic West Indian island, Walsh set up an experiment at a site receiving an average annual precipitation of almost 5500 mm. The slopes exceed 25°, but neither Hortonian overland flow nor widespread saturation overland flow were observed, even during the heaviest and most prolonged rainstorms. The extremely high infiltration capacities of the allophane soils prevent the occurrence of Hortonian overland flow, while the absence of saturation overland flow on the slopes can be linked to (a) the high water storage capacity of the very deep (20–30 cm) and very porous organic topsoils, (b) the relatively permeable subsoil, which allows much water to percolate instead of being deflected laterally as throughflow, even in very intense rainfalls, (c) the high *lateral* permeability of the topsoil, and (d) the high slope angle (in excess of 25°). The last two factors together render topsoil throughflow relatively efficient at transporting water down slope and hence prevent *in situ* saturation of the soil.

However, in this area, Walsh did observe saturation overland flow along channels – in very narrow floodplains along the steeply incised valleys. Nevertheless, more important as source areas of saturation overland flow in central Dominica are ridge-top flats. These flat interfluve areas are essentially remnants of the original constructional volcanic surface. Overland flow from these areas provided almost all the storm runoff and (since the channel is reduced to a trickle between storms) most of the total runoff of West Stream.

Three possible reasons for these differences need consideration: slope angles, rainfall intensities and soils. Slope angles range from 17 to 25° across the four areas. Soils have different parent materials, but all tend to be red-yellow latosols with kaolinitic or similar clays. Rainfalls vary both in annual totals and in storm intensity, but the two wettest areas, north Queensland and Dominica, have somewhat similar rainfall regimes. These two have very different parent materials and this might well account for their differences in hydrologic behaviour. The Amazonas soils are sandier than the others, and the stability of the sand particles probably leads to the very high porosity.

Throughflow. All four experiments show the considerable importance of throughflow. In Malaysia most throughflow occurs between about 20 and 80 cm depth. In Queensland the bulk of the flow is in the top 25 cm of the soil. In Amazonas, the bulk of the throughflow is around 50–100 cm depth, while in Dominica it occurs in the top 50 cm. All sites have important throughflow contributions. Actual pathways depended on the character of the soils.

Pipeflow. Subsurface piping is recognised as a fairly widespread feature in many humid tropical situations. In Malaysia, piping derived from root pores is probably significant in the pathways taken by throughflow in the Pasoh area. Macropores create a type of pipeflow in the Babinda, Queensland, experiment. Pipeflow also occurs in Dominica, but is not widespread.

Seasonal fluctuations of ground water and solutes in springs

The roles of seasonal water table fluctuations in weathering mantles and of storm events in solute removal are well illustrated by investigations on the Fouta Djalon plateau in Guinea (Mamedov *et al.* 1983). In the dry season only two hydrogeological zones may be distinguished in the weathering mantle, a permanently aerated zone and a permanently waterlogged zone. At this time the discharge of springs on the upland slopes is 0.01–0.1 l s^{-1} and the pH of subsurface water varies from 5.0 to 6.5. The principal solutes present are HCO_3, with admixtures of Cl and, less frequently, NO_3 and SO_4, Ca, Mg, NH, some Na and K, and H_4SiO_4, with minor amounts of Al and Fe.

In the wet season, four hydrogeological zones develop rapidly with the onset of the rains. The permanently aerated zone is a narrow layer close to the surface, below which occurs a zone of subsurface water fluctuation with waterlogging for less than 20% of the time. This latter zone gives way to a similar zone where saturated conditions prevail for more than 20% of the time, with below that the permanently waterlogged zone, now much nearer the surface than in the dry season. During this rainy season a multitude of springs develop on the slopes with discharges of 0.1 to 4 l s^{-1} with localised surface pools and saturated swampy areas. The chemical composition of subsurface water changes, pH decreasing by 0.3 to 0.8, with a sharp increase in HCO_3, the virtual disappearance of Cl, more carbonates, less NH_4 and Fe, and more SiO_2 and O_2.

Such a seasonal fluctuation has profound effects on weathering processes, gibbsite being formed only in the uppermost part of the weathering mantle where saturation occurs less than 2% of the time. Kaolinite is formed below, where the duration of saturation is up to 150 days, while montmorillonite is only found in zones that are saturated for more than 200 days.

Between individual storms on the Fouta Djalon plateau, the water table fluctuates rapidly, falling up to 4 m before the next storm, at a rate of 2 to 6 cm h^{-1}. The fluctuations occur within the gibbsite–geothite–kaolinite zone of the weathering mantle. These vertical changes in weathering conditions produce variations in the solutes released to streams and are crucial for the release of nutrients to ecosystems. Any climatic change altering the hydrological regime would change the patterns of weathering, clay mineral genesis and water quality in springs.

3.4 Sediment and solute transport

Transport of sediment by surface wash

A variety of techniques, including erosion pin, Gerlach trough and runoff plot experiments, have been used to assess the transport of sediment by surface wash in the tropics (Table 3.5). Estimates of soil loss range through several orders of magnitude, from 25 to 19 000 kg ha^{-1} yr^{-1}. Slope angle is a major source of variation in the values given, but, as the data from northern Thailand indicate, the nature of soils and parent material is also important.

In peninsular Malaysia, the rate of loss on soils derived from shales at Pasoh is higher (0.7809 cm^3 cm^{-1} yr^{-1}) than on granitic soils at Bukit Lagong (0.2992 cm^3 cm^{-1} yr^{-1})(Peh 1978). The higher rates of runoff and of soil detachment and movement at Pasoh reflect the lower porosity and clay content and higher silt content than in the soils at Bukit Lagong. The contrast between

Table 3.5 Rates of denudation and sediment transport by surface wash in tropical areas.

Location	Vegetation	Angle of slope (deg)	Technique	Ground lowering (mm yr^{-1})	Loss per unit area (kg ha^{-1} yr^{-1})	Volumetric transport (cm^3 cm^{-1} yr^{-1})	Source
Pasoh,	lowland dipterocarp	8–30	traps	–	–	0.781	Peh (1978)
Negeri Sembilan	rainforest	8–30	pins	2.6	–	–	
Northern Thailand	teak forest on shale	all plots on same angle	plots	–	19240	–	
	teak forest on limestone	but angle not defined	plots	–	6090	–	Sa-ard Boonkird (1968)
	dry dipterocarp forest		plots	–	6160	–	
SE Mindanao	rainforest	14	plots	–	46	–	Kellman (1969)
	logged-over forest	14	plots	–	836	–	
Tawau, Sabah	clear-felled	32	stakes	35.6 (6 months)	–	–	Liew (1974)
Ivory Coast	rainforest	4–50	stakes	1.5–3.5	–	–	Rougerie (1960)
Adiopodoume, Ivory Coast	rainforest	4	plots	–	30	–	
		12.5	plots	–	200	–	Roose (1971)
		24	plots	–	1000		
Adiopodoume, Ivory Coast	secondary rainforest	7–8.5	plots	–	25	–	Fournier (1967)
Senegal	rainforest	37	plots	–	76	–	Roose (1971)
		1	plots	–	20–220	–	Fournier (1967)
Azaguie, Ivory Coast	rainforest	14	plots		75		Roose (1980)
Rupununi,	rainforest G	10	plots		7290		Kesel (1977)
Guyana	H	15	plots		8550		Kesel (1977)
Divo, Ivory Coast	semideciduous forest	5	plots	–	500	–	Roose (1971)
Northern Territory,	savanna on granite		traps	0.05	–	36.34	
Australia		1–15	traps	0.06	–	22.86	Williams (1973)
Gonse, Upper Volta	tree savanna	0.5–1	plots		84		Roose (1980)
Rupununi, Guyana	savanna B	8–10	plots		3310		Kesel (1977)
	D	8–10	plots		7710		Kesel (1977)

undisturbed and regenerated logged forest on granite appears to exert little influence. Soil loss from a regenerated forest with a single litter layer of freshly fallen leaves at Bukit Mersawa close to Bukit Lagong is 0.3134 $cm^3\,cm^{-1}\,yr^{-1}$. Differences in rates of sediment transport at different trap sites are not related to slope angle. At Pasoh, however, there is a significant positive correlation between amount of sediment collected and distance from the crest. Here, also, structural properties of the soil and the characteristics of the litter cover appear to be important in accounting for differences in the rate of sediment transport.

The figures for soil loss in Pasoh obtained by Peh have a larger range and higher maximum values than those obtained in the same locality by Leigh (1982), 0.208–2.626 $cm^3\,cm^{-1}\,yr^{-1}$, compared to 0.058–0.735 $cm^3\,cm^{-1}\,yr^{-1}$. The range in values probably reflects differences in site characteristics and variation in rainfall totals and frequency–duration characteristics during the different collection periods. Leigh notes a relationship between rates of soil loss and the thickness and completeness of the litter and humus layers immediately above overland flow traps. Thus if a lowland tropical rainforest has the mosaic structure of regeneration, building and mature stages described in the previous chapter with consequent variations in litter and humus accumulation, local variations in surface soil loss are inevitable, regardless of the physical and chemical character of the soil and regolith.

Slope, parent material, topography, vegetation and soil cover by litter all appear significant in soil erosion by slopewash under rainforest. Nortcliff and Thornes (1981) have rightly emphasised the importance of soil permeability and porosity in overland flow/throughflow relationships. Rainfall quantity and frequency–duration characteristics are also important in the differences between sites, but clearly there is no simple climatic explanation of the variability of this type of geomorphic process. The impact of climatic change on such processes is thus likely to be expressed through the ability of the plant–soil–water regime to withstand a stress, such as a longer dry season or occasional drought period.

Solute transport by humid tropical rivers

These pathways are vital components of the denudation systems. Solute transport by lateral eluviation at Pasoh, Malaysia, accounts for 27.4% of the total sediment transported by surface work and throughflow at pit 1 and 15.5% of the total at pit 2. In Dominica the throughflow discharge contains more solutes than the saturated overland flow because of the contact with the soil particles. Although solute concentrations are low throughout the year, more solutes are moved in the wet season when flows are high in Amazonas.

Such considerations lead us to examine the relationships between soils, stream catchment conditions and solutes. As already demonstrated, weathering in the humid tropics is effective. Yet, strangely, solute concentrations in rivers are low (Table 3.6). South American rivers like the Amazon are particularly dilute. Although the Amazon supplies 18% of the world's fresh water

Table 3.6 Dissolved and solid loads of major tropical rivers (after Meybeck 1976, 1979, Milliman & Meade 1983).

River	Area (km^2) $\times 10^3$	Runoff (mm)	Dissolved load ($m^3\ km^2\ yr^{-1}$)	Suspended load ($m^3\ km^2\ yr^{-1}$)		Chemical composition ($mg\ l^{-1}$)							
				Meyb.	M. & M.	SiO_2	Ca	Mg	Na	K	Cl	SO_4	HCO_3
Amazon	6300	840	17.5	29.8	58.5	11.2	6.5	1.0	3.1	1.0	3.9	3.0	22.5
Brahmaputra	580	990	49.1	517.0	[a]	–	22.5	4.1	2.9	3.7	5.9	4.0	84.1
Congo (Zaire)	4000	294	4.4	5.0	4.5	9.8	2.4	1.3	1.7	1.1	2.85	2.95	11.2
Ganges	975	360	29.4	202.6	[a]	12.8	24.5	5.0	4.9	3.1	3.4	8.5	105.2
Irrawaddy	430	936	–	264.0	246.5	–	–	–	–	–	–	–	–
Magdalena	240	930	44.1	377.4	366.7	12.6	15.0	3.3	8.3	1.85	13.4	14.4	49.3
Mekong	795	690	28.3	164.1	81.0	8.9	14.2	3.2	3.6	2.0	5.3	3.8	57.9
Niger	1125	165	3.4	22.6	13.2	15.0	4.1	2.6	3.5	2.4	1.3	1.0	36.0
Orinoco	950	948	19.6	34.3	83.2	11.5	3.3	1.0	1.5	0.7	2.9	3.4	11.0
Parana	2800	192	7.5	15.1	–	14.3	5.4	2.4	5.5	1.8	5.9	3.2	30.7
São Francisco	470	186	–	–	3.8	–	–	–	–	–	–	–	–
Tocantins	900	366	–	–	–	–	–	–	–	–	–	–	–
Zambezi	1340	159	4.3	28.3	6.7	12.3	9.7	2.2	4.0	1.2	1.0	3.0	24.9

[a]Milliman and Meade estimate that the joint sediment discharge of the Ganges and Brahmaputra together is 451 $m^3\ km^2\ yr^{-1}$.

to the oceans, it only yields 8% of the dissolved salts to the world's oceans (Gibbs 1972).

Even so, a closer examination of the Amazon data shows that 86% of the solutes discharged by the river are supplied from the 12% of the total area of the basin comprising the mountain zones, and 82% of the suspended solids discharged come from this environment also (Gibbs 1967) (Table 3.7). In the lowlands of the Amazon, the character of river waters can be divided into 'white', 'clear' and 'black' waters.

The white waters, such as those of the Solimões/Amazonas and Madeira described earlier, are turbid, carrying suspended sediments. It was thought that this sediment came from the Andes, as Gibbs (1967) suggested, but now it is believed that the silt more probably comes from the foothill zone of the Andes where the rivers leaving the mountains flow over soft Tertiary sediments. Channel migration across these alluvial sediments provides a ready sediment supply, with surface runoff playing only a minor role in the origin of this suspended sediment (see Ch.10). These white, turbid waters have relatively high concentrations of calcium, magnesium, potassium and other cations, while the clear waters contain extremely low concentrations of these elements, but are richer in alkalis, aluminium and iron (Klinge *et al*, 1981). The black waters, such as those of the Rio Negro, are clear waters containing dissolved humic substances. They usually have low pH values and are often associated with tropical podsols.

Not only do these solute concentrations vary spatially, but they also vary in time. The Amazon rises to a seasonal flood peak, with the different chemical consitituents changing in concentration as water levels rise. While calcium concentrations fall with increasing discharge, silica and potassium concentrations alter less rapidly, the lowest potassium concentrations occurring well after the high-water period. During individual storm runoff events, similar fluctuations in solute concentrations occur. In the Barron River at Picnic Crossing in north Queensland during the passage of cyclone Flora in December 1964, the total dissolved solids concentration fell as discharge increased, but the concentrations of individual ions changed in differing ways, potassium levels increasing while sodium and calcium concentrations fell. Silica concentrations dropped initially as discharge rose, but then increased, perhaps as silica was lost from particulate matter carried into the stream (Douglas 1972). Similar patterns of variation in solute concentrations have been observed in many other tropical rivers, indicating that solutes are derived from differing sources and enter into solution at different rates and through different chemical reactions.

When waters flood low-lying areas, substances may be released from solution, particularly in groundwater and soil-water bodies. Such exchanges probably occur in floodplain areas along major rivers, but are less likely to be important in steep headwater tributaries. Such processes lead to the reformation of clay minerals and to the accumulation of less mobile elements such as aluminium and iron, perhaps leading to the formation of iron concretions, or even ferricrete of the type described by Lamotte and Rougerie (1962).

Table 3.7 Sediment yields in tropical environments.

Region/Country	Station	Catchment area (km^2)	Runoff ($mm\,yr^{-1}$)	Natural vegetation cover (%)	Suspended sediment yield ($m^3\,km^2\,yr^{-1}$)	Source
Papua New Guinea	Ok Ningi	4.56	7208	95	2980–4050	Pickup *et al.* (1981)
	Ok Tedi	420	5695	95	1720–2960	
	Ok Menga	240	6660	95	370–460	
	Alice	3900	5870	95	300–560	
	Fly at Kiunga	6300	5360	95	260–350	
	Aure	4360	2220	95	4190	
	Purari at Wabo	26300	2830	95	790	
	Ei Creek	16.25	2625	85	36.3	Turvey (1974)
Philippines	Bararo	191		10	366	De Vera (1981)
	Bauang	353		19	773	
	Aringay	274		9	536	
	Ambayaon	312		9	477	
	Caranglan	258		13	79	
	Talavera	285		6	835	
	Patalan	347		16	74	
	Rio Chico	1177		26	69	
	Penaranda	512		22	204	
Indonesia	Konto	185	1333	65	3190	Brabben (1978)
	Kwayangan	53		70	3810	
	Brantas at Karangkates	2050	1037	15	3410	Brabben (1979)
	Mondo	0.187	3460	100[a]	209–313	Bruijnzeel (1982)
	Pelus	13.2			270–400	
	Rambut	45.0			420–630	
	Sanggreman	0.629		1	1355–2551	Van der Linden (1978)
	Serayu	666			410–615	Bruijnzeel (1982)
	Serayu at Banyumas	2665			3800	Snowy Mts Eng. Corp. (1974)

	Bengawan Solo	2890			4146	Van der Linden (1978)
	Rambut	45	2740		200	Van Dijk and Ehrecron (1949)
	Cacaban	79	1285		2500	
	Cilulung	600	1352		8000	Snowy Mts Eng. Corp. (1974)
	Lower Serang at Godong	3047			2400	
	Lusi at Purwodadi	1966			2500	
Tanzania	Ikowa	640			191	Stromquist (1981)
	Matumbula	15			581	
	Msatatu	8.7			556	
	Imagi	2.2			610	
	Kisongo	9.3			481	
	Mogoro	19.1	1200	44	147	Rapp *et al.* (1972)
	Rufiji	156000	1142	63[b]	60	Temple and Sundborg (1972)
Nigeria	Rima	35370	45.2		58	Oyebande (1981)
	Sokoto	12590	62		148	
	Gagere	5670	276		110	
	Bunsuru	5900	232		165	
	Hadejia	17400	116		134	
	Kano	6980	173		83	
	Challawa	6890	100		279	
	Watari	1450	83		182	
	Jamaari	7980	261		173	
	Benue at Makurdi	304300	330		16	
	Katsina Ala	22000	1045		24	
	Cross at Ikom	16900	2009		27	
Kenya	Iuini	11.3	81		223	Thomas *et al.* (1981)
Cambodia	Mekong at Stung Treng	534000			130	Carbonnel (1965)
	Kampong Lan at Thnot Chum	420			19.8	Carbonnel and Guiscafré (1965)
	Sen at Kampong Thom	13670			16.5	Carbonnel and Guiscafré (1965)
	Pursat at Pursat	4480			14.8	Carbonnel and Guiscafré (1965)
	Krakor at Krakor	138			8.0	Carbonnel and Guiscafré (1965)

Table continued

Table 3.7 (*Continued*)

Region/Country	Station	Catchment area (km^2)	Runoff ($mm\,yr^{-1}$)	Natural vegetation cover (%)	Suspended sediment yield ($m^3\,km^2\,yr^{-1}$)	Source
Malaysia	Perak at Kenering	5500			144	Douglas (1970)
	Kial, Cameron Highlands	21			111	
	Bertam, Cameron Highlands	73			103	
	Perak, Bersia	3600			88	
	Perak, Temangor	34000			88	
	Gombak, $12\frac{1}{2}$ milestone	41.3	1	100	24	Douglas (1978a)
	Telom, Cameron Highlands	77			21	Douglas (1970)
Cameroons	Sanaga at Nachtigal	77000		30	11.3	Nouvelot (1969)
	Tsanaga at Bogo	1526	167		80.6	
	Mbam at Gowa	42300	559	25	34	
	Djerem at Mbakaou	20390	634	Savanna	22.3	
	Rao at Foumban	1345	846	Savanna	6.4	
Australia	Davies Ck, rainforest	5.59	–	99	9.34	Douglas (1973)
	Freshwater, Copperlode	44.03	968	95	6.30	
	Behana, Aloomba	82.88	2325	85	29.78	
	Barron, Crater	11.91	–		6.22	
	Barron, Picnic Crossing	225.33	657		14.96	
	North Babinda, Boulders	15.36	–		190.51	
	Millstream, Ravenshoe	91.95	621		14.09	
	Wild, Recorder	585.34	421		18.76	
Kenya						
Lake Victoria Basin	Nzoia	1470			5.1	Edwards (1979)
	Kipkarren	808			1.4	
	Nzoia	8620			8.6	
	Yala	2390			23.9	
Rift Valley	Lelgel	108			6.0	

Athi River Basin	Ruiruaka	101	38.1
	Tigoni	17	0.7
	Ruiruaka	68	0.6
	Nairobi	161	5.3
	Riara	38	21.8
	Kiu	38	17.8
	Ruiru	346	13.3
	Athi	5590	15.7
	Mzima Springs	306	3.1
Tana River	Sagana	90	13.5
	Nairobi	119	15.2
	Sagana	501	33.8
	Amboni	473	24.0
	Chania	517	47.5
	Thika	331	146.2
	Thiba	1970	29.1
	Tana/Kambura	9520	19.3
	Tana/Grand Falls	17400	793.6
	Tana/Garissa	31700	143.6
	Kalundu	25	206.0
Ewaso Ngiro Basin	Ew. Narok	58	10.3
	Equator	157	5.9
	Pesi	135	3.6
	Ew. Narok	878	0.7
	Ew. Ngiro	405	23.2
	Burgaret	98	6.8
	Ngobit	256	14.4
	Nanyuki	68	20.9
	Ontulili	61	33.4
	Kongone	14	44.0
	Sirimon	62	4.9
	Teleswani	36	46.1
	Timau	64	75.4

Table continued

Table 3.7 (*Continued*)

Region/Country	Station	Catchment area (km^2)	Runoff ($mm\,yr^{-1}$)	Natural vegetation cover (%)	Suspended sediment yield ($m^3\,km^2\,yr^{-1}$)	Source
	Lili	184			5.7	
	Ew. Ngiro	15300			72.3	
Amazonia	Amazon at mouth	6300000	–		29.8	Gibbs (1967)
Mountainous headwaters	Ucayali	406000	741		115.9	
	Maranon	407000	842		94.9	
Mixed	Napo	122000	1188		69.4	
	Ica	148000	1216		23.3	
	Japura	289000	1214		45.3	
	Madeira	1380000	719		59.4	
Lowland	Javari	106000	1094		25.8	
	Jutai	74000	1027		15.4	
	Jurua	217000	907		18.6	
	Tefe	24400	1024		0.8	
	Coari	55500	973		0.8	
	Purus	372000	917		16.3	
	Negro	755000	1863		3.8	
	Tapajos	500000	448		0.5	
	Xingu	540000	450		0.3	
	Araguari	45200	1128		2.6	

[a] *Agathis* forest plantation up to 40 years old.
[b] Mostly woodland offering poor protection to the soil.

Sediment transport in humid tropical rivers

In his 1960 text on the physical geography of the intertropical zone, Birot wrote: 'The action of watercourses is relatively inefficient, in the humid tropics, by comparison with other climates.' In 1981, Pickup, Higgins and Warner wrote that in the Ok Tedi River: 'Overall denudation rates calculated from the suspended sediment load data are high by world standards but not unusual for Papua New Guinea' (Table 3.7). These two statements encompass the great contrasts that are present in the humid tropical areas.

Birot's arguments about the nature of fluvial processes in the humid tropics start from geomorphological criteria – from the character of the long profiles of the rivers. He writes that it is surprising to find on major rivers rapids only a few kilometres from the base level, or in other words not far upstream from the sea. Although one might be tempted to suggest that such rapids relate to recent tectonics, they can be found even on rivers in old land surfaces of low relief, such as the Ivory Coast, where elevations rarely exceed 200m. Here rapids are found not far upstream from the coast. Birot argues that this is a fundamental characteristic of the climatic geomorphology of the humid tropics.

Furthermore, Birot would argue that rarely have waterfalls been cut back along gorges, but rather that many of these rapids in humid tropical rivers coincide with the escarpments formed by the resistant rock layers that cause the waterfalls. The reason for this, according to Birot, is due to the absence of large pebbles in tropical rivers, and thus a shortage of coarse fragments to abrade the bed and create potholes in the hard rock. The material supplied by the slopes, he argues, is fine, as a result of the deep weathering, and is thus unable to attack resistant rocks.

Studies of a waterfall complex in Surinam led Bakker (1957) to note that, although the erosive action of the river broke fragments of rock away from the bed of the channel, such fragments were quickly worn away and the pebbles so formed disappeared by about 6 km down stream. It is this rapid rate of chemical decomposition which, as studies of weathering have shown, leads to a rapid diminution in size of any material supplied to a river. Birot argues that this would explain the relative absence of incised meanders and gorges in the humid tropics.

In his later book on the influence of climate on continental sedimentation, Birot (1970) begins to qualify his earlier generalisations about tropical rivers. While the generalisations would seem to apply to the ancient shield areas, such as Brazil and Africa, they are less applicable in the beds of steep streams incised into the west of the massif of Madagascar, and even less to rivers in crystalline rocks in the great Tertiary fold mountains. At the foot of the eastern flank of the Peruvian Andes, in the Amazon headwaters, several pebbles up to 50 cm in diameter, which are moved by major floods, can be found. These larger pebbles are made not only of rhyolites and similar very resistant material but also of grey granites and crystalline limestone. Both pink granites with a high feldspar content and schists decompose rapidly. However, even in this situation, as we saw earlier with the suspended sediment supply, a considerable portion of the pebbles are derived from erosion of an old terrace which may have been formed in a drier climate when chemical weathering was less effective.

In Puerto Rico, stream channels in the centre of the island, Birot notes, have large quantities of pebbles, with some over 50 cm in diameter being found in the reservoirs built on these mountain streams. Rounded, volcanic pebbles are found 15 to 20 km down stream of the last rock outcrops. Here again, though, some of the pebbles are derived from river terraces which could have formed in drier climates.

For Birot, this essential difference between the ancient shield areas and the mountainous zones stems from the greater depth of weathering on shield areas. In Puerto Rico, the rivers have cut down to bedrock along practically the whole of their length, giving the impression that the river has fixed its bed at the base of the regolith, at the boundary between weathered and unweathered rock. The pebbles come then from the bed itself rather than from the slopes.

These comments on Birot's views are important, because they reflect the type of generalisation carried forward into many textbooks. Büdel (1982), for example, notes that in humid tropical *lowland* rivers bedload consists mostly of fine sand around 20 to 200 μm in diameter. He implies that fluvial erosion in the humid tropics is weak. Büdel (1980) also notes that mountain valleys in the humid tropics rarely have gravel blankets and that, lacking erosional tools, humid tropical rivers are not capable of vertical erosion. Where they happen to flow across fresh bedrock they are not able to cut through it, forming rapids and waterfalls instead. These rivers serve only as *passive* base levels and conveyors, but they are not active in the sense of linear erosion. In general, Büdel says, they cannot work ahead of the areal denudation; instead they are fully integrated in the overall process of land surface lowering.

Garner (1974) develops these ideas about weathering and sediment supply to rivers to suggest that the forested humid tropical ridge and ravine topography might be at an equilibrium in terms of the processes operating, forming what he called a *selva landscape* of *humid climax topography*. He notes that here the streams are capable of transporting far more material than they actually carry and that erosion by a stream at the foot of the slope is often necessary to trigger off a landslip upslope. Garner suggests that rarely is there sufficient solid sediment in transit in such areas to form extensive point bars on the inside of stream bends, and it is doubtful if such a stream could maintain a delta even on a near-zero energy coast.

3.5 Specific environments: New Guinea and Malaysia

The evidence from New Guinea

In his book on New Guinea, Löffler (1977) notes quite different types of fluvial processes, refuting the climatic contrasts in river power suggested by authors such as Birot and Büdel. Fluvial erosion is undoubtedly the most important process operating in the mountainous landscape of Papua New Guinea. Because of the great tectonic relief there is a huge amount of energy available for the rivers to erode deeply into the bedrock. The rivers contain large amounts of coarse gravel and boulders which travel considerable distances down stream. In the Strickland, for example, coarse gravel, mainly quartzite and limestone, forming extensive gravel banks and bars, is to be found up to 100 km down stream from where the stream leaves the mountains.

The New Guinea river beds are narrow, mostly cut into bedrock with the river occupying nearly the entire width of the bed. Side slopes are very steep and straight and there are few lower slope concavities, as most of the debris transported to the foot of the slopes is quickly removed by the rivers. Strongly oversteepened profiles are also rare as slope processes keep pace with the rate of incision.

Quantitative data on river loads in New Guinea are few. There were virtually no data on bedload in 1977, although it is undoubtedly the major form of fluvial transport in the mountains. Löffler (1977) estimates that 2.5×10^6 m^3 of sediment accumulated in a dam in the Bewani Mountains with a catchment of 78 km^2 in 30 years – a rate of 2831 $t\,km^2\,yr^{-1}$. Many New Guinea rivers have braided channels where they emerge from the mountains. Bedload transport in the braiding floodplains of the Markham River and its northern tributaries appears to be particularly great. Buried tree trunks bear witness to these rapid processes. Partially buried trees in some locations indicate that up to 2.7 m of gravel was deposited in a two-year period. One river, the Leron River, had its bed raised by 0.5 m in three months. Löffler (1977) points out that these are extraordinary rates and should not be regarded as representative, yet they give an indication of some maximum rates of bedload movement.

The calibre of the bedload can also be truly impressive. Boulders of several metres in diameter have been observed in many large rivers. Although most have not moved, some have travelled several kilometres downstream.

More recently a thorough study has been made of the headwaters of the Fly River (Pickup *et al.* 1981). The Fly must be regarded as one of the most spectacular rivers of the world. It rises in the Hindenberg Range in an area of karst and ridge and ravine topography with mountains reaching altitudes of over 3000 m. However, the mountains only occupy a small part of the catchment and the river mainly cuts across a low alluvial plateau in an incised valley zone 5 to 15 km wide. A braided gravel channel extends as far as Konkonda where it gives way to a meandering channel which extends down to the tidal zone and the Gulf of Papua. In the headwaters mean annual rainfall may exceed 10 000 mm, but in the lowlands it is about 2000 mm.

In the mountain areas there are three major sources of sediment: reworked stream bed and bank sediment; landslides; and material contributed by slopewash. For much of the time, rivers carry very small loads because the bed material is too coarse to transport at low and medium flows and the depth is insufficient to cause significant bank erosion. The bed of the river is therefore quickly swept clean of small sediment particles. Sediment concentrations increase rapidly once the river flow becomes great enough to erode the banks or where surface runoff brings in material from sources outside the channel such as recent slides or landslide debris. However, the greatest increases are likely to occur where the flow is sufficient to erode the bed material and expose the finer material trapped interstitially below the surface layer.

The two or three new landslides a year which occur in the Upper Ok Tedi catchment may each involve tens of millions of tonnes of debris. Such huge mass movements are a major source of fluvial sediment.

In the Fly the mountains are the main source of sediment and, as the river proceeds down stream, the sediment derived from the mountains gets diluted by clearer water from the lowland tributaries. Sediment concentrations vary both in space and time as a result of the irregular occurrence of landslides. Although Pickup *et al.* (1981) calculated denudation rates from suspended load data, without considering bedload, their estimates (Table 3.7) are among the highest sediment yields for any tropical rivers. While the Javanese rivers (Table 3.7) reflect erosion accelerated by human activity, the New Guinea results indicate the power of erosion in steep mountainous terrain in the humid tropics. Such rates of denudation are, however, not peculiarly tropical and similar erosion rates are also found in the relatively easily eroded rocks of the alpine terrain of the west coast of South Island, New Zealand (Walling 1982).

Rivers in the monsoonal humid tropics of SE Asia

Malaysian rivers in their headwaters have much gravel material in their beds. However, the pebbles are eroded selectively, those decomposing easily being worn away before the more resistant quartz and quartzite fragments. Yet even well down the courses of major rivers, substantial quantities of gravel occur. The Pahang River is dredged for gravel at Temerloh. Terraces along the river are sometimes cut by very high flows revealing banks of coarse gravel. Nearby there are no obvious sources of sediment, so the gravel beneath the bed must either be transported a considerable distance down stream or be derived from erosion of terraces which could perhaps have accumulated in a drier past climate when pebble decomposition would not have been so rapid. One difficulty is that many plant ecologists argue that this part of Malaysia has much evidence that rainforest has prevailed over the area for a long time and that Quaternary climatic changes were not great. Sediment loads from landslips or surface wash under rainforest, as described by Leigh (1978), can be large, but in virtually all storms the sediment carried by rivers is far less than the ability of the stream to carry debris. This is shown by the way in which sediment concentrations reach a maximum on the rising stage of the hydrograph, well before the peak runoff.

Former channels of these Malaysian rivers graded down to Quaternary low sea levels are recognisable by sand and gravel lenses that are now exploited for alluvial tin offshore. At the present time, the rivers flowing to both the east and west coasts of the peninsula carry no coarse material to the sea, the sediment load being almost entirely composed of fine clay. However, it is important to recognise the role of lithology. As mentioned above, the amount of sediment is limited by the supply. If hard resistant rocks are supplied, they will persist further downstream, as illustrated by the quartz fragments from the Kelang Gates Ridge in the Sungai Gombak near Kuala Lumpur (Douglas 1981).

Rates of denudation on older Mesozoic and Palaeozoic terrain such as that of peninsular Malaysia (Table 3.7) are two or three orders of magnitude less than those of the Fly in New Guinea or the Serayu in Java. Relief, tectonics, volcanicity and lithology clearly play a major role in the dynamics of tropical landforms.

Milliman and Meade (1983) have made useful calculations of regional sediment yields (Table 3.6) which show the dramatic contrasts between tropical areas. For example, Central America, south of Mexico, has a yield of 400 $t\,km^2\,yr^{-1}$, and the Magdalena catchment one of 900 $t\,km^2\,yr^{-1}$, while northern South America, which includes the Orinoco and Amazon basins, provides only 150 $t\,km^2\,yr^{-1}$. SE Asia and the Himalayas yield 796 $t\,km^2\,yr^{-1}$, while the rest of India provides 154 $t\,km^2\,yr^{-1}$. New Guinea yields are of the order of 1000 $t\,km^2\,yr^{-1}$, while African rivers yield between 17 and 80 $t\,km^2\,yr^{-1}$, except that the Tana River in Kenya yields 1000 $t\,km^2\,yr^{-1}$, largely as a result of human activity. These comparisons show the importance of the tectonically active mountain areas and confirm the low rates of denudation associated with old land surfaces, especially in Africa and eastern South America.

3.6 Response of geomorphic processes to environmental change

Records of environmental change in the tropics are increasing in availability but relatively few have been closely related to landform development. Later chapters in this volume will illustrate instances where this has been done successfully, but our task is now to illustrate how the fluvial processes just discussed would have varied with climatic change. One area where such an analysis is possible, largely through the careful pollen analyses by Kershaw (1975, 1978, 1980), is north-east Queensland. Here at the base of the Cape York Peninsula the Gilbert and Mitchell rivers rise in the highlands close to the east coast and flow out on to the Gulf of Carpentaria plains. Kershaw has obtained a 130 000-year pollen record from Lynch's Crater in the highlands not far east of the Mitchell headwaters (Fig. 2.5). A sequence of five stages of alluvial fan development by the westward-flowing rivers since the late Pliocene has been documented by Grimes and Doutch (1978). The last three of these stages are believed to have occurred in the last 150 000 years and the two earlier stages are probably composites of many more stages, of which little evidence remains.

Concentration on the last 150 000 years will show how the climatic oscillations affected processes (Table 3.8). Throughout the period rainfall appears to

Table 3.8 Environmental change and fluvial processes in north Queensland.

Years BP	Lynch's Crater		Carpentaria Plains
	Rainfall (mm yr^{-1})	Vegetation	River development
c. 8000–present	1500–3500	SNVF–CMVF	stage 5 fans formed after 6000 BP
c. 26 000–8000	750–900	sclerophyll	low sea level 20 000 BP; dry, possible calcrete formation on then exposed Gulf of Carpentaria
c. 38 000–26 000	800–950	LMVF + araucaria	
c. 50 000–38 000	800–1000	LMVF + araucaria	dry conditions, possible aeolian deflation of fluvial deposits; rare floods, short-distance tranport by braided channels
c. 63 000–50 000	1000–1500	CNVF + araucaria	warmer, possibly more seasonal discharge, trend to aggradation of channels; ferruginisation of existing fluvial deposits
c. 79 000–63 000	1000–1500	LMVF + araucaria	cooler, dissection of fans
c. 86 000–79 000	2500–3000	SNVF–CMVF	Humid, continued fan development by seasonal floods associated with cyclones
c. 116 000–86 000	2000–2500	SMVFR	migration of anastomosing distributary channels, e.g. Mitchell River fan
c. 123 000–116 000	2500–3300	SNVF–CMVF	stage 3 fans formed
c. 132 000–123 000	800–2500	CNVF + araucaria	

SNVF, simple notophyll vine forest; CMVF, complex mesophyll vine forest; LMUVF,
CNVF, complex notophyll vine forest; SMVFR, simple microphyll vine/Fern forest.

have fluctuated, but the pollen record shows climatic oscillations of a similar magnitude to those recorded in deep-sea cores. Not all the vegetational changes are due to climate alone and fluvial processes could have been affected by tectonism as well as the sea-level changes associated with climatic change. Quantitatively, the lack of knowledge of the seasonality and erosivity of past rainfalls makes extrapolation of empirical relationships between present-day hydrometeorological parameters and sediment yields hazardous (Douglas 1980a). However, the type of channel metamorphosis described by Schumm (1969, 1977) for the Murrumbidgee River in southeastern Australia would be expected to have occurred. Building of fans or inland deltas is probably likely to have occurred during periods of high sediment yield and runoff, with lateral migration and erosion of fans by meandering streams possibly during periods of high runoff but low sediment yield. Dry periods might well have seen single straight channels and wind erosion of interchannel areas, with any excess runoff discharging into a brackish or possibly fresh Lake Carpentaria surrounded by low plains linking Australia to New Guinea (Torgersen *et al.* 1983).

It is likely that the lowering of the sea level during glacial periods would have reduced the frequency of tropical cyclones in the present north Queensland land area. Less available latent heat, less disturbed conditions and decreased migration of the Intertropical Convergence Zone, reduced depth of the tropical easterlies, fewer incursions of upper tropospheric level cold pools and anticyclones would all provide an assemblage of conditions less favourable to tropical cyclones (Oliver 1980). As cyclones are the chief mechanisms producing runoff and sediment yield in the rivers draining westwards to the Gulf of Carpentaria (Douglas 1973), any reduction in their frequency would lead to a decrease in fluvial sediment transport and thus in the development of alluvial fans. Although examination of the sediment yield–mean annual precipitation relationship (Langbein & Schumm 1958, Schumm 1968) suggests that sediment yields would increase with decreasing vegetation cover below about 1000 mm mean annual precipitation, the most effective periods of runoff and sediment yield would be the periods of increasing sea level and return of cyclonic activity at the end of a dry phase. Such a sequence of events would conform to the suggestion that geomorphic work done would be at a maximum immediately on a change towards more humid conditions (Knox 1972, Douglas 1980b). In the periods before dense ground cover became established even in the headwater areas, more frequent heavy rains probably would have evacuated the sediments mantling the pediments and footslopes of the ranges. Rivers are likely to have developed wide shallow braiding channels which frequently overflowed in the low-angle slopes towards the Gulf of Carpentaria, developing anabranches and distributary channels. Abundant evidence of prior channels can be traced on these Pleistocene fans. Streams like the Gilbert River still have wide beds choked with sand which is only shifted when cyclones cause huge floods that turn much of the Carpentaria plains into a temporary lake or virtual extension of the Gulf.

Off the western coast of Africa, near the mouth of the Zaire River, evidence of fluvial deposition in the period *c*.40 000–33 000 BP during the interstadial characterised locally by the Inchirian transgression, reveals fine material carried by frequent high flows from abundant and regular rains (Giresse 1978). Vegetation at this time was 80% forest (Caratini & Giresse 1979) and probably the fluvial sediment loads were low by world standards and of similar magnitude to those of the Zaire River today (Table 3.6).

During the subsequent lowering of sea level, from *c*.30 000 to 12 000 BP, vegetation changed to a savanna form (Caratini & Giresse 1979) and seasonally active rivers flowed occasionally with sufficient energy to dissect the estuarine accumulation of the Inchirian transgression and to carry coarse debris out to sea (Giresse 1978). However, further north with the expanded Sahara, the Senegal River ceased to flow into the sea and, at the transition from the arid late Pleistocene to the wetter early Holocene, the River Niger was blocked by aeolian and fluvial sediments (Michel 1980, Diester-Haass 1976, Talbot 1980).

Sedimentation off the mouths of the Congo and Niger rivers was particularly rapid between about 12 000 and 10 500 BP, when deposits accumulated at 1.6 m/1000 years (Giresse *et al.* 1982). This rapid activity corresponds to the phase of rapid erosion in the Simen Mountains of Ethiopia (Table B.1; Hurni 1982). A core from the outer Niger Delta suggests that peaks of river discharge occurred at 13 000–11 800 and 11 500–4000 BP. Furthermore, sedimentation rates increased 20-fold between 11 500 and 10 900 BP (Pastouret *et al.* 1978); Street (1981) has suggested that this was due to the reintegration of the drainage network, to lake overflow in continental Africa and to intense runoff from still partly unvegetated slopes.

The Holocene transgression which followed saw the re-establishment of forest vegetation, renewed development of coastal mangroves and the gradual infilling of the deep channels, now drowned, that had developed during the preceding low sea level. By about 8000 BP the rivers were depositing sediments similar to those carried today, but a slight lowering of sea level at about 4000 BP saw some coarse deposits accumulating and some wind action on coastal deposits (Giresse & Moguidet 1980). A later rise in sea level saw the infilling of coastal lagoons, evidence of fluvial transport of coarse debris during floods, perhaps reflecting bank erosion of coarse material accumulated in a more arid phase when fluvial transport was less efficient, and accumulation of organic matter. Scanning electron microscope studies of quartz grains show that fluvial energy and silica solution were active at this time, implying high runoff and high temperatures (Giresse & Le Ribault 1981). Thus by about 2000 BP the present-day forests were fully established and the fluvial processes were akin to those of today. Despite the evidence of the coastal and offshore deposits, it remains difficult to interpret sediment sources and to know whether the material deposited at the river mouth relates to that being supplied to the stream at the present time, or merely that which the river is reworking from older deposits accumulated under a past set of environmental conditions.

At least in the Amazon and Orinoco basins there is evidence to suggest that much of what is being supplied to the rivers comes from the erosion of old river terraces. As Tricart describes in Chapter 10, much of the sand in present-day rivers is reworked from such older deposits as the relict dunes of the Llanos of Venezuela (Tricart 1974b). Deposits coarser than those of the present day are found both in great low-angle alluvial fans developed by major rivers and in the piedmont fans along the eastern flank of the Andes (Tricart & Michel 1965, Tricart & Millies-Lacroix 1962). Sinuous channels now cross fossil parallel dune ridges. Short reaches of river run parallel with the ridges. Where the channels cut through ridges, their incised channels derive sediment from the old dunes. Present-day fluvial processes and landforms in this part of the humid tropics are as much a reflection of the legacies from the drier periods of the Quaternary as of the dynamics of rainforest ecosystems. Thus, while the study of present-day processes provides an indication of what may happen in environmental change, rates of denudation measured today often show the ability of rivers to evacuate debris left from past climates as much as the work of the present climate on the slopes. When this is added to the differentiating role of tectonic activity described earlier, the task of explaining how the present climate affects the landscape is difficult on anything more than a small watershed study of an area of less than about 10 km^2. Nevertheless, at least the study of present-day processes does show the diversity of geomorphic work rates in the tropics and warns the students of environmental change of the difficulty of accurate environmental reconstruction.

4
Aspects of present-day processes in the seasonally wet Tropics of West Africa

J-M. Avenard & P. Michel

4.1 Introduction

In French-speaking West Africa, over the last 25 years, research on present-day geomorphic processes has progressed on two fronts:

(a) generally qualitative studies by geomorphologists, either regional (Michel 1973) or thematic (Avenard 1969) in character;
(b) applied agronomic studies carried out by pedologists concerned with soil conservation and crop protection.

Indeed, with the reduction in length of the fallow period, through combined demographic and socio-economic pressures, accelerated erosion phenomena have appeared in many places. Furthermore, under the effect of the growth in demand for primary products, a rapid extension of land clearance by mechanical means has been considered beneficial. Faced with all-too-often proven setbacks, agronomists have blamed the fragility of tropical soils and the effects of intense, highly erosive rainfall and high temperatures, before even attempting to adapt cultivation practices that have been well proven in temperate latitudes.

In order to confront this pressing problem, the Office de la Recherche Scientifique et Technique Outre-Mer (ORSTOM) and the Instituts Français de Recherches Appliquées, at the instigation of Professor Fournier, have, since 1954, set up a network of experimental field plots. The aims of this programme are to evaluate the magnitude and causes of erosion, the factors that alter its effects and soil conservation measures suitable for African conditions (Roose 1977). These investigations rely on quantitative assessment of the factors affecting erosion and runoff,* in the general framework of the Wischmeier

* Without going into details, it may be useful to define the terms 'runoff' and 'erosion' ('*ruissellement*' and '*érosion*' in the original French version – Eds). Hydrologists and soil scientists use 'runoff' only to define a term in the water balance (the water that flows over the surface of

equation. Thus, the ORSTOM pedologist, Roose, has undertaken a series of experiments on the processes of erosion, soil creep and impoverishment, and vertical and lateral eluviation in different ecological situations from extremely dry ferallitic soils to humid tropical ferruginous soils (Roose 1967, 1968, 1971, 1972a, b, Roose & Bertrand 1971).

Although the aims of these two types of research, geomorphological and pedological, are not identical, it is to be regretted that more fruitful collaborative efforts have not been made, despite one or two attempts (Avenard & Roose 1972). Geomorphologists in particular have neglected the results that can be obtained from the observation of examples of erosion and the calculation of rainfall erosivity.

4.2 Morphoclimatic conditions

The seasonally wet intertropical area of West Africa (Fig. 4.1) forms a transition zone between the humid tropics and the deserts. Attention may be directed to the general climatic features which in turn determine the importance and nature of present-day processes. Rainfall generally decreases northwards (from an average of 1300 mm yr^{-1} in the Guinean zone to about 600 mm yr^{-1} on the dry margin). Towards the north, aeolian action begins to dominate over water action. However, this generalisation needs to be modified in the light of the special climatic characteristics of this zone: first, irregular rainfall with, as a corollary, exceptional intensities that set in motion rare but highly effective processes; and, secondly, the importance of inherited morphoclimatic features that determine the general character of the land surface (*modèle*) on which present-day processes work either in consort or in opposition. Other factors that intervene are the nature of the substratum and superficial deposits and the protective role of savanna vegetation.

This transitional nature also affects the relationship between soil and landform development in a twofold manner: in humid periods, conditions approach those of the equatorial zone and weathering dominates; whereas in dry periods, conditions resemble more closely those of the semi-arid or arid zones. One can perceive here that morphogenetically critical conditions can occur at the passage from one set of conditions to the other, such as at the onset of the rains when the dry, but unprotected, soil is particularly susceptible to surface runoff, just at the moment when the dry vegetation provides the least efficient protection for the soil (Tricart 1965). However, at the same time, the

---

the soil). For geomorphologists, 'runoff' comprises the effects of water flow through the superficial layers of the soil (pedologists refer to this as selective surficial layer erosion)(*érosion sélective en nappe*). In the same way, for pedologists, 'erosion' defines the total solid matter lost (both fine material carried long distances in suspension and coarser material travelling short distances), while they use 'chemical erosion' to cover the migration of solutes in surface runoff and drainage waters. Geographers talk more readily of 'surface lowering' (*ablation* = removal – Eds), 'erosion' being reserved for the complete surface lowering–transport–deposition complex.

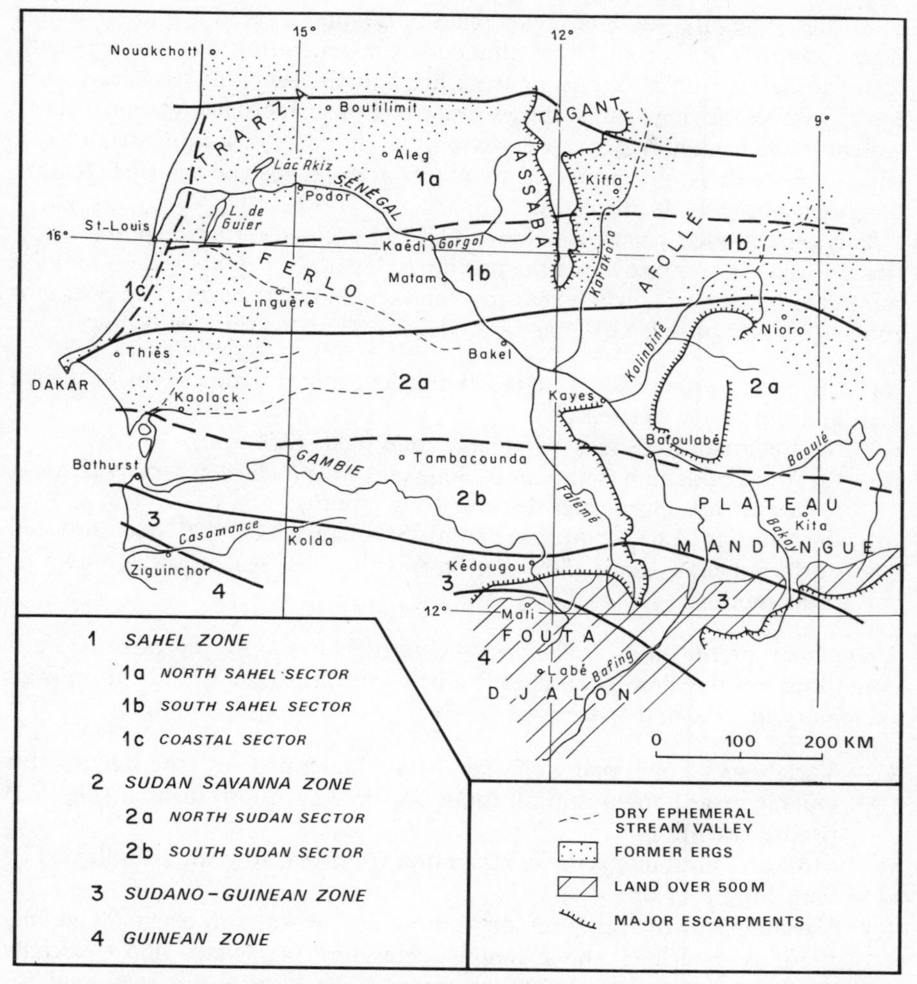

Figure 4.1 Principal bioclimatic zones of the West African seasonally wet tropics.

landforms themselves create differences: humid depressions and hollows allow soil moisture to be retained, giving rise to weathering conditions close to those of the humid tropics, whereas the drier interfluves experience conditions resembling those of arid areas.

The climate

General information. The rainfall regime depends above all on the existence of two principal tropical air masses, continental and maritime, which confront each other along the Intertropical Convergence Zone (ITCZ). The annual migration of the ITCZ leads to the appearance of two contrasting seasons: the dry season and the wet season (winter) (Mietton 1980).

As the generally subdued relief causes no major modifications to this regime, rains result from either thunderstorm-generating disturbances with convectional or frontal origins or from line-squalls, generally associated with easterly waves moving against the general trend of surface airflow. Particularly violent at the beginning of the wet season, the thunderstorm rains striking soil which is bare or poorly protected by plants are perhaps the principal feature of present-day fluvial processes.

This general mechanism, however, does not exclude certain regional variations. From the south towards the north, the length of the wet season and the rainfall total decrease, while year-to-year variability increases. The principal zones thus recognised, as defined by isohyets, (Fig. 4.1) are as follows:

(a) the Sudano–Guinean zone (1250 mm and more) with a tropical regime and dominant wet season;
(b) the Sudanian (or south Sudanian) zone (900–1250 mm);
(c) the north Sudanian (or Sudano–Sahelian) zone (600–900 mm) with a wet season of not more than three or four months;
(d) the Sahelian zone, comprising south Sahelian sector (400–600 mm) and north Sahelian sector (200–400 mm) with a wet season of only two or three months.

Irregularity of the rains. The zones defined above only represent average conditions as the dominant characteristic of this zone is one of rainfall variability in time and space.

(a) Variability from year to year is as illustrated by the record for Ouahigouya; annual rainfall totals plot in a sawtooth fashion (Fig. 4.2) (Pallier 1978).
(b) Monthly variability is no less than that from year to year, as indicated by data for Pô (Table 4.1).
(c) Variability of the rainy season and the dry season also occurs. The fluctuations described above mainly determine the length and especially the commencement of the wet season. The latter event is crucial for agriculture as it determines the success of the sowings. Baldy (1977) has pointed out that, although the normally adequate May rains at Bobo-Dioulasso average 97 mm, in one year out of ten they are only 36 mm, and in one out of four only 64 mm, which clearly would not lead to a successful sowing. The distribution of rain from day to day in the month is also important: it would be comforting to know that, three years out of four, there would be enough rain in the last ten days of June if one had finished a sowing in the first ten days of the month.

However, this variability has important consequences on the moisture status of the soil and thus on the quantity of runoff. It is known that the infiltration rate in an originally dry soil decreases in time as a result of the progress of a wetting front and the formation of a sealed splash surface (Roose 1977).

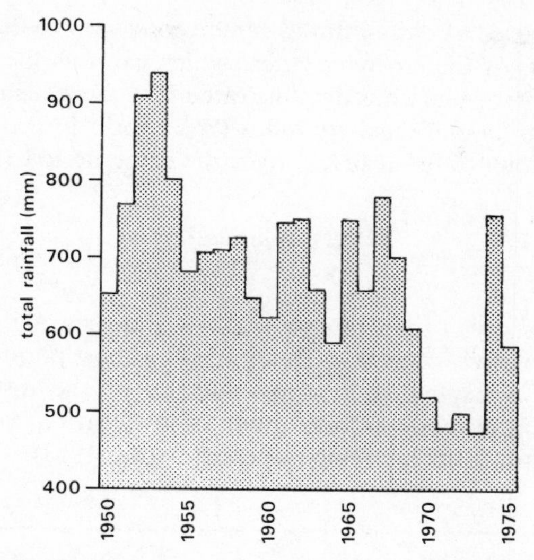

Figure 4.2 Variability of annual precipitation of Ouahigouya.

The same sort of variability occurs from one rain gauge to another as a result of the localised cellular nature of thunderstorm rainfall. Such characteristics have important repercussions for geomorphic processes as they give rise to a special kind of erosivity or climatic aggressivity.

Climatic aggressivity. Several authors have attempted to define a climatic aggressivity index and have established correlations between sediment transport and rainfall, or good linear regressions between the kinetic energy and the quantity of rainfall (Fournier 1960, 1962, Charreau 1969, CTFT/HV 1973).

At present, it seems that the research of Roose, using the equation developed by Wischmeier and Smith (1960), provides the best approach to this problem. By converting successive daily rainfall totals for experimental sites to the erosivity index using this regression, it has been found that the average error over five years is not more than 5% of the mean index value. Thus long

Table 4.1 Variability of annual and monthly rainfall at Pô, Upper Volta (after Mietton 1980).

	Maximum value	Year of occurrence	Minimum value	Year of occurrence
annual total rainfall	1429	1950	503.3	1977
May monthly total rainfall (mm)	209.4	1958	25.5	1965
August monthly total rainfall (mm)	568	1950	86.8	1976
number of rain days	84	1967	39	1953

20- to 50-year records of daily rainfall can be converted to determine monthly and annual means of the erosivity index, which can then be mapped (Roose, *et al.* 1974). This procedure has demonstrated that there is simple relationship between the mean annual erosivity index (R_{am}) for a five- or ten-year period and the mean annual rainfall (H_{am}) over the same period (Roose 1977):

$$\frac{R_{am}}{H_{am}} = 0.50 \pm 0.05$$

The preparation of a map of the erosivity index, using this relationship, has shown that erosivity decreases in a manner almost parallel to the isohyets (Fig. 4.3). This is explained by the similarity of the distributions of the intensity–duration curves, the magnitude of once in ten-year rains and the mean annual rainfalls in the region (Brunet-Moret 1963).

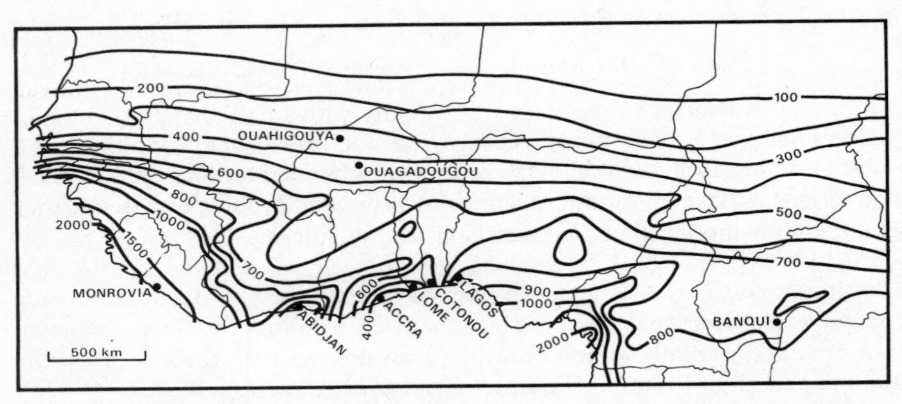

Figure 4.3 Map of climate aggressivity based on Roose's mean annual erosivity index.

The substrate.
Even though contrasts in rock type cause the operation of processes to vary, so creating phenomena of differential erosion, several factors tend to make present-day processes fairly uniform.

Weathering produces much clay. The resistance to erosion of these largely kaolinitic clays is low. The few soil mechanics tests that have been made give liquidity limits above 30% moisture, often up to 50% moisture, while plastic limits are between 20 and 35% and shrinkage limits are of the order of 15 or 20%, with values sometimes higher (for example 27.1% in weathered slate). These properties influence the nature of present-day processes because:

(a) shrinkage cracks are virtually non-existent, even when drought prevails, thereby limiting the amount of infiltration during the first rains of the wet season;

(b) the high plastic limit reduces the role of landslips;
(c) the very high liquid limit practically prevents all solifluction, superfical rotational slip or mudflow phenomena (Avenard 1962).

Rock types vary little; granites and schists underlie nearly all the area, while slates strikingly dominate the landscape as buttes rising 300 to 400 m above the general level of the land surface. Indeed the role of rock type is often expressed indirectly depending on whether the surface is duricrusted, a relict slope or other feature. These factors determine whether surface runoff or infiltration dominates.

Morphoclimatic legacies

Present-day processes operate on pre-existing landforms, either perpetuating or interrupting the evolution of those features. The inherited forms often play a dominant role in the action of present-day processes, as a result of the slopes that they have created and through the influence of relict duricrusts. Such influences, in terms of slope angle, length and shape, have long been studied by Roose (1977) who, for example, has observed at Sefa, Senegal, that removal of material and surface runoff increase rapidly with small changes in slope.

Processes thus act in a framework of renewed erosion from the edges of granitic domes, inselbergs and ancient duricrusted surfaces, relics of which still protect the landscape. Under the morphogenetic processes of the present climate, these more or less reduced and stripped duricrusted buttes supply debris to long, gentle slopes which develop at their base as extended glacis.

The vegetation cover

Even though the typical vegetation of the seasonally wet tropics is savanna, in reality this very general category covers a range of plant formations which protect the soil with varying degrees of effectiveness. On the whole, the relative ineffectiveness of the vegetation screen has two consequences. The first is poor protection against variations in temperature and runoff, although this varies with the physiognomy of the savanna, wood savanna offering better protection than treeless savanna. The second is the irregular, uneven protection of the soil. The tussocky character of the grass leaves part of the soil bare, allowing runoff to develop without hindrance. Once started in this manner, such erosion is self-reinforcing. On the other hand, seasonal variations in plant metabolism are important, because vegetation differs in its protective efficiency from season to season. Dense wet-season vegetation intercepts the rainfall and protects the soil, while the poor protection offered by sparse dry-season vegetation can be further reduced by bush fires. Again it must be noted that the first rains of the wet season fall on poorly protected soil.

Bush fires can be important, for their occurrence can lead to much erosion. Measurements at three stations (Avenard & Roose 1972) have shown that:

(a) in the Guinean zone (Bouake, Ivory Coast), soil loss was greatly reduced, falling from 182.4 kg ha^{-1} yr^{-1} in 1967 to approximately 50 kg ha^{-1} yr^{-1} in 1968–9 and to less than 2 kg ha^{-1} yr^{-1} in 1970–1, once the plot had been protected from bush fires;

(b) in the south Sudanian zone (Korhogo, Ivory Coast), surface soil loss ranged from 100 to 200 kg ha^{-1} yr^{-1} according to the importance of bush fires;

(c) in the Sudanian zone (Gonse, Upper Volta), this soil loss increased from 50 kg ha^{-1} yr^{-1}, when the savanna was not burnt, to between 150 and 200 kg ha^{-1} yr^{-1} when bush fires occurred.

Protection against fire and the development of an increasingly dense vegetation lead to better absorption of the kinetic energy of falling rain and a reduction in the effectiveness of runoff. The excellent qualitative description provided by Monnier (1968) is thus confirmed by precise measurements.

Many trends are leading to further vegetation degradation especially around towns. At Ouagadougou, the rapidly expanding capital of Upper Volta, the search for wood to make charcoal for urban domestic use has led to the replacement of the tree savanna by a dominantly grassland vegetation. Even the latter has been affected, for a comparison of aerial photographs taken in 1956, 1961 and 1980 has shown a great increase in the number and size of patches of bare soil.

4.3 Processes and their effects on slopes

The general environmental conditions described above determine the nature of present-day processes: water, derived from rainfall, and wind, of increasing importance towards the north, provide the energy to initiate sediment transport.

The impact of falling rain on the soil

Varying according to the protective role of the vegetation and the erodibility of the soil, the impact of raindrops is the first stage in the erosion of surface materials. Thunderstorm rains at the end of the dry season are particularly effective as the soil is poorly protected: the energy of the raindrops breaks up aggregates, little craters being formed by the rebounding of fine droplets loaded with suspended matter. As Mietton (1980) notes, the best way to appreciate the power of raindrop splash is to observe it at maximum rainfall intensities. Soil aggregates are broken apart and material is set in motion by the sheet of surface water that begins to run off. It is, however, difficult to separate this splash effect from the action of runoff once any slope, however slight, is involved. In effect, a discontinuous layer of water is formed, which displaces the splashed particles, depositing them a little further down slope. The slope surface is thus subject to an irregular glazing with clay which dries out in the dry season to form a crust, or splash skin, of the same composition as the underly-

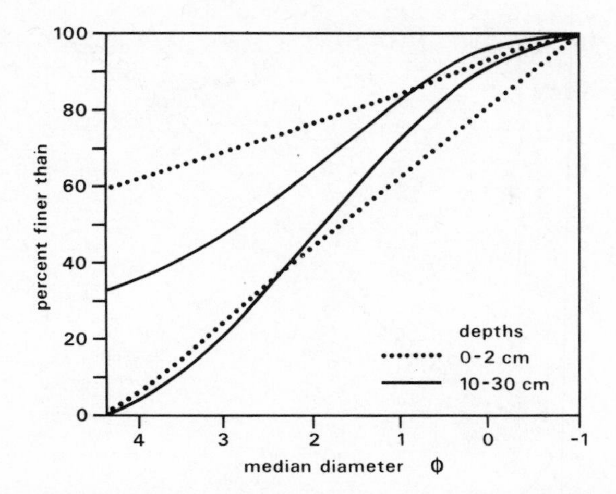

Figure 4.4 Mean particle size distribution for surface soil material and underlying material at two sites.

ing material (Fig. 4.4.) Antecedent soil moisture and rainfall intensity and duration are important factors acting against each other in this process.

Towards a typology of runoff

Studies currently in progress in Africa are likely to modify the terminology adopted here. The roles of the splash skin (*croûte de battance* or *pellicule de glaçage*) and of subsurface runoff (*ruissellement en nappe*) have not yet been adequately defined.

The following suggestions are based on previously published descriptions and some further more recent observations: a statistical study using edge punch cards on which are summarised environmental conditions and granulometric analyses is under way.

Incipient runoff (ruissellement aérolaire). Mietton (1980) suggests that, whatever the nature of the runoff which eventually occurs, it begins by the formation of small patches of localised water movement which end at little deposits of debris below which water seeps into the soil. This initial stage varies in duration, as the state of the soil surface, the antecedent soil moisture conditions and the nature of the rain collectively determine the lag before runoff commencement. Incipient runoff involves the washing of small fragments of debris down slope to the next obstacle, up slope of which a small accomodation of either sand or plant debris develops (ULP-CGA 1977). This selective surface degradation, while small in magnitude, may, through its repeated occurrence, help to explain the formation of splash skins.

Embryonic runoff (ruissellement pelliculaire ou embryonnaire). The flow of small threads of water on slopes of 2 to 3° for short distances down slope can

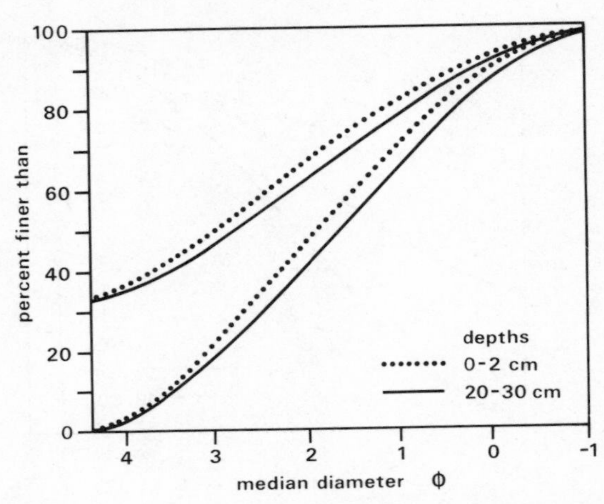

Figure 4.5 Mean particle size distributions for surface and subsurface materials affected by embryonic runoff.

transport fine material (ULP-CGA 1977). The granulometric curves of slope materials affected by this type of runoff characteristically show a loss of fine material (Fig. 4.5). Similarly, the undermining of vegetation tussocks, or the development of little steps on bare slopes, indicates the role of this type of runoff in slope development.

Diffuse runoff. This type of runoff accentuates the characteristics of the previous type – small threads of water flow become little streams, capable of transporting coarser, sand-sized debris over relatively long distances by repeated short movements during each rain event. This process can expose the roots of tussocks and create large slabs of bare soil on which vegetation has difficulty in becoming re-established. The crust that develops after rain becomes impermeable and is broken up into a series of miniature steps by subsequent storms (Fig. 4.6).

Concentrated runoff. Small rills, not easily eradicated by normal ploughing, tend to persist from one year to the next. The rills form a connected drainage network ending in small fan-like spreads of debris further down slope. In certain cases these rill systems allow the transport of debris down slope through a series of erosion–deposition sequences. Steeper (3 to 4°) slopes are incised by these rills, which give way to spreads of debris on gentler slopes. On these debris spreads, the water loses its competence and develops anastomosing channels, which come together again where the slope steepens once more.

A consequence: generalised slope stripping

These different types of runoff are generally interlinked, either in time or in space, giving rise to a general, insidious degradation of the whole slope, posing

Figure 4.6 Formation of miniature steps by dissection of the surface wash crust.

serious difficulties that are not always perceived by anyone attempting to grow crops.

Concentration of water: linear incisions

Linear gully incisions, found in many localities, result from present-day processes, but develop under conditions that are difficult to predict. They are not due to base-level change or to exceptional climatic erosivity. Except in exceptional circumstances, it seems that degradation of the vegetation by human intervention is the main cause: the instability caused in this way sets in motion a series of morphogenetic changes which threaten the stability of the whole denudation system (Fig. 4.7).

Wind action

Although difficult to measure, the effects of wind action are so clearly expressed in the landscape that they cannot be ignored. Deflection carries away fine particles leaving behind sand and ferruginous pellets which form a surface paving. This paving greatly affects runoff processes. Acting predominantly in the dry season, wind action works to prepare material for wet-season runoff. Accumulation features, or nebkhas, are most frequent in the Sahel zone, but localised wind-blown deposits behind tussocks of vegetation or crops can occur further south. Thus removal of vegetation can also augment the effect of wind action.

Figure 4.7 Linear gully incision in the Pô region, with retreat of gully walls by as much as several metres in a single wet season.

Before the recent Sahel drought, wind action on the fixed Ogolian (15 000 to 25 000 years BP) dunes along the lower Senegal valley was slight. It was only in the places where the grass cover had been disrupted by overgrazing or by trampling by people and animals that surface sand was entrained by the wind (Michel 1968a). Along paths followed by herds of domestic animals converging on the valley, mobile sand was often arranged in a series of ripples.

Wind erosion has always been much more important in the south, on the undulating sandy terrain of the former Cayor erg, between Thies and Linguère (Fig. 4.1). These unconsolidated materials had been cultivated for a long time and the fallow periods had been becoming progressively shorter. Cultivation lowered the organic content of these sandy soils (called dior) with the result that the uppermost horizons lost their structure and became extremely susceptible to wind erosion. During the long dry season, the Trade winds now sweep across these denuded lands, carrying away silts and fine sands. Gradually the surface is becoming covered with a coarse sand layer which the wind cannot transport. Systematic ground-nut cultivation by the pioneer settlers of the Mourides colonies on the south-west part of Ferlo has greatly accelerated this deflation.

Wind action has been further enhanced by the Sahel drought, which has greatly impoverished the vegetation cover. In the Sahelian savanna of northern Ferlo, where the average annual rainfall is 350 mm, only 33 mm fell in 1972.

ORSTOM botanists observed at that time that the herbaceous stratum had entirely disappeared and the tree layer had been greatly reduced, 53% of the *Acacia senegal* being dead. Trees were greatly damaged by pastoralists who lopped branches in an effort to feed their herds.

The fixed Ogolian dunes began to become reworked. On the edge of the former Cayor erg, the semi-fixed coastal dunes of Lompoul became completely gullied, at a latitude with an average rainfall of 450 mm yr^{-1} (Toupet & Michel 1979). Wind erosion has increased as a result of the rapid growth of the Dakar urban area, caused by rural migration. Around the satellite town of Pikine, constructed in the semi-fixed dunes, all the vegetation was utterly destroyed within a few years, making sandstorms and dust storms, almost unheard of before 1970, more and more frequent in Dakar.

Such wind erosion must be studied as carefully as gully erosion and fluvial processes, as it has in western Senegal since 1970 by Sall (1971, 1978, 1979, 1982). He has complemented field observations and measurements at different scales with the interpretation of aerial photographs and satellite imagery.

4.4 Fluvial processes

Stream action in West Africa, as throughout the tropics, is equally dependent on bioclimatic environments. One of us (P.M.) has had the opportunity of studying fluvial processes in the Senegal and Gambia basins during applied geomorphological surveys of alluvium for mineral prospecting and rural development (Michel 1959, 1962, 1966, 1968b, 1973).

The Sudano–Guinean zone

In this zone, where biochemical weathering is so important, the transport of material on the interfluves is largely in solution. Erosion only supplies a small amount of coarse material, which rarely reaches the rivers.

Streams descend from mountainous areas by a series of waterfalls and rapids across resistant rock bands, resulting in extremely irregular long profiles. When crossing resistant layers of quartzitic sandstone or dolerite, channels subdivide into many branches around rocky outcrops and islets. Some sectors of river courses, especially in quartzitic sandstones, follow joint networks and thus changes in direction frequently occur. Channel cross sections are, therefore, extremely varied in character (Michel 1973, vol. 2, Fig. 164).

The rivers are generally fringed with a gallery forest of luxuriant, moisture-loving plants. Such trees protect the banks; their branches reach down to the water level and reduce the velocity of flood flows. The convex profile of the banks suggests that there is little erosion. Forest trees also grow on the mud channel islets and rock outcrops.

The transport of solid material by the rivers is thus greatly reduced. No coarse debris is supplied from the interfluves. Bank erosion is highly localised below rapids and waterfalls and yields only fine particles. This lack of debris results in virtually no abrasion of the rock bars and ledges. Furthermore, these ledges and bars are protected by a black ferromanganese rind covering all rock types, including quartz. Some rock bars, however, are pitted by many small potholes. The potholes appear to be relict features as some contain pebbles cemented into a conglomerate by oxides of iron (Michel 1962); they may have been formed during dry periods when the rivers carried pebbles.

At the present day, deposits are not found and banks of sands and fine gravel are rare, being found only below rock outcrops. Despite high peak discharges, sand transport is not observed; only fine particles are carried long distances in suspension. Thus each rock bar or ledge evolves independently, so maintaining the irregular long profiles of the major rivers.

The Sudanian zone

The mechanical disintegration of rocks and their transport by runoff is more active in this zone. Rivers often flow through sandy–clayey alluvia. Gallery forests along the rivers are discontinuous in the southern part of this zone and disappear altogether towards the north. Moreover, the vegetation has often been cleared to allow for cultivation after rain or following a flood. Rivers can thus erode their banks by undercutting during high flows, attacking both the fine material of recent terrace deposits and the underlying gravels (*graviers sous berge*), which are often cemented by iron oxides. Banks thus have a concave cross-sectional profile.

Rivers in flood carry a considerable sediment load, which may be transported over long distances if the material is fine. However, observations on the Falémé, the principal tributary of the Senegal (Fig. 4.1), and on the middle Gambia and certain of its tributaries have shown that pebbles are only moved short distances. Gravel bars are found immediately down stream from exposures of gravel in the river banks or from outcrops of conglomerates on the older terraces. Petrographic and granulometric similarities between bank and gravel bar materials confirm that such sediment is only carried a short distance (Michel 1959). On the other hand, major floods carry coarse sand considerable distances. Thus at low flows sand and gravel bars are exposed in the bed of the lower Falémé, enabling the river to be forded; much of this bed material is trapped by the Sénoudébou rock bar (Fig. 4.8). In the major flood of 1958, the Senegal deposited considerable quantities of well sorted coarse sand, containing layers of heavy minerals, especially ilmenite, in several places between Bafoulabé and Kayes (Fig. 4.1).

Thus in the Sudan zone, in contrast to more humid regions, rivers have some debris with which to abrade the rock ledges and outcrops at waterfalls and rapids. The Bakoy and the Senegal continue to deepen and enlarge huge potholes in the Billy and Felou rock bars (Tricart 1956b). Channels are enlarged by the cutting of pothole walls which then collect all the flow in dry periods. The surface of the rocks and the walls of the potholes are always covered by a ferromanganese patina. Incision appears to be extremely slow although the

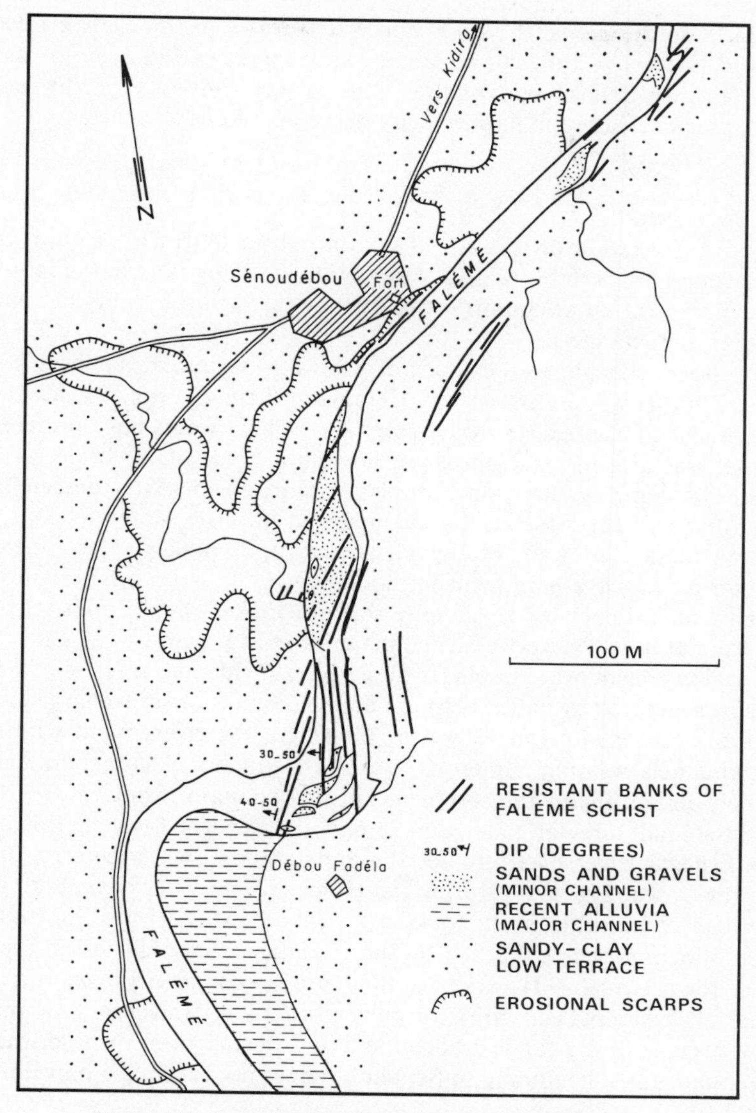

Figure 4.8 Rock bar at Sénoudébou, lower Falémé River,

rate depends on lithology. Little erosion occurs on resistant quartzitic sandstones for example; on this lithology, only small potholes are found on rock bars and waterfall recession is slight.

The major rivers have well developed minor channels through banks of sand and gravel which shift during floods. These are rivers with mobile beds, like those of temperate regions. The Baoulé and the Falémé possess large meanders which first appear in the north of the Sudan zone when the ratio of flow to sediment load decreases: the slope and water inflow to the channels decrease while the load grows through undercutting of banks and gullying of terraces. The rivers then deposit part of their load at any convexity on the inside of meander bends where the flow velocity decreases. Thus a succession of growing meanders develops.

The Sahel zone

The sparse vegetation of this zone offers no obstacle to river action. Intense surface runoff leads to brutal, but often brief, flooding. Discharge falls rapidly, but there is often an important underflow through the sandy bed material occupying the channels.

The Assaba plateau, in the south of central Mauritania, has been strongly dissected in this way by the upper tributaries of the Gorgol, the lowest right bank affluent of the Senegal River (Fig. 4.1). These wadis have preferentially exploited fractures in the sandstones, flowing in some places in deep, narrow gorges. The slope of their long profiles is quite steep. After descending the western flank of the Assaba, the channels cut into the sandy deposits of the piedmont plain. The wadi Bounguel, for example, makes several meanders, undercutting concave banks during floods when it carries a large amount of sand and silt (Michel 1973). Despite the violence of these floods, the major rivers are not able to remove all the material fed to them by the dense, branching drainage networks. In places these rivers divide into a series of unstable channels, separated by natural levées and great banks of sand characterised by bushes of *Acacia nilotica*. However, elsewhere the rivers wind within their major channels creating numerous irregular meanders. Channel gradients are extremely gentle. Below the confluence of its two major branches, the Gorgol flows in a small alluvial valley and rejoins the Senegal River at Kaedi.

The Senegal River itself crosses the Sahel zone down stream of Bakel to rejoin the ocean (Fig. 4.1). During the Holocene, the river created an alluvial valley 10 to 25 km wide, in the form of a great arc (Michel 1973, pp. 560–605). It is an allochthonous river, fed by the abundant rains that fall in the upper parts of the catchment. However, annual floods vary greatly from one year to another. The waters begin to rise at the end of June. However, flow velocities are low because of the gentle gradient and thus the height of the floodwave that passes Bakel at the beginning of September does not reach St Louis until early November.

At the beginning of the seasonal flood, turbulent water undercuts the concave banks of meanders usually of post-Nouakchottian (1800–5000 years BP) sands and silts. The river drops the coarser fraction, the medium sands, as soon as its velocity falls on the convex bank of the next meander. Thus the meanders are either continually enlarged, or redeveloped following cutoffs. The silts carried in suspension may eventually reach the river mouth.

The effectiveness of this lateral erosion depends on the material of the river banks. Silts and sands are easily attacked, but clay layers are more resistant and tend to stand out. The most rapid erosion occurs on the unconsolidated sands of fixed Ogolian dunes. Such erosion can be catastrophic during major floods, as the majority of villages are built on the highest parts of levées on concave banks. Thus, for example, many huts were washed away when the river bank at Mouderi was eroded back by 5 m during the 1964 floods (Michel 1966). Observations suggest that concave river banks are cut back by an average of 1.0–1.5 $m\,yr^{-1}$

The flooding of the major channel occurs initially through exploitation of crevasses across levées. As the water level rises, the waters pass through bigger crevasses. Silts and fine sands are deposited on the convex banks of meanders. Backswamps become progressively submerged, remaining under water for 4–12 weeks, depending upon the size of the flood. With virtually stagnant waters, the suspended particles gradually settle out and are deposited.

Water levels fall as a result of evaporation or drainage as the flood recedes: some infiltration takes place, temporarily saturating the clay soils. However, in some parts of the most poorly drained depressions, water stands for much longer and leads to a much greater deposition of clay and higher moisture levels. Many swamps thus persist until the end of the dry season despite the very high evaporation rate.

Upstream of Podor (Fig. 4.1) many sandbanks appear in the minor channel of the river at baseflow, in places occupying two-thirds of the river bed. The surface of these sands may be fashioned into dune bedforms, with the steep face of the dune on the downstream side. Certain sand layers contain heavy minerals and their high ilmenite content indicates the transport of fine sand from much further upstream (Michel 1973). Where the river touches the eastern edge of its valley, several rock ledges of Lutetian quartzitic sandstone outcrop in the channel bed. The river has not eroded these bars and it seems to be unable to incise its channel into the outcrops of resistant rocks down stream of Bakel, although these strata rocks do seem to divert fluvial energy towards lateral erosion at times of flood.

When the water recedes from the major channel, the vertisols and hydromorphic soils of the depressions crack and develop gilgai features. Saline depressions (sebkhas) in the delta area are often subjected to wind erosion, particularly at the end of the dry season when the clay material has been well broken up by salt crystallisation (Tricart 1954, 1955). At this time of the year two types of strong wind, the maritime trades and dry 'tornadoes' blow. Some of the windblown material accumulates on the southeastern edge of the sebkhas as lunettes of saline silt aggregates. The high, fluviodeltaic, post-Nouakchottian levées of the lower valley and the delta, which have a sparse vegetation cover, are also swept by dry-season winds. Sandy silt accumulations form successions of flattened nebkas fixed by small herbaceous plants. This aeolian microrelief often develops around villages where vegetation has been destroyed by overgrazing and trampling.

Throughout this area, population growth, the consequential reduction in bush fallows, and the general geomorphic instability caused by the destruction of vegetation has led to an accentuation of erosional processes, and in particular mechanical action. These effects of human activity illustrate the instability and fragile nature of the seasonally wet tropical zone; every disruption of an ecological equilibrium initiates an acceleration of present-day geomorphic processes. Landform instability is thus becoming more widespread in these complex environments where management is difficult.

5

The influence of climate, lithology and time on drainage density and relief development in the tropical volcanic terrain of the Windward Islands

R. P. D. Walsh

5.1 Introduction

Although much work has been done on drainage density and drainage networks since the classic studies of Horton (1945) and Melton (1957), remarkably few exist on (a) drainage densities of high rainfall areas, particularly in the tropics, and (b) rates of drainage network development and adjustment to changed environmental conditions. Thus, of 44 studies cited by Gregory (1976) in a recent review of world patterns of drainage density, only five came from the humid tropics and only one covered areas with annual rainfall exceeding 3750 mm. With the exceptions of the short-term study of Schumm (1956) of waste tips in New Jersey and the less detailed studies of North American tills (Ruhe 1952, Leopold *et al.* 1964) and upraised lake floors (Morisawa 1964), there has been very little study of real-world (as opposed to theoretical) drainage network development. Furthermore, almost nothing is known about the response times of drainage networks (as indeed of other geomorphological variables and soil and vegetation) to changes in climate.

The volcanic landscapes of the Windward Islands, which range from Miocene to Historic in age and where annual rainfall ranges from 1000 to 10 000 mm, provide an opportunity for both these questions to be investigated. This study presents drainage density maps of the Windward Island group and attempts to examine the influence of climatic, soil, lithological and time factors in accounting for variations in drainage density and relief development within the islands. First, however, the nature and range of environments found within the Windward Islands are briefly described.

The Windward Islands

The Windward Islands, which comprise the formerly British islands of Dominica, St Lucia, St Vincent, the Grenadines and Grenada, lie between 12 and 16°N in the Eastern Caribbean (Fig. 5.1). The islands form part of the

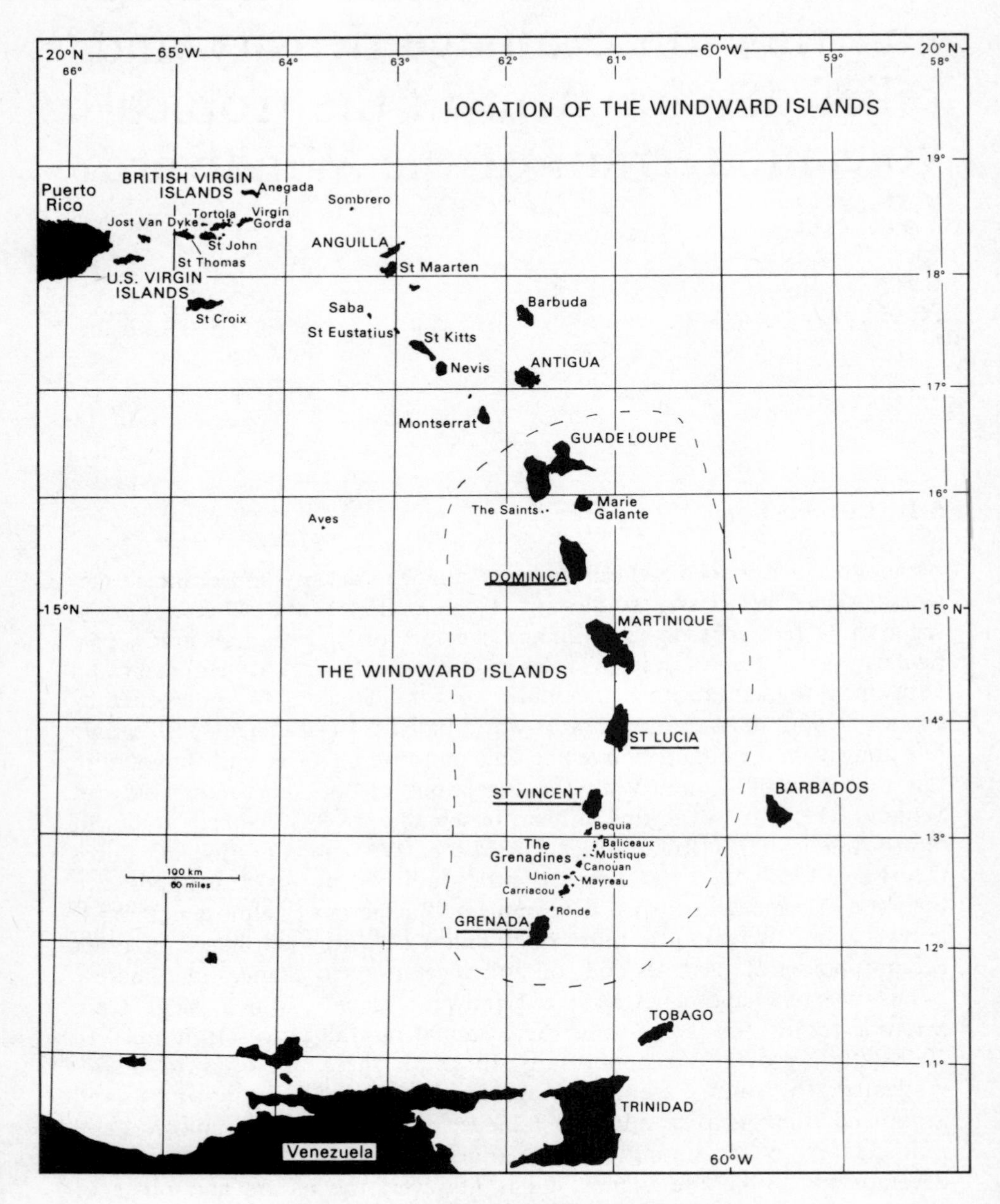

Figure 5.1 Location of the Windward Islands.

Lesser Antillean double island-arc system and are composed almost entirely of volcanic deposits ranging from Miocene to Historic in age (Arculus 1976, Nagle *et al.* 1976). Grenada and St Vincent are composed of basalts and basaltic andesites, whereas in the more northerly islands of St Lucia and Dominica acid andesite and dacite rocks predominate. There is also a contrast in the physical nature of the rocks between the islands. In Grenada and St Vincent lava flows commonly outcrop in the mountains, and ash (and in the case of Grenada reworked ash) deposits cover much of the surface. In St Lucia and Dominica, on the other hand, permeable pyroclast flow deposits predominate and bouldery lava forms the highest peaks; fine ash and lava flows are comparatively rare. All the islands are mountainous, the highest peaks being Diablotins (1422 m) and Troisè Pitons (1406 m) in Dominica, Morne Gimie (950 m) in St Lucia, Soufriere (1220 m) in St Vincent, and Mt St Catharine (849 m) in Grenada. Steep slopes, often in excess of 40°, and rugged relief are predominant throughout the island group.

The islands lie in the trade wind belt and are characterised by a humid tropical climate. Mean annual temperature at sea level is around 26°C and the annual range is very small, invariably less than 4°C. Rainfall amounts and seasonality vary greatly with altitude and aspect in the islands. In the coastlands, rainfall occurs chiefly in summer and autumn, when the Intertropical Convergence Zone lies close to the islands and easterly waves and other tropical disturbances affect the region. The winter months are dry because of the proximity of the Azores subtropical anticyclone with its low-level inversion and descending stable air. However, in the mountainous interiors of the islands, orographic uplift and orographic-induced instability result in substantial winter rainfall and enhanced summer rainfall. Rainfall statistics for a representative number of stations in the islands are given in Table 5.1. In Dominica, climates range from highly seasonal on the sheltered leeward coast and weakly seasonal on the windward (eastern) coast to perennially very wet in the central mountains, with mean annual rainfalls ranging from 1000 mm on the leeward coast to as high as 10 000 mm in the central mountains. In St Lucia, annual rainfall ranges from 1250 mm in the southern and northern penisulas to 3750 mm in the west-central mountains. In St Vincent, annual rainfall varies from 1600 mm in the extreme south to an estimated 6250 mm over the northern mountains, although the absence of mountain stations in St Vincent makes annual rainfall maps less reliable than for the other islands. Grenada is the least mountainous island and annual rainfalls are more moderate, ranging from 1100 mm in the south-west and north-east peninsulas to 3750 mm in the western mountains.

Heavy daily rainfall magnitude–frequency likewise varies considerably with altitude and aspect within the islands and strong positive linear relationships were found to exist between heavy daily rainfall parameters and annual rainfall. In Dominica, for example, correlation coefficients obtained for the relationships between mean annual rainfall, on the one hand, and the frequencies of daily rainfalls of at least 25.4, 50.8, 76.2 and 127 mm and mean annual

Table 5.1 Rainfall data for some Windward Island locations.

Station	Island	Altitude (m)	Length of record (yr)	Mean annual rainfall (mm)	Rain days per annum		Mean annual maximum daily rain fall (mm)	Highest recorded daily rain fall (mm)
					$\geqslant$76 mm	$\geqslant$127 mm		
Trafalgar	Dominica	260	15	5920	–[a]	–	–	349
Brantridge Can.	Dominica	440	2	5788	11.5	2.00	(149)	–
Wet Area	Dominica	335	3	5432	7.0	1.00	(139)	–
Quilesse Nurs.	St Lucia	305	11	4045	–	–	157	383
Belvidere	Grenada	350	31	4026	–	–	136	260
Annandale	Grenada	210	38	3541	1.4	0.40	123	233
Copt Hall	Dominica	105	8	3445	3.5	1.10	138	–
Les Avocats	Grenada	335	34	3440	–	–	151	383
Bath Nursery	St Lucia	305	30	2605	2.1	0.30	106	297
Samaritan	Grenada	90	50	2226	1.6	0.33	106	287
Camden Park	St Vincent	30	24	2158	1.6	0.17	102	165
Morne Bruce	Dominica	120	33	2151	1.6	0.40	104	191
Botanic Gardens	Dominica	12	53	1922	1.4	0.30	106	217
Lower Marli	Grenada	70	48	1720	1.3	0.11	89	198
La Fargue	St Lucia	70	16	1547	1.0	0.20	92	195
Moule à Chique	St Lucia	150	27	1278	–	–	82	144
Colihaut	Dominica	60	6	1269	0.8	0.00	86	–
Calliste	Grenada	30	17	1089	0.6	0.10	71	169

[a]The dash indicates insufficient data.

maximum daily rainfall on the other were +0.998, +0.98, +0.95, +0.88 and +0.88 respectively. Mean annual maximum daily falls (Table 5.1) range from 71 mm at Calliste (mean annual rainfall 1032 mm) in south-west Grenada to 157 mm at Quilesse (mean annual rainfall 4045 mm) in the St Lucia mountains. The mean annual frequency of daily falls of 76 mm and more likewise ranges from less than once per year at stations with less than 1500 mm annual rainfall to as high as 11.5 at Brantridge (mean annual rainfall 5788 mm) in central Dominica. Extreme falls of between 250 and 380 mm have been recorded at many mountain stations, usually as a result of tropical cyclones, but at coastal stations, daily falls rarely exceed 200 mm. Fifty-six tropical cyclones passed through the island group between 1900 and 1975, but only 13 of these were of hurricane intensity.

Natural vegetation, of which very little remains except in Dominica, varies with annual rainfall and seasonality and has been mapped in detail by Beard (1949). Deciduous seasonal forest is the natural vegetation of the driest areas with less than 1250 mm annual rainfall. Semi-evergreen seasonal forest covers the rest of the areas with a marked dry season (3–5 months) and annual rainfalls of 1250–2000 mm, notably large areas of St Lucia and Grenada and the leeward coastlands of St Vincent and Dominica. Rainforest is the natural vegetation of the wetter areas with either a very short or no dry season and an annual rainfall of at least 2000 mm; in Dominica, undisturbed rainforest covers large areas of the interior, but in the other islands most of the original rainforest has been cleared for agriculture. On exposed mountain ridges and peaks and on excessively steep slopes, rainforest in turn gives way to stunted montane vegetation, although the area covered by these formations (montane thicket, elfin woodland and palm brake) is small. Further details of natural vegetation distribution and land-use history can be found in Walsh (1980a).

Hydrologically important changes in soil type accompany these climatic and vegetational differences and they have been described in detail for Dominica by Lang (1967). Impermeable smectoid clays, rich in the swelling 2 : 1 lattice clay mineral montmorillonite, predominate in the strongly seasonal zone with less than 2000 mm annual rainfall. More permeable kandoid (kaolin-rich) clays characterise the less wet rainforest areas with a weak dry season, and these are succeeded by extremely permeable allophane latosolic soils in areas of above 3750 mm annual rainfall. Finally, allophane podzolic soils occur in the wettest areas with more than 7000 mm annual rainfall.

5.2 Measurement of drainage density and relief development

The definition of drainage density used in this study is based on Horton's original definition (Horton 1932, 1945) as modified by Melton (1957) for use in the field. It is simply 'The sum of stream channel lengths per unit area, where the channels are subject to concentrated water flow at least in times of maximum runoff under current climatic conditions' (Melton 1957) and it is clearly a geomorphological definition with the emphasis on channels as important and identifiable landscape features. The drainage networks from which drainage density values were measured were derived from the DOS 1 : 25 000 topographic map series for the Windward Islands. These maps had been constructed direct from 1950s aerial photographs of similar or only slightly different scale. As with most topographic maps, the blue-line network considerably understates the real stream-channel network and hence the contour crenulation method of extending the blue-line network was used, followed by field checking (in the summer of 1973) of 30 sample drainage basins throughout the island group. The rugged nature of the relief meant that the channel network was for the most part accurately indicated by the contour crenulations. Even in rainforest areas, where the vegetation canopy can obscure channels on aerial photographs, correspondence between map-derived and field-checked networks was high, apart from a few short valley-side gullies in areas of low relative relief.

Drainage density was calculated and mapped on a grid square basis rather than using drainage basins for a number of reasons. First, the use of grid squares of standardised area avoids the problem of correlations (spurious or caused by downstream changes in environment) between drainage density and basin area and order (Ferguson 1978). Secondly, a drainage basin approach is particularly unsuited to the Windward Islands where environment changes so rapidly with increasing altitude within short distances of the coast; the elongated basins associated with the youthful radial drainage pattern tend to extend over too wide a range of rainfall, vegetation and soil. A grid square approach, however, allows a more complete and precise coverage of the range of environments within the islands. Thirdly, it avoids the need to delimit and measure basin areas and lends itself more readily to mapping (Newson 1978).

In choosing the grid mesh size of 0.65 km^2 (quarter square mile), the desirability of as fine a grid mesh as possible in view of the intense spatial variability in environment across the islands had to be balanced against the introduction of excessive 'noise' if grid squares were too small to be representative of their environment. The high drainage densities of the islands allowed the relatively fine mesh size to be selected. The total length of stream channels in each grid square was measured using a chartometer and then converted to drainage density by division by the grid square area. Drainage density maps of the islands were produced using these values. The mean annual rainfall of each grid square was also recorded from mean annual rainfall maps of the islands.

Also calculated on a grid square basis was source frequency (sometimes termed drainage texture (Morgan 1976)), defined as the number of stream sources per unit area. Source/network ratios were calculated by dividing source frequency by drainage density. It expresses the number of sources (or first-order streams) per unit length of stream-channel network and appears to be a useful indicator of the stage of drainage network development.

Finally, in order to assess relief development, a very simple measure was employed, termed the 'ruggedness index'. This measure attempts to summarise the 'up-and-downedness' of the landscape on a grid square basis. For 10.36 km^2 squares, the number of contours crossed by two north–south lines, two west–east lines and the two diagonals (NW–SE and SW–NE) of the square were counted. Multiplying by the contour interval and dividing by the total length of the six lines, a (directionally unbiased) measure of altitudinal change per unit distance within the square was obtained. These values were derived for land surfaces of ten volcanic centres of different ages in Dominica.

Age-of-landscape maps for the Windward Islands

An attempt was made to map the ages of the original constructional land surfaces upon which and since which the present drainage networks of the islands have developed. This effectively involved mapping the age of the last major volcanic deposits laid down in each area. The data used in the construction of the maps were derived from the existing geological literature. Five main lines of evidence were used by these studies in dating deposits:

(a) a limited number of ^{14}C and K–Ar dates (Tomblin 1971, Sigurdsson 1972, Wills 1974, Briden *et al.* 1979);
(b) foraminiferal evidence from limestone horizons;
(c) the existence of limestone and marine deposits at heights well above any possible interglacial high stands of the sea, indicating that deposits were of at least mid-Pleistocene age, as regional uplift ceased at the end of the mid-Pleistocene (Wills 1974);
(d) metamorphism, folding and faulting and igneous intrusions, regarded as indicating Mio–Pliocene orogenic activity and hence that the rocks were of at least Miocene age;
(e) standard stratigraphical principles.

Because of the complex history of most of the volcanoes, the availability as yet of comparatively few absolute dates and the lack of detailed knowledge of the spatial extent of deposits from particular volcanoes and particular eruptions over an island's landscape, the maps (Figs 5.2–5.5) must be regarded as tentative and very approximate. Because of this, the rough timescale categories given below were adopted in classifying and mapping the volcanic centres of the islands.

	Years BP
Historic	<300
Recent	300–10 000
late Upper Pleistocene	10 000–50 000
Upper Pleistocene	50 000–400 000
mid-Pleistocene	400 000–1000 000
Lower Pleistocene	1000 000–1800 000
pre-Pleistocene	>1800 000

The volcanic landscapes of the islands vary immensely in age. Grenada (Fig. 5.2) is comparatively old with volcanic activity ceasing towards the end of the mid-Pleistocene (Arculus 1976). The southern and eastern centres are considerably older. Most of St Lucia is likewise old (Fig. 5.3); younger volcanics, of Upper Pleistocene and late Upper Pleistocene age, are confined to the south-west of the island, where a series of young domes have been mapped and dated by Tomblin (1964, 1965, 1971). The most recent pyroclast flow has been dated by ^{14}C methods as 39 050 ± 1500 years BP and the most recent lava domes to emerge are Terre Blanche and Belfond, both of which are still completely undissected (Tomblin 1971).

The geology of southern Dominica has been investigated in some detail by Wills (1974) and he has ranked the volcanic centres in terms of the age of the youngest primary deposits and mapped the approximate boundaries of their respective deposits (Fig. 5.4). Four areas are regarded by Wills as being particularly young: Patates in extreme south, Micotrin and Trois Pitons in the centre-south, and part of the large Diablotins centre in the north. ^{14}C work by Sigurdsson (1972) on the Micotrin centre has shown that the most recent thick and extensive pyroclast flows occurred between 46 000 ± 4500 years BP and 34 000 years BP. Smaller flows occurred around 28 400 years BP, at which time the northern part of the dome was extruded. The Trois Pitons and north-west Diablotins areas are both considered by Wills as less than 50 000 years old and the Patates centre is dated as being as late as 10 000 years BP.

St Vincent is by far the youngest of the islands and major eruptions of Soufrière have occurred in 1717, 1812, 1902–3 and 1978 as well as throughout the Upper Pleistocene. For the pre-1978 situation, three mapping units of land surface age of the Soufriere volcano have been identified (Hay 1959a,b, Robson 1965) (Fig. 5.5):

(a) a late Upper Pleistocene surface of the northeastern slopes, protected from later outpourings by an old crater rim and dated at 11 000 to 14 000 years BP (Hay 1959a);
(b) a Recent surface dated by ^{14}C methods at around 4000 years BP (Hay 1959a);
(c) the 1902–3 deposits which covered the areas close to the current crater rim and which as glowing avalanche deposits followed the major valley systems (Hay 1959b).

Older rocks of volcanoes which ceased activity by the Upper Pleistocene underlie the centre and south of the island (Robson 1965), but the whole area was later covered by thick ash deposits from Soufriere during the Upper Pleistocene (Hay 1959a) and has therefore been mapped as that age.

Drainage density patterns in the islands

Maps of drainage density for the islands are given in Figures 5.2 to 5.5. In the older volcanic islands of Grenada (Fig. 5.2) and St Lucia (Fig. 5.3), roughly

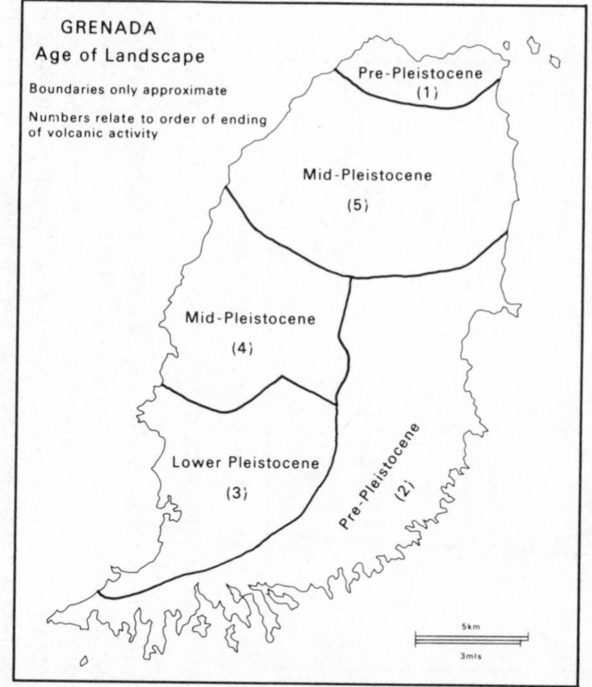

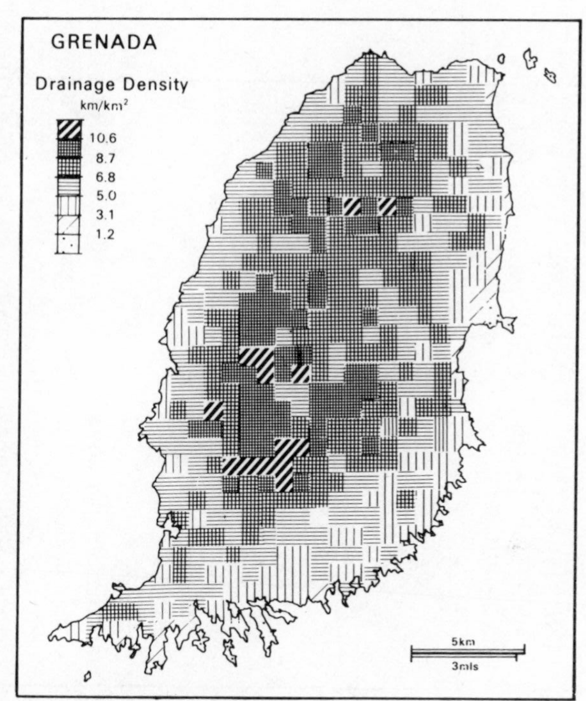

Figure 5.2 Grenada: age of landscape and drainage density.

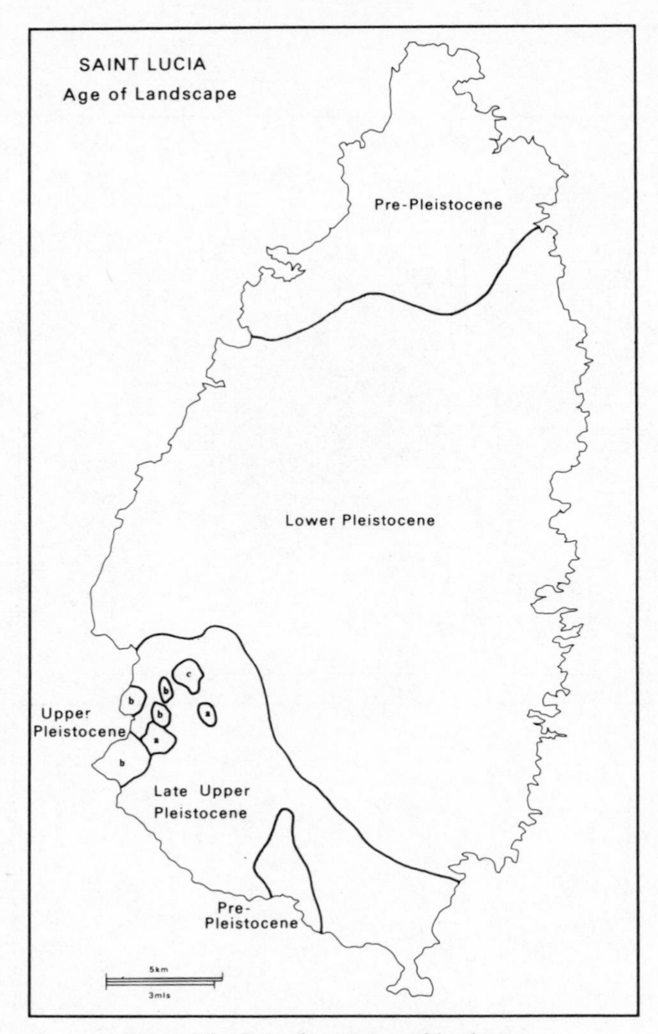

Figure 5.3 St Lucia. (a) Age of landscape.

concentric spatial patterns of drainage density are apparent, with drainage density rising inland to a maximum over the mountains of the centre-west. In Grenada, drainage densities increase from around 4 km km^{-2} in the dry coastlands to 10 km km^{-2} and higher in the central mountains. In St Lucia, drainage densities are somewhat lower, ranging from 3–4 km km^{-2} in the dry northern and southern peninsulas to 7–10 km km^{-2} over large areas of the centre-west of the island. In the south-west of the island lie some Upper Pleistocene domes, and densities in parts of this area are very low (0–3 km km^{-2}).

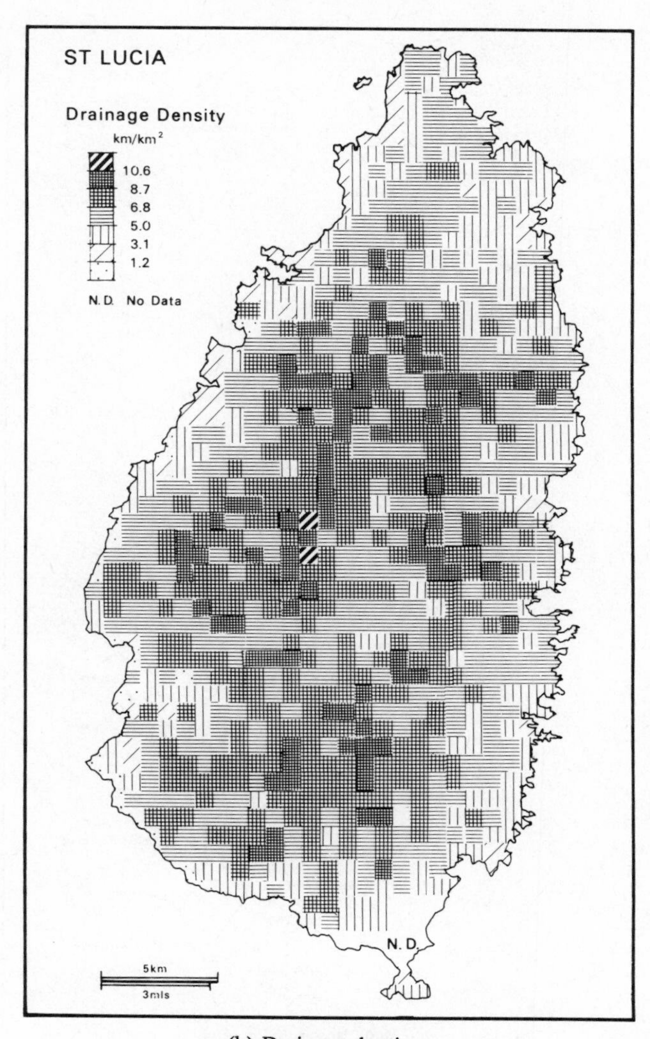

(b) Drainage density.

In St Vincent and Dominica, simple concentric patterns of drainage density do not exist. In St Vincent (Fig. 5.5), drainage densities are considerably higher than in the other islands. The mean drainage density for the island is 8.2 km km^{-2}, compared with means of 6.6, 6.2 and 6.3 km km^{-2} respectively for Grenada, St Lucia and Dominica. Over quite extensive areas on the eastern side of Soufrière in the north of the island, drainage densities range between 10.7 and 13.6 km km^{-2}. There are no extensive areas with drainage density less than 5 km km^{-2}, although the southern coastlands have rather lower drainage densities than the south-central interior and the north.

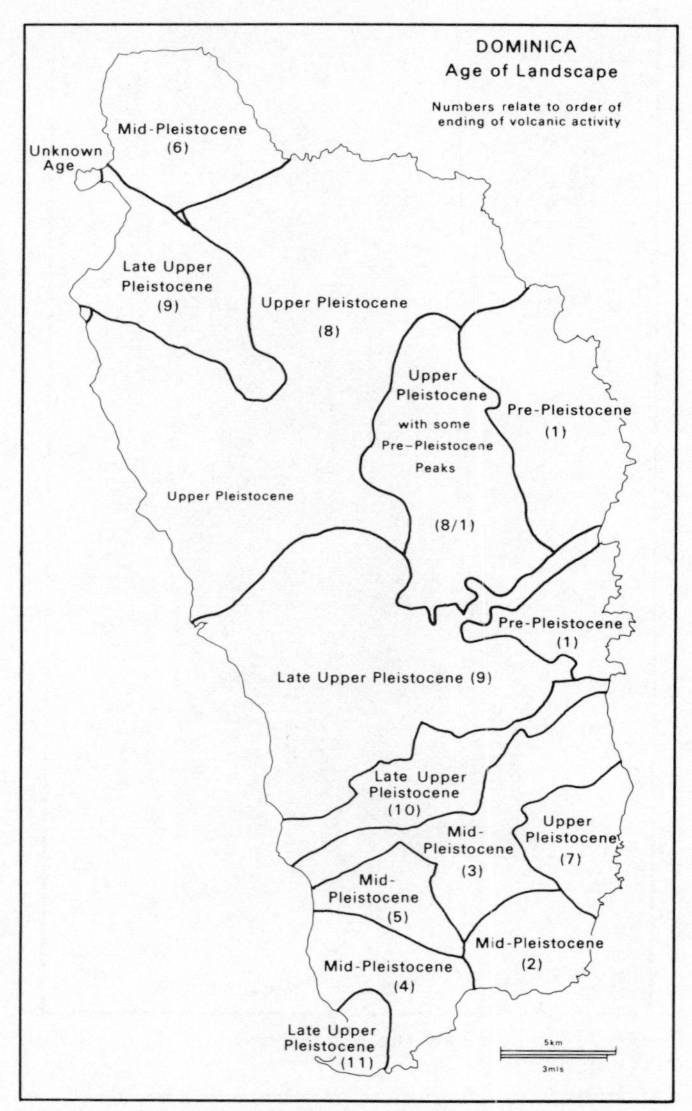

Figure 5.4 Dominica. (a) Age of landscape.

Dominica (Fig. 5.4) has the most complex pattern. The only extensive area of high drainage density is the southeastern mountain interior, where densities are generally over 8.7 $km\,km^{-2}$ and in places exceed 10.7 $km\,km^{-2}$. However, extensive areas of anomalously low drainage density occur not only in the comparatively dry south-west but also over parts of the very wet interior,

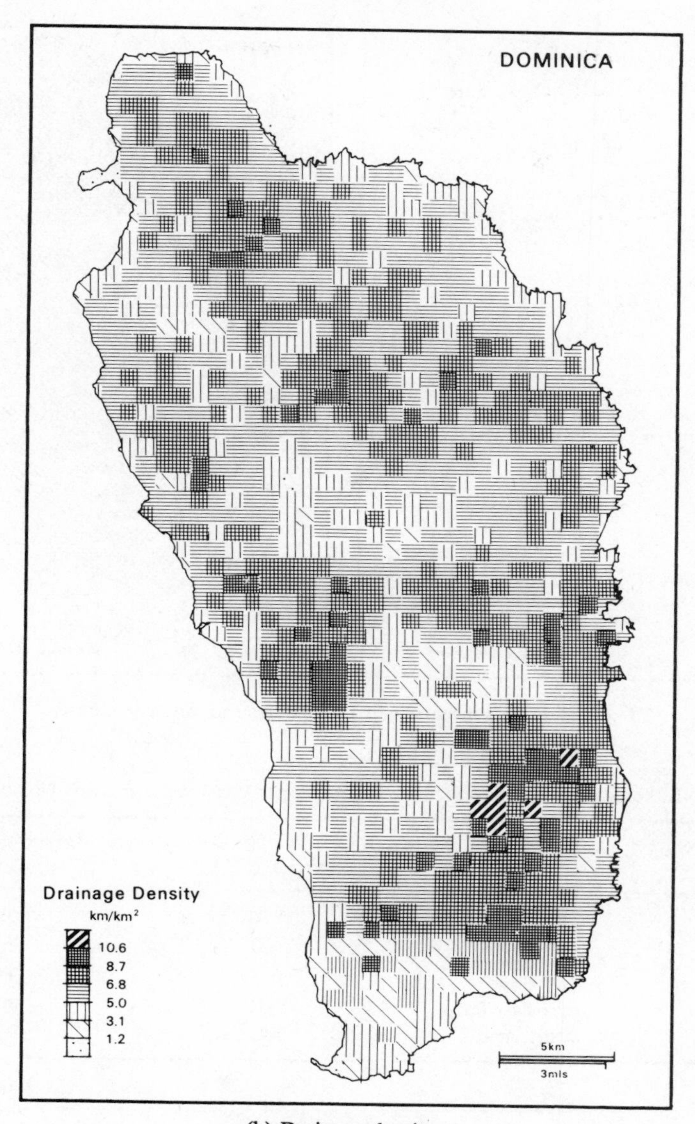

(b) Drainage density.

notably (a) north-west Diablotins, (b) an area between Diablotins and Trois Pitons, (c) the dome and flanks of Trois Pitons itself, and (d) the Micotrin dome – the latter two in the south-centre of the island. In all these areas, drainage densities are less than 5 km km^{-2} and are locally less than 3.1 km km^{-2} (Table 5.2)

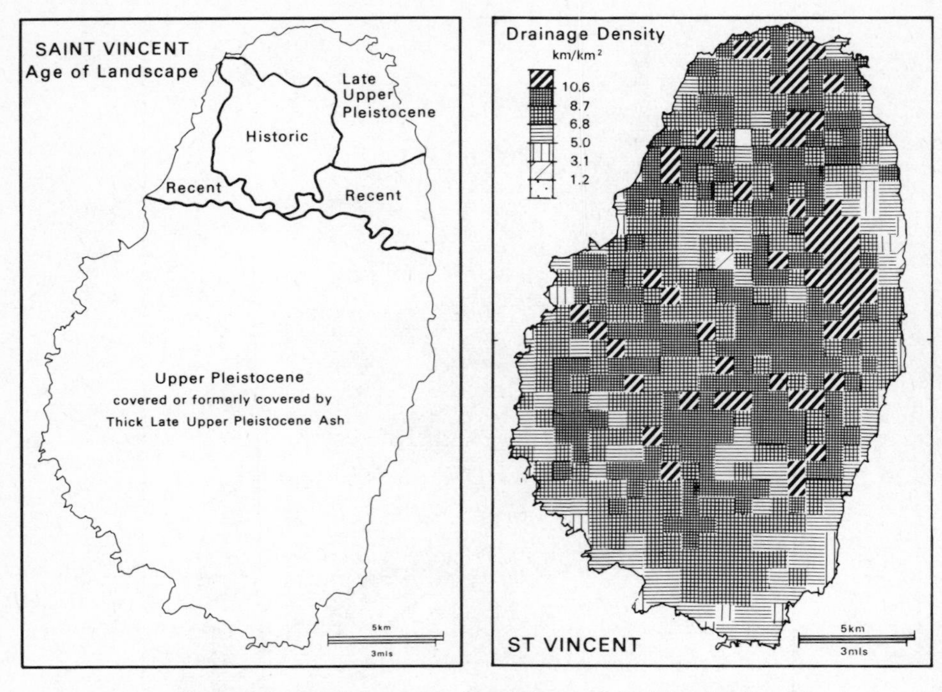

Figure 5.5 St Vincent: age of landscape and drainage density.

Table 5.2 Lithology and drainage density in late Upper Pleistocene areas of Dominica.

Volcanic centre	Lithology	Mean drainage density ($km\ km^{-2}$)	Mean annual rainfall (mm)
Micotrin	pyroclast flows	5.80	5385
	dome lava	3.85	9017
Trois Pitons	lava flows	7.11	4321
	pyroclast flows	5.63	4932
	dome lava	4.71	9440

5.3 Analysis and interpretation of drainage density

Tentative explanations of the variations in drainage density within the Windward Islands involve the interplay of climatic, bioclimatic and lithological factors acting *through* their influence on regolith hydrology and regolith erodibility and *within* the framework of time (as measured by age of landscape). As a first step in the analysis of the drainage density results, scattergrams of drainage density against annual rainfall were produced for each island and for each volcanic centre or age-of-landscape class within each island; scattergrams for 12 of the areas are given in Figures 5.6 and 5.7. (Annual rainfall, although a very crude parameter, was selected because it is strongly related to heavy rainfall magnitude–frequency and vegetation and soil characteristics in the islands.) In volcanic areas of roughly mid-Pleistocene age and older, broad linear or curvilinear relationships between the two variables are apparent (Fig. 5.6), but in younger volcanic areas, scatters are much wider and generally unsystematic (Fig. 5.7). Results from these two areas are now analysed in detail separately.

Older volcanic landscapes

In the older volcanic areas, drainage density increases with annual rainfall, but the rate of increase (a) is relatively small and (b) decreases considerably once annual rainfall exceeds around 2700–3000 mm. This is indicated not only by the scatters of Figure 5.6, particularly those for Lower Pleistocene areas of St Lucia and the Morne Plat Pays/Morne Watt centres of Dominica, but also by the fact that slopes of calculated linear regression lines for the scatters decline as the range and magnitude of mean annual rainfall covered by the grid square data increase. Thus, in Grenada, regression slopes (which indicate the *rate* at which drainage density increases with annual rainfall) range from 0.264 ($km\,km^{-2}$ rise per 100 mm rise in annual rainfall) in Lower Pleistocene and pre-Pleistocene areas, where annual rainfalls range only from 1067 to 3683 mm, down to 0.135 in mid-Pleistocene areas, where the annual rainfall range is 1321–4318 mm. Similarly, in St Lucia, the regression slope in the dry pre-Pleistocene north (annual rainfall range 1143–3073 mm) is 0.161 compared with a slope of 0.111 in the wetter Lower Pleistocene areas of the island (annual rainfall range 1473–5080 mm). In the wettest island, Dominica, where the range of annual rainfall covered by most volcanic centre data sets is from 1905 to 7000 mm and higher, slopes are less than 0.100 $km\,km^{-2}$ per 100 mm.

The regression slope evidence, together with the general form of the scattergrams covering both low and high annual rainfalls (notably St Lucia Lower Pleistocene and Dominica Morne Plat Pays/Morne Watt – see Fig. 5.6), suggests either (a) a convex curvilinear relationship, or (b) a dog-legged relationship composed of two straight lines: a steeper line covering lower annual rainfalls from 1000 to 3000 mm and a less steep line applying at higher annual rainfalls. The wide scatters prevent any firm conclusions on this point.

Clearly annual rainfall itself does not control drainage density, and consequently it is variables that are themselves correlated with annual rainfall which are responsible for the form of the drainage density–annual rainfall relationship. Three variables, all of which vary systematically with increasing annual rainfall and decreasing seasonality, appear to be important, namely:

(a) extreme rainfall magnitude–frequency,
(b) soil hydrological properties,
(c) ground erodibility.

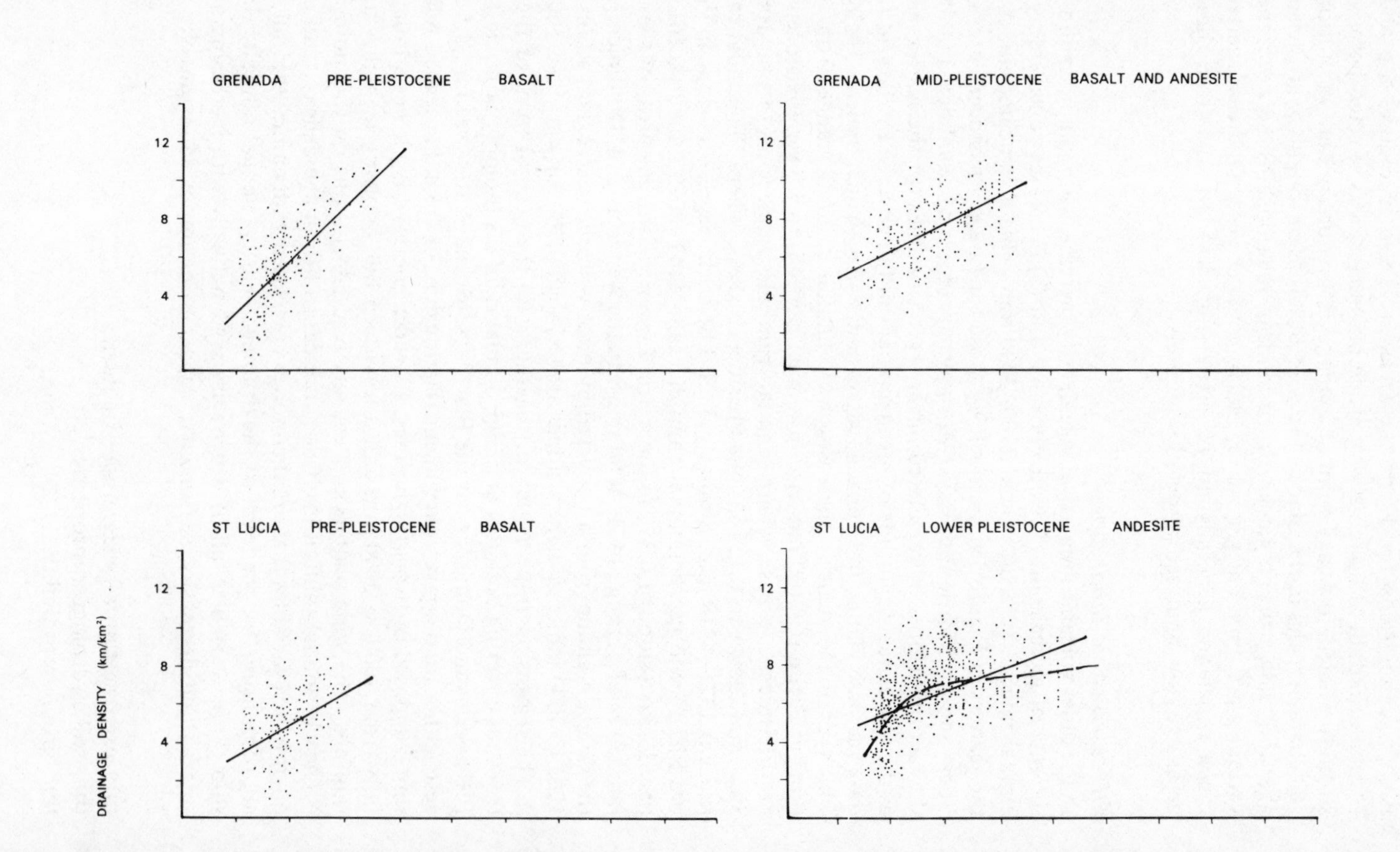
GRENADA PRE-PLEISTOCENE BASALT
12
8
4
GRENADA MID-PLEISTOCENE BASALT AND ANDESITE
12
8
4
ST LUCIA PRE-PLEISTOCENE BASALT
12
8
4
ST LUCIA LOWER PLEISTOCENE ANDESITE
12
8
4
DRAINAGE DENSITY (km/km²)

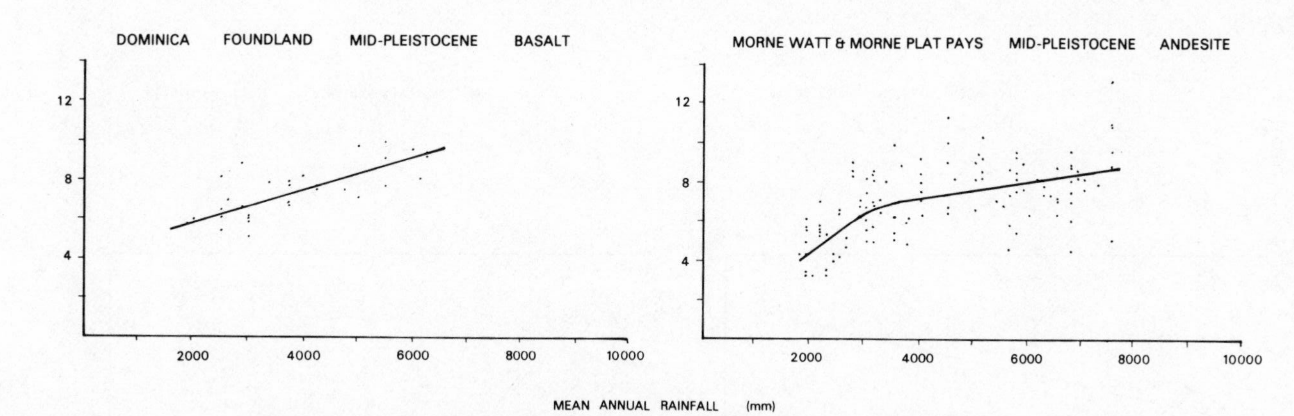

Figure 5.6 Drainage density and mean annual rainfall in the older volcanic areas.

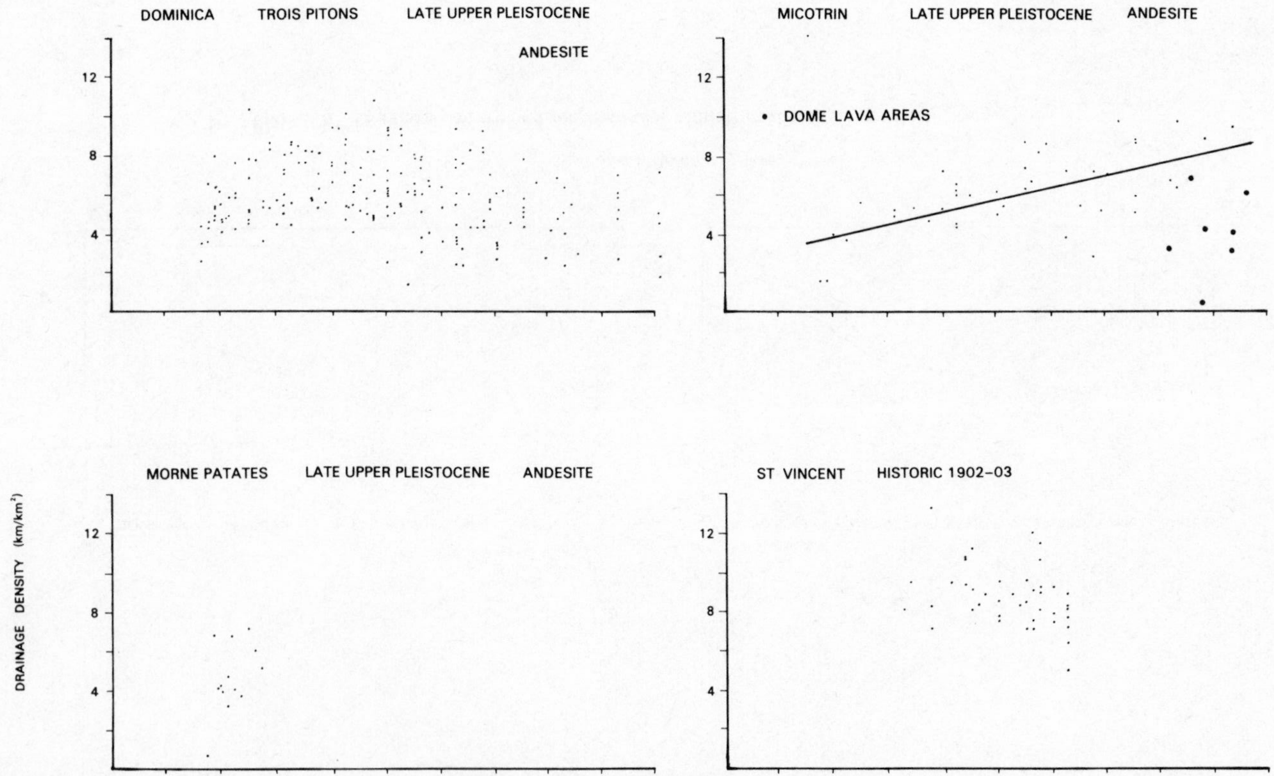

DOMINICA TROIS PITONS LATE UPPER PLEISTOCENE ANDESITE

MICOTRIN LATE UPPER PLEISTOCENE ANDESITE

• DOME LAVA AREAS

MORNE PATATES LATE UPPER PLEISTOCENE ANDESITE

ST VINCENT HISTORIC 1902–03

DRAINAGE DENSITY (km/km²)

12

8

4

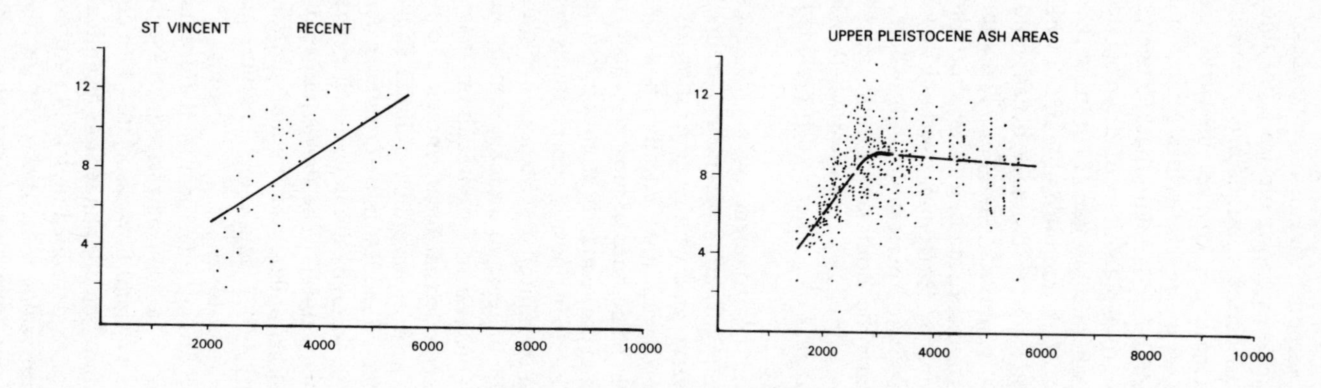

Figure 5.7 Drainage density and mean annual rainfall in the younger volcanic areas.

Although it is generally accepted that in most areas of the world (areas where peak runoff is generated by snowmelt are a possible exception) infrequent, high-magnitude, short-period rainfalls are responsible for generating the runoffs that form and maintain channel networks, there is very little knowledge or consensus as to the precise return period and duration of the rainfall events involved. Melton (1957) found that the five-year one-hour rainfall contributed significantly to explanations of drainage density in the semi-arid American south-west, whereas Morgan (1976) used the return periods of 50 mm and 100 mm daily rainfalls and the daily rainfall with a ten-year return period as variables in accounting for drainage density and drainage texture variations in humid tropical peninsular Malaysia. The appropriate duration, therefore, varies with environment, tending to be short (hourly rainfalls) in impermeable areas dominated by Hortonian overland flow and longer (daily rainfalls) in vegetated areas with permeable soils where throughflow and saturation overland flow are the dominant runoff processes. However, it should be stressed that it is still largely unknown whether it is events of one year, five year, ten year, 50 year or greater return periods that are critical in controlling drainage density.

As the Windward Island environment is one of thick vegetation and permeable soils, longer-duration rainfall parameters are indicated as being of more relevance. Field observations of the generation of overland flow in Dominica (Walsh 1980a) suggest that ideally data on 6–12 h rainfall magnitude–frequency would be perhaps of most relevance, but such data were non-existent in the islands and consequently attention focused on the relatively good daily rainfall data. The strong positive correlations between heavy daily rainfall parameters and annual rainfall described earlier in this chapter suggest that the main reason for the increase in drainage density in the mountainous interiors of the islands is an increase in extreme rainfall magnitude–frequency. However, it is interesting to note that mean annual maximum daily rainfall increases relatively slowly with mean annual rainfall (Table 5.1). This may partly help to explain the overall modest rate of drainage density increase with annual rainfall in the islands, whereby drainage densities less than double with a six-fold increase in annual rainfall. The *linear* nature of the relationships between heavy daily rainfall parameters and annual rainfall, however, fails to explain the curvilinearity (or dog-legged nature) of the drainage density–annual rainfall relationship.

The main reason for the attenuating form of the drainage density-annual rainfall curve is considered to be the progressive increase in permeability and decrease in susceptibility to overland flow associated with changes from montmorillonite to first kaolin and then allophane soils as annual rainfall increases towards the island interiors. As described in detail in a previous paper (Walsh 1980a), the smectoid (montmorillonite-rich) clays of the highly seasonal areas with less than 2500 mm annual rainfall are characterised by shallow topsoils, low topsoil water-holding capacity and very low subsoil permeability; consequently, in heavy rains saturation overland flow is generated quickly and tends to be widespread over slopes (see Ch. 3). In the kandoid (kaolin-rich) clay areas, permeabilities of both topsoil and subsoil are much higher and the deeper topsoils have a greater water-holding capacity; the soil is consequently less easy to saturate and widespread saturation overland flow over slopes is very rare and probably confined to cyclone rainfalls. In the allophane latosolic areas, the soil is of exceptionally low bulk density and is extremely permeable; despite the frequent heavy rainfalls, widespread saturation overland flow has never been observed. In both the kaolin and allophane latosolic soil zones, throughflow is the dominant runoff process and overland flow is confined to localised ridge-top flats and valley-bottom locations.

Thus, it is argued, the increase in drainage density that rainfall factors tend to produce is progressively lessened by the marked increase in soil permeability and depth (D. G. Day 1980). This argument is given additional weight by the fact that the lower rate of increase in drainage density with annual rainfall appears to start at around the 3000 mm isohyet, which is within the kaolin soil zone.

A third possible factor concerns soil erodibility. Ground vegetation cover is sparse under seasonal forest types and rainforest, but there are important changes in litter cover and surface root density between the forest types. The perennial, relatively complete (albeit thin) leaf litter cover and dense surface and near-surface root mat of the rainforest afford more protection to the ground surface against erosion by overland flow than do the sparse, seasonal litter and deeper roots of the seasonal forest formations. How important these differences are remains unassessed, however.

In contrast to the young volcanic centres, the role of lithological differences in causing variations in drainage density in the older volcanic areas appears to be relatively small. Drainage densities of old domes such as Morne Watt in Dominica and Mornes Tabac and Gimie in St Lucia are as high as on surrounding pyroclast flow and lava flow deposits. It may be that, once weathering and soil development in the bouldery lava domes become more advanced, the anomalously high permeability associated with the initial bouldery surface is replaced by a lower permeability of weathered mantle and soil. Certainly, however, some of the scatter of the graphs of Figure 5.6 is due to lithology. At a general level, drainage densities in Grenada (and St Vincent), where lava flows are more prominent, are somewhat higher than at similar annual rainfalls in St Lucia and Dominica, where more permeable pyroclast flow deposits predominate.

Younger volcanic areas

Simple relationships between drainage density and annual rainfall are not generally apparent in the younger volcanic areas (Fig. 5.7). Scatters are wide and unsystematic (with one or two exceptions) and fitted regression lines and correlation coefficients have no statistical validity. Great contrasts in drainage density (and by implication therefore in rates of drainage network

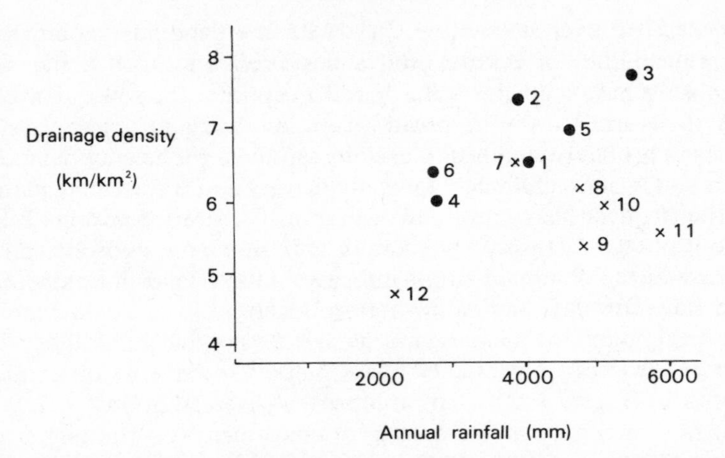

Figure 5.8 Mean drainage density and mean annual rainfall for volcanic centres in Dominica.

development also) are evident both between and within the different volcanic areas and these contrasts appear to be principally related to differences in lithology.

In Dominica (Figs 5.7 and 5.8), mean drainage densities of the four youngest volcanic centres of late Upper Pleistocene age are markedly lower than for the older volcanic centres of corresponding mean annual rainfall. Furthermore, more detailed analysis of the Micotrin and Trois Pitons data (Table 5.2) shows that, within these centres, drainage densities are highest on lava flows, intermediate on pyroclast flows, and least on the extremely permeable bouldery dome lava areas. In both cases drainage densities on the dome lava are remarkably low (3.85 and 4.71 km km^{-2}) despite their extremely high annual rainfalls (9017 and 9440 mm respectively). In the case of Micotrin, if the dome lava data are excluded from the scatter, a consistent increase in drainage density with increasing annual rainfall can be perceived for the pyroclast flow deposits (Fig. 5.7). The correlation coefficient of a linear regression fit to the data increases to $+0.66$ ($n = 45$) for the pyroclast flow data subset, compared with only $+0.28$ ($n = 52$) for the original undivided data set. Drainage densities on the pyroclast flows are nevertheless well below those of the adjacent Foundland and Plat Pays/Watt centres of mid-Pleistocene age (Fig. 5.6) – the regression lines being about 3 km km^{-2} apart.

Within Dominica, rates of drainage network development appear to be (a) comparatively slow, since mean drainage densities of the late Upper Pleistocene centres (age 10 000–50 000 years) are still 3 km km^{-2} less than those of mid-Pleistocene age and earlier, and (b) different for different types of volcanic deposit, being slowest on dome lava, intermediate on pyroclast flows and fastest on lava flows. Both these features can perhaps be explained in terms of the permeability characteristics of the deposits. The dome lava areas are mostly composed of large boulders with only small patches of highly organic soil; permeabilities are extremely high with water readily disappearing underground between the boulders. Pyroclast flow deposits are predominantly coarse grained with numerous small boulders, and weathering and soil development are much more rapid than on the dome lava; permeabilities are not as extreme as on the dome lava, but they are still very high. Lava flows, on the other hand, tend to be more massive and less permeable and overland flow and channel formation are likely to occur earlier and at a faster rate than on the other types of deposit. However, all three types of deposit are permeable by world standards and this is thought to be the reason for the slow rates of drainage network development despite the very high rainfall intensity environment. Drainage networks on the Upper Pleistocene and late Upper Pleistocene domes and flows of the south-west of St Lucia have likewise been very slow to develop.

The lack of drainage channels on the young bouldery domes of Dominica and St Lucia contrasts sharply with the high drainage densities and rapid drainage development of the predominantly ash deposits of the historically active volcano of Soufrière in St Vincent (Fig. 5.7). Mean drainage densities on the 1902–3 surface (8.77 km km^{-2}) are similar to the mean densities for the older Upper Pleistocene, late Upper Pleistocene and Recent surfaces of St Vincent (8.11, 8.90 and 7.85 km km^{-2} respectively) and are amongst the highest in the whole Windward group. Values for individual grid squares reach as high as 13.27 km km^{-2}. The reasons for this are lithological. The relatively fine-grained ash deposits of St Vincent are extremely erodible, especially when unvegetated, as they are in immediate post-eruptional times. Then high-intensity rainfalls would reduce infiltration capacities through rainbeat and result in widespread overland flow and rapid gully erosion. (The susceptibility of the ash to gullying can be seen currently in the form of incised gullies up to 3 m deep which have developed from footpaths along some of the ridges of southern St Vincent.) Although initial drainage development may be extremely rapid, once a protective vegetation cover becomes established and soil profiles develop, further development may be much slower.

Although for the 1902–3 surface there is no discernible relationship between annual rainfall and drainage density, on the Recent surface (around 4000 years old) there appears to be already a distinct curvilinear relationship (Fig. 5.7). This suggests that a rough 'equilibrium drainage network' may be achieved much faster on the ash areas of St Vincent than on the pyroclast flow and dome lava deposits of St Lucia and Dominica. This suggestion is perhaps given additional weight by the fact that a curvilinear relationship also seems to apply on the Upper Pleistocene ash areas of the south and centre of the island (Fig. 5.7).

However, the latter ash deposits present problems as regards interpretation of the drainage networks developed on them. Hay (1959a) has shown that these ash deposits were probably the result of intermittent light falls of ash spread over a long period, as there is evidence in the form of root holes throughout the 7–14 m thick deposits that the vegetation maintained itself during deposition. It would seem likely, therefore, that the pre-existing drainage network already developed on the older underlying volcanic deposits would easily be maintained throughout the ash deposition. It also means that in this case any attempt to correlate drainage density with the age of the *latest* volcanic deposits (i.e. the ash) would be invalid.

5.4 Factors affecting drainage density and relief evolution

Age of landscape and the degree of drainage and relief evolution

A provisional investigation was made into how two simple parameters of drainage network organisation and relief vary with age of landscape. Variations in source/network ratio (defined earlier) between 17 volcanic centres were examined and mean values ranged between 1.75 and 2.87 sources per kilometre. Although source/network ratios in general tend to increase with drainage density (Fig. 5.9), values for young centres tend to be lower than those for older centres of similar drainage density. The reason for this may be that the older the volcanic landscape, the less the drainage network is constrained by the initial conical or dome form of a volcano, which is normally characterised by radial drainage dominated by a relatively small number of long, subparallel first-order streams (and hence low source/network ratios).

Using the index described earlier, relief ruggedness was assessed for ten areas of contrasting age in Dominica (Table 5.3). Values are relatively low (305–577 m km^{-1}) for the late Upper Pleistocene areas, reach a maximum (900–1005 m km^{-1}) in the mid-Pleistocene centres and fall to 660 m km^{-1} in the pre-Pleistocene Eastern Volcanics area. This would suggest that dissection and ruggedness increase for at least 400 000 years and reach a peak sometime

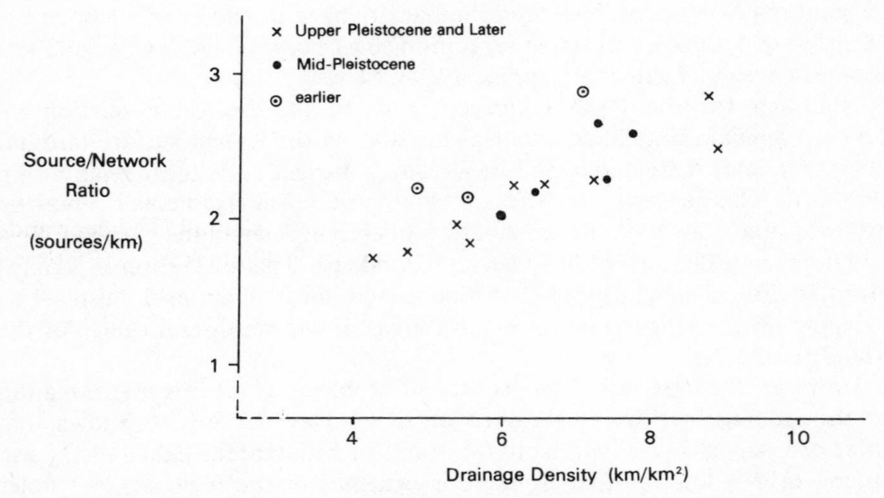

Figure 5.9 The relationship between means source/network ratio and mean drainage density for 17 volcanic areas of the Windward Islands.

Table 5.3 Ruggedness index values for volcanic centres of different ages in Dominica.

Volcanic area	Age of land surface	Ruggedness index[a] (m km^{-1})
Eastern Volcanics	pre-Pleistocene	660
Foundland	mid-Pleistocene	928
Morne Plat Pays	mid-Pleistocene	900
Morne au Diable	mid-Pleistocene	1005
Grand Soufrière Hills	Upper Pleistocene	824
Morne Diablotins	Upper Pleistocene	774
Trois Pitons		
older western area	late Upper Pleistocene (?)	611
newer dome and pyroclast flows	late Upper Pleistocene	305
North-West Diablotins	late Upper Pleistocene	557
Micotrin	late Upper Pleistocene	577

[a]For explanation of index see text.

before 1800 000 years, after which time (according to the Eastern Volcanics data) relief contrasts are being gradually reduced by valley widening, slope decline and ridge-top reduction. However, the occurrence of major uplift in the mid-Pleistocene and the varying original sizes of the volcanoes are complicating factors and, until more detailed work is done, no firm conclusions can be drawn.

The interrelated problems of climatic change, relief changes and tectonic activity

Interpretations and explanations of the drainage density and network development data presented for the Windwards are rendered even more difficult by a history of climatic change and relief instability in the islands. There is mounting evidence of major long-term and short-term climatic fluctuations in the region. Table 5.4 summarises some of the changes of the last few hundred years.

Mean annual rainfall in the islands has fluctuated considerably over the past hundred years and at most stations with long records the late 19th century was 15–25% wetter than at present (Table 5.4a). More significant as regards drainage density are the parallel changes in heavy daily rainfall frequency and in the tracks and frequency of tropical cyclones. At St Thomas Police Station, Barbados, daily falls of 76 mm and higher were over three times more frequent in the late 19th century than in the early 20th century and again in recent years (Table 5.4b); and at Roseau, Dominica (for which daily data were only available from 1921), such falls were twice as frequent during the wetter 1929–58 period than since 1959.

Exceptionally heavy rainfalls are often associated with the passage of tropical cyclones, and marked fluctuations in cyclone tracks and frequency in the eastern Caribbean have been recorded over the last 325 years (Walsh 1977). As Table 5.4c shows, peak frequency over the southern Windwards and

Table 5.4 Recent climatic change in the Windward Islands.

(a) Mean annual rainfall (from Walsh 1980b).

Station and island	Mean annual rainfall (mm) pre-1899	1959–76	Change (%)
Roseau, Dominica	2114	1781	−16
Codrington/Seawell, Barbados	1491	1111	−25
Belmont, St Kitts	1961	1509	−23

(b) Heavy daily rainfall magnitude–frequency (from Walsh 1980b).

Station	Period	Mean annual rainfall (mm)	F_{76}	F_{127}	Mean annual maximum daily rainfall (mm)
St Thomas	1889–1906	2166	2.3	0.56	128
Barbados	1907–1925	1574	0.6	0.16	89
	1926–1958	1776	1.1	0.13	90
	1959–1972	1570	0.7	0.20	97
Roseau	1929–1958	2039	1.7	0.37	111
Dominica	1959–1976	1781	0.8	0.22	96

F_{76} and F_{127} = mean annual frequencies of daily rainfalls exceeding 76 and 127 mm, respectively.

(c) Mean annual tropical cyclone frequency (from Walsh 1977).

Period	Guadeloupe, Dominica and Martinique	Grenada, St Lucia, St Vincent and Barbados
1650–1764	0.12 (p.a.)	0.11 (p.a.)
1765–1792	0.68	0.22
1793–1805	0.00	0.08
1806–1837	0.72	0.41
1838–1875	0.18	0.13
1876–1901	0.35	0.81
1902–1927	0.50	0.31
1928–1958	0.32	0.52
1959–1975	0.44	0.22

Barbados occurred in the period 1876–1901 (0.81 per annum) compared with only 0.22 per annum in recent years and even lower frequencies during the periods 1650–1764, 1793–1805 and 1838–75. Over Dominica and the French islands of Guadeloupe and Martinique, peak frequencies occurred in the late 18th (1765–92) and early 19th (1806–37) centuries, with very few cyclones occuring in the periods 1650–1764, 1793–1805 and 1838–75. The question therefore arises, particularly in view of the slow rates of drainage network development noted for the pyroclast flow and dome lava deposits of Dominica and St Lucia, as to whether drainage densities in the older volcanic areas are adjusted to present or past climatic epochs or whether they are permanently in a state of disequilibrium because climatic fluctuations occur so often.

Over a much longer timescale, much larger climatic fluctuations are indicated. Relatively unweathered sands of last glacial age found in deep-sea cores between 20°N and 10°S along the Atlantic seaboard of the Americas and the Caribbean have been interpreted by Damuth and Fairbridge (1970) as indicating arid or semi-arid climates in the region during the last glacial phase. This would be a logical consequence of the 8°C drop in sea surface temperature for the Caribbean and a 4°C drop in the equatorial Atlantic estimated for the last glacial age by Emiliani (1966, 1971) using oxygen-isotope techniques and foraminiferal evidence. Such a sea surface temperature drop would also presumably result in an absence of tropical cyclones during glacials, since cyclones require a sea surface temperature of at least 28°C to develop (Palmer 1951).

Finally, as rainfall in the islands is highly dependent on relief and aspect, the changes in island relief which have undoubtedly resulted from (a) glacio-eustatic sea-level changes, (b) volcanic eruptions and sub-aerial erosion (leading to relief creation and reduction respectively) and (c) tectonic uplift, which was important particularly in the mid-Pleistocene (Wills 1974), would presumably have had major effects on spatial patterns of rainfall in the islands. Again, the question arises of how long drainage networks take to respond to such changes.

Comparisons

Method and map scale influence drainage density estimation and make comparisons between different studies problematical. Bearing this in mind, the drainage densities of the older volcanic areas of the Windwards, although high compared with those of humid temperate areas, are somewhat low in relation to mean annual rainfall compared with values reported from other tropical areas (Table 5.5). In peninsular Malaysia, drainage densities at 2250–2540 mm annual rainfall range between 7.5 and 12.4 km km^{-2} on granitic, arenaceous an calcareous rocks (Eyles 1966, Morgan 1976) compared with densities of 5–7 km km^{-2} at similar annual rainfalls in Grenada, St Lucia and Dominica. Only on the volcanic ash deposits of St Vincent are densities (7.8 km km^{-2}) at 2500 mm annual rainfall of a similar order to those of Malaysia. The reason is probably lithological: volcanic deposits (excepting fine ash, which is also highly erodible) are highly permeable compared with most other lithologies.

Drainage densities in the granitic highlands of Sri Lanka range from around 5 km km^{-2} at 1346 mm annual rainfall to 15 km km^{-2} at 5000–5500 mm annual rainfall (Madduma Bandara 1974a,b). These values also are much higher than those in the Windwards and furthermore the *rate* of increase in drainage density as annual rainfall rises is much higher too. These differences may again be the result of lithology. Granite is less permeable than volcanic

Table 5.5 Drainage density data from tropical areas.

Territory	Lithology	Vegetation/Land use	Annual rainfall (mm)	Drainage density (km km^{-2})	Source
Windward Islands	volcanics[a]	seasonal–rainforest	1000–8000	3–11	Walsh (this chapter)
Sri Lanka	granite	tea plantations	1346–5436	5–15	Madduma Bandara (1974a, b)
Queensland	various	semi-arid–rainforest	889–3988	4–14[b]	Abrahams (1972)
Malaysia	granite	rainforest/cultivated	*c.*2250–2540	7.5	Eyles (1966)
Malaysia	arenaceous	rainforest/cultivated	*c.*2250–2540	9.3	Eyles (1966)
Malaysia	calcareous	rainforest/cultivated	*c.*2250–2540	9.0	Eyles (1966)
Malaysia	various	rainforest/cultivated	*c.*2250–2540	7.5–12.4	Morgan (1976)
Fiji	granite	grass and scrub	1778–2540	15.5–18.6	Wright (1973)
Brazil	granite	ND	1270–1524	6.5	Cunha *et al.* (1975)
Brazil	sandstones/clays	ND	1270–1524	4.0	Cunha *et al.* (1975)
Uganda	gneiss, micaschists and granite	savanna	1016–1270	1.2–2.2	Doornkamp and King (1971)
Uganda	quartzites and conglomerates	savanna	1016–1270	3.0–4.0	Doornkamp and King (1971)
India	rhyolite	sparse	356	1.9	Ghose *et al.* (1967)
India	granite	sparse	356	1.3	Ghose *et al.* (1967)
India	sandy Alluvium	sparse	356	0.8	Ghose *et al.* (1967)

[a]Windward Islands data are for older volcanics.
[b]Highest drainage densities in Abrahams' study were in semi-arid areas, which were the only areas with maps of adequately large scale for accuracy.

deposits and furthermore the exceptionally permeable allophane soils of the wettest parts of the Windwards do not develop on granitic rocks, and this latter fact may help to account for the sustained higher rate of increase in drainage density on granite than on volcanics.

The few available data on rates of drainage network development from other parts of the world display disparities in rates of a similar order to those found in the Windwards. Thus, in unvegetated areas such as the waste tips of Perth Amboy, New Jersey, in the humid temperate north-east of the United States (Schumm 1956), drainage and relief development are extremely rapid, as on the ash deposits of St Vincent. On the other hand, the evidence from a series of progressively older glacial tills in the Mid-West of the United States suggests that at least 20 000 years were required there for drainage density to approach an equilibrium level (the data of Ruhe 1952, as analysed by Leopold *et al.* 1964). The fact that drainage networks in the Windward Islands on pyroclast flows and dome lava need at least 50 000 years (and perhaps considerably more time) to approach equilibrium levels is of interest, particularly in view of the much higher rainfall input of its environment compared with humid temperate areas.

5.5 Conclusions

Variations in drainage density within the Windward Islands are strongly related to climate, lithology and age of landscape. The influence of climate in the islands is both direct (through variations in heavy rainfall magnitude–frequency) and indirect (via its strong zonal influence on vegetation and soil type). The relationship between drainage density and annual rainfall in the older volcanic areas appears to be an attenuating curvilinear rather than a linear one. The principal reason for this is thought to be the marked increase in permeability of soils as one moves from the seasonal coastlands into the kaolin and then allophane soil zones of the rainforest interiors of the islands. In the older volcanic areas, densities increase from 3–4 $km\,km^{-2}$ at 1000 mm annual rainfall to around 10 $km\,km^{-2}$ at 7000 mm annual rainfall. Because of the relatively high permeability of the volcanics, these values are a little low compared with other areas of the tropics.

In the younger volcanic areas (Upper Pleistocene and later), drainage density and the rate of drainage network development appear to vary strikingly with lithology. On fine ash deposits, drainage densities reach high values within a few years of an eruption, whereas on the more permeable pyroclast flow and dome lava deposits of Dominica and St Lucia, drainage densities are still well below equilibrium values after 50 000 years. No single timescale for drainage development in the islands is applicable. The slow rates of drainage network evolution on the pyroclast flow deposits, despite a very high rainfall environment and high available relief, raise major questions about degrees and rates of adjustment of landscapes to climatic change or tectonic uplift. It seems unlikely that drainage networks are completely in equilibrium with the prevailing environmental conditions even in the older volcanic areas in view of the evidence of recent climatic change and uplift, together with the apparently slow rates of channel adjustment.

6

Surface and underground fluvial activity in the Gunung Mulu National Park, Sarawak

P. L. Smart, P. A. Bull, J. Rose, M. Laverty, H. Friederich & M. Noel

6.1 Introduction

Relatively little is known about the Quaternary palaeoclimate and evolution of the Malaysia area. Much of the evidence is derived from the interpretation of the alluvial placer deposits in peninsular Malaya and the Sunda shelf, which have been worked for tin (for instance Aleva *et al.* 1973, Stauffer 1973, Batchelor 1979). Some of these deposits have been dated using the radiocarbon method (Haile & Ayob 1968), while others have been shown to have reversed palaeomagnetism, implying an age in excess of 700 000 years (Haile & Watkins 1972). The remains of a volcanic ash, dated at 30 000 years by the fission track method using zircons, are present at some localities (Stauffer *et al.* 1980). A second ash dated at 70 000 years is also reported from deep-sea cores by Ninkovich *et al.* (1978). In Borneo the details of fluvial terrace deposits are not so well known, suggested chronologies are very tentative, and no palaeoclimatic interpretation of the deposits has yet been attempted (Leichti *et al.* 1960, Wilford 1961, Wall 1967, Woodroffe 1980). However, in Sabah, Mount Kinabalu (4100 m) is known to have been glaciated at least once during cold phases of the Quaternary (Koopmans & Stauffer 1968), and work is in progress to investigate altitudinal variations in the vegetation zones during this time (Flenley Ch. 7 this volume.).

Palaeoclimatic inferences have also been derived from studies of the evolution and distribution of the fauna (Molengraaff & Weber 1920, Medway 1972, Verstappen 1975), whose migration was frequently made possible by glacial lowering of sea levels. Much of the Sunda shelf area was exposed during these times, and Biswas (1973) reports the occurrence of mangrove swamps, freshwater marshes and coal swamps at depths up to 70 m below present sea level. A radiocarbon date at one site indicated an 11 000 year age for lignites at 67 m depth, but two or possibly three other recessions were also observed.

Batchelor (1979) gives evidence to show these may have been up to 130 m below present sea level. In postglacial times Haile (1969) suggests that sea levels have reached 6 m higher than present, but he refutes earlier suggestions of extensive Quaternary submergence.

The object of the present study was to interpret the origin and development of the extensive fossil and active cave systems found in the limestone hills of the Gunung Mulu National Park, Sarawak, and to relate these, and their sedimentary sequences, with the terraces of the Melinau valley. It was hoped that palaeoclimatic interpretation of the terrace-forming mechanisms, combined with absolute dates from speleothems found in the caves by the $^{230}Th/^{234}U$ disequilibrium method (Harmon *et al.* 1975), could substantially advance our understanding of the Quaternary geology of Borneo.

6.2 The study area

The Gunung Mulu National Park, some 500 km^2 in area, is located close to the border of Brunei in northern Sarawak at a latitude of 4°N (Fig. 6.1). The area has recently been studied by a Royal Geographical Society expedition, although the geology had previously been described by survey and oil company workers (Leichti *et al.* 1960, Wilford 1961). The climate is equatorial with consistent, high temperatures averaging 27°C. Rainfall is in excess of 5000 mm yr^{-1} but there are marked monthly variations. Over most of the Park, a dense tropical rainforest vegetation is found.

The region can be divided into two topographical zones (Fig. 6.2), an eastern upland area rising to 2400 m on the summit of Gunung Mulu (Mount Mulu) and a western lowland, generally below 200 m. Much of the upland is developed on the quartzitic sandstones and interbedded shales of the Mulu Formation. This terrain has a high drainage density with sharp-crested ridges separating deep V-shaped valleys with steep, straight, weathering-limited slopes. Landslides are common, particularly on the more shaly beds, and are generally shallow and ribbon-like translational slides, developed at the junction of the regolith and bedrock (M. J. Day 1980). Stratigraphically above the Mulu Formation is the dense, fine-grained and exceptionally massive Melinau Limestone Formation, a biothermal limestone. This forms a spectacular series of karstified steep-sided mountains up to 1700 m high (G. Api), through which rivers from the Mulu uplands pass in either impressive deep gorges, such as that of the Sungei Melinau (Melinau River) or by underground drainage, such as that from Hidden Valley via the Nasib Bagus cave system. The autogenic drainage of the limestone uplands is also underground, and combines with runoff from the adjacent lowlands to resurge at major risings, such as that of the Clearwater Cave complex (48 km long), which drains a major part of G. Api (for detailed survey see Eavis 1981). To the West of this mountain in the Melinau valley there is an abrupt transition to lowlands with

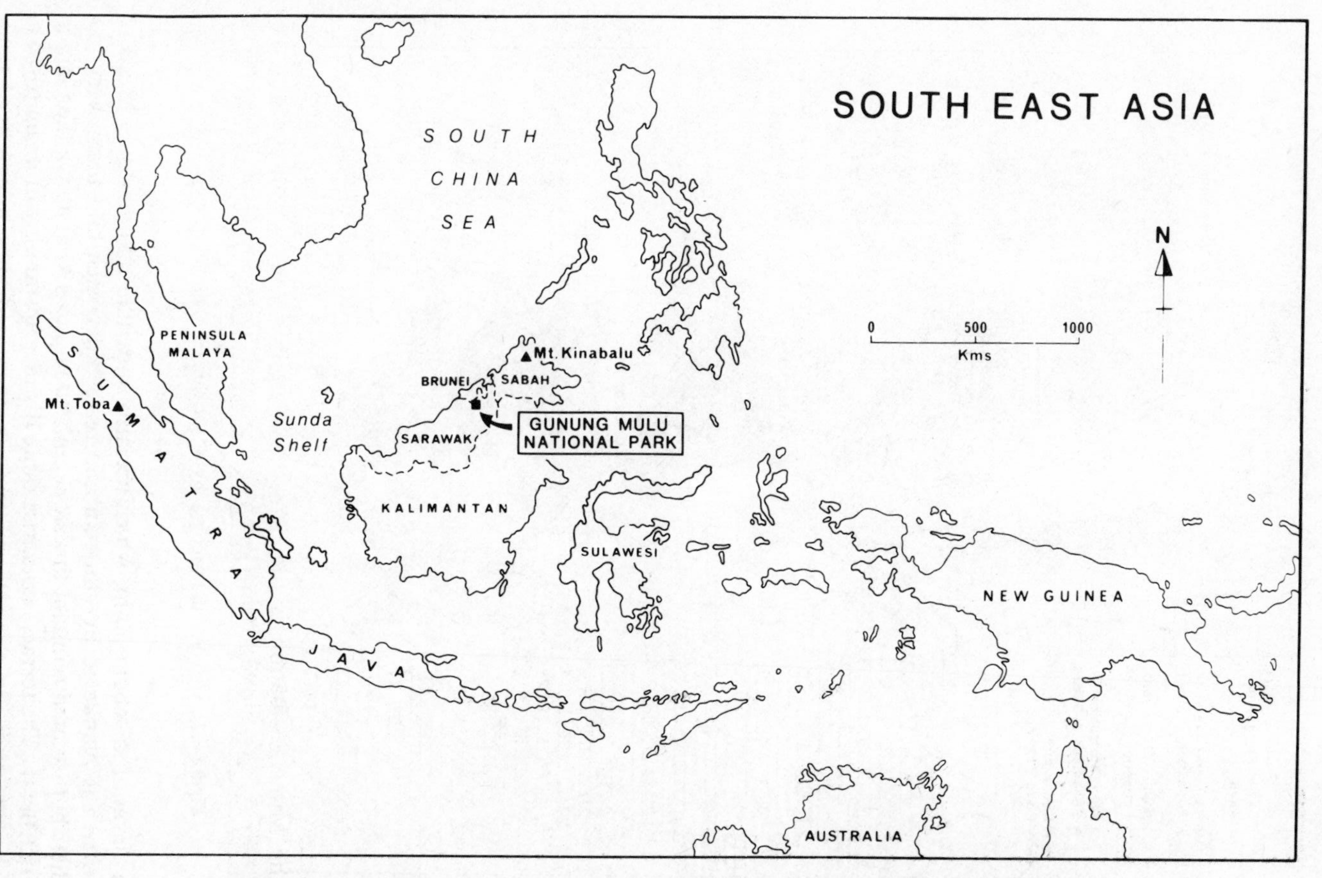

Figure 6.1 Location of the Gunung Mulu National Park study area in south-east Asia.

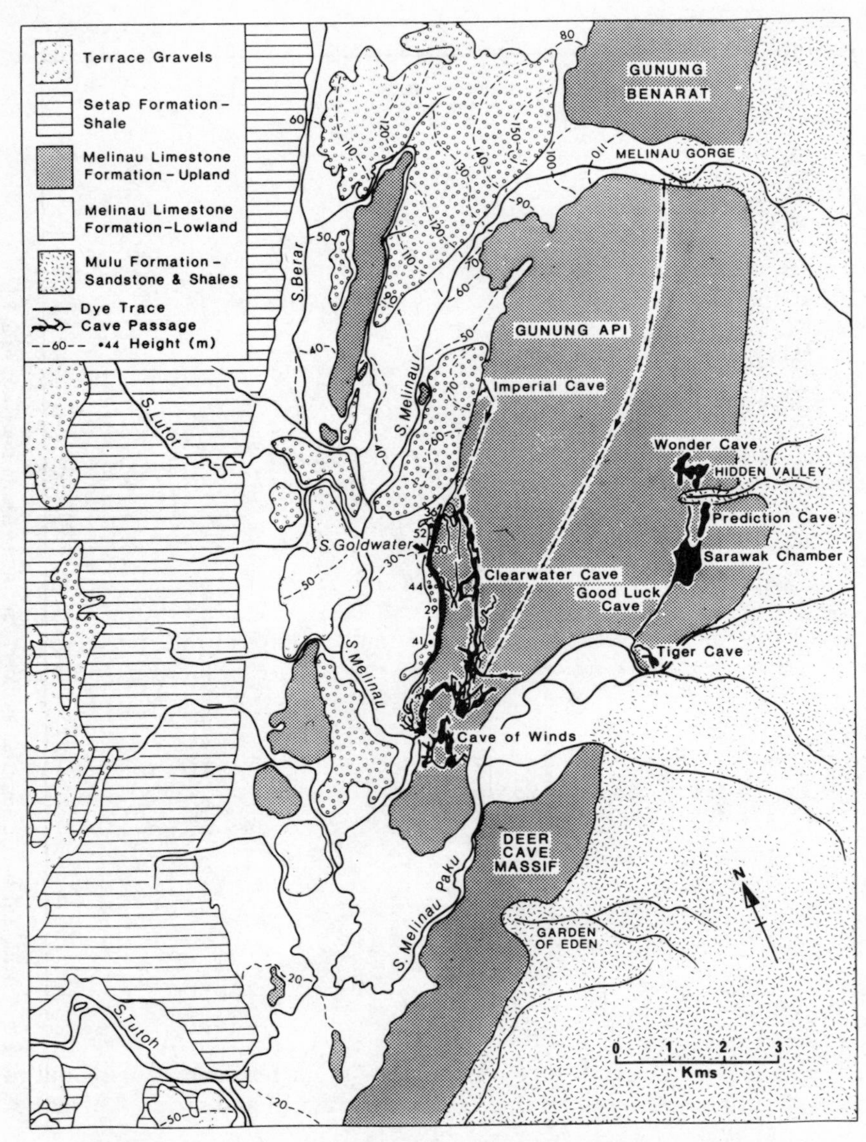

Figure 6.2 Geology and topography of the Sungei Melinau drainage basin.

an elevation of about 30 m, which are also underlain by limestone. Much of the area is blanketed by alluvium and terrace deposits, and has a very low relief, but isolated remnant limestone hills rise up to 330 m above the plain at Bukit Berar. The terraces comprise quite distinct upstanding units up to 70 m high with steep marginal slopes, and are composed predominantly of sandstone boulders set in a sandy matrix. The limestones pass eastwards under the younger Setap Formation, a sandy shale giving rise to low hills and ridges at about 150 m elevation. Terrace remnants are also found in this area, and on the sandstones of the Belait Formation, which forms an eastward-facing scarp along the topographic divide which is the Sarawak/Brunei border.

6.3 The development of the regional and limestone drainage

The north-west Borneo area was first exposed sub-aerially after folding and uplift in the late Pliocene. A regional drainage pattern to the north-west developed in response to the tilting of the accreted sedimentary prism towards the subduction zone north-west of the present Borneo coast (Hamilton 1979). However, this pattern was modified by particularly active uplift along a plunging north-east/south-west axis passing through G. Mulu, and forming an elongate dome. The drainage pattern therefore shows two major northwestward-flowing rivers (Fig. 6.3), the S. Limbang to the north of the Mulu uplift, and the S. Tutoh to the south. In between, there is a zone of radial drainage centred on G. Mulu, and including an element of southeasterly flow to the Tutoh headwaters. To the north-west of G. Mulu, the earliest drainage may well have been into the present Belait basin, but a combination of subsidence along the Melinau syncline, and the strong north-east/south-west structural grain, would appear to have caused diversion of this drainage, forming the present trellised pattern, particularly apparent near Batu Ulu Tutong on the Belait Formation, but also evident on the limestones and shales. The present Melinau valley must date from this time, and may have drained into either of the regional river systems depending on rates of incision and differential uplift. The Melinau valley is therefore a feature of some antiquity, whose drainage history may be very complex.

The northwestward flow of aggressive water from the impermeable Mulu Formation onto the Melinau Limestone gave considerable potential for the development of major cave systems. But this potential could not be realised until an outlet for ground water could develop. Two factors controlled this outlet: the first is the structure of the Melinau Limestone unit in relation to the regional drainage pattern, and the second the presence of an overlying Setap Shale cover. Owing to both a reduction in thickness and wedging out of individual beds, the limestones thin markedly to north and south, where they are replaced by Setap Shales (Fig. 6.3). This lenticular structure is accentuated by the development of the Melinau syncline (Fig. 6.4), an elongate basin with maximum depression just north of the S. Berar junction with the S. Melinau. The lowest points on the margin of this basin were in the north and south, where the regional rivers cut the axis of the Melinau syncline. These were therefore preferred outlets for ground water, resulting in a strong strike component in the resulting regional circulation. This was accentuated by the strongly developed and laterally continuous nature of particular bedding partings in the very massive limestone (Webb 1982).

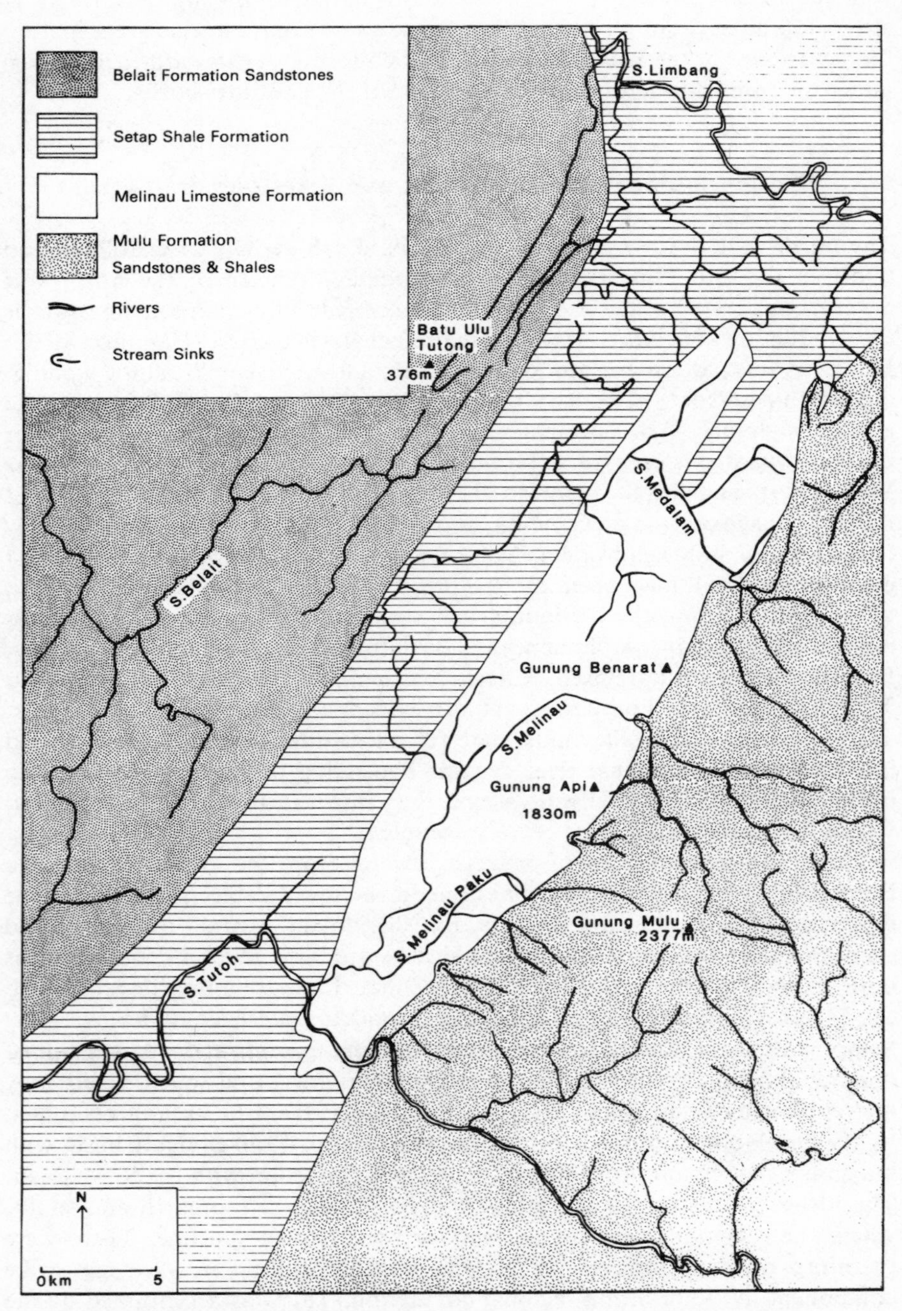

Figure 6.3 Regional drainage pattern of the Gunung Mulu National Park Area.

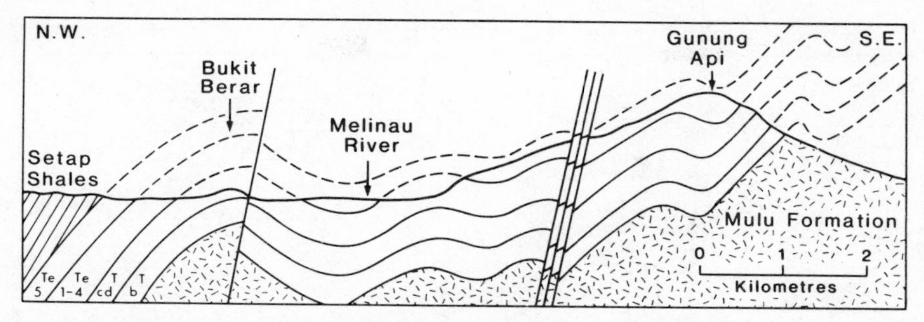

Figure 6.4 Geological section north-west to south-east through the summit of G. Api (after Webb in Eavis 1981).

The distribution and form of many limestone landforms on G. Api can be explained by the progressive westward stripping of the Setap Shale cover. The section through G. Api (Fig. 6.4) shows that relatively little limestone has been removed from extensive areas on the upper part of the hills, which must have been shale-capped until relatively recent geological times. These areas are characterised by small, shallow heterogeneous depressions, which contrast markedly with the linear steep-sided tower karst forms seen to the east between Hidden Valley and the Melinau Paku. There is also strong evidence that the dissected valley complex west of Hidden Valley was formed by superposition of surface drainage from this overlying shale cover. Thus water entering the limestones along downdip routes from the east was not able to discharge upwards to the surface through this impermeable formation, and was conducted into major strike conduits. This is best seen in the High Level Series of Clearwater Cave (Fig. 6.5) where downdip flow in Ronnie's Delight derived from Hidden Valley transfers into a dominantly strike conduit in the Scumring Series.

The S. Tutoh was therefore the major base level for cave development in G. Api, and the greatest proportion of passages in the Clearwater Cave complex are developed along the strike with flow to the south (Fig. 6.2). As the S. Tutoh incised into the Mulu uplift and the base level for cave development fell, a vertical sequence of cave passages was formed in the limestone (Fig. 6.6). Some of these passages were developed under water-filled (phreatic) conditions, such as the High Levels passages, which have a pressure flow segment rising vertically over 100 m (Fig. 6.5). As the water level in the limestone fell, circulation through the highest levels was no longer possible and lower routes developed. The High Levels passages were abandoned as circulation developed in the Revival level for instance. The higher routes may well have functioned initially as flood overflows, or had isolated segments of free surface (vadose) flow, such as the isolated trench found downstream of the Volcano, but soon they became completely fossilised. Some passages, although initially phreatic, developed a free air surface as incision occurred. Providing passage downcutting occurred at a rate comparable to that of base-level lowering, a canyon

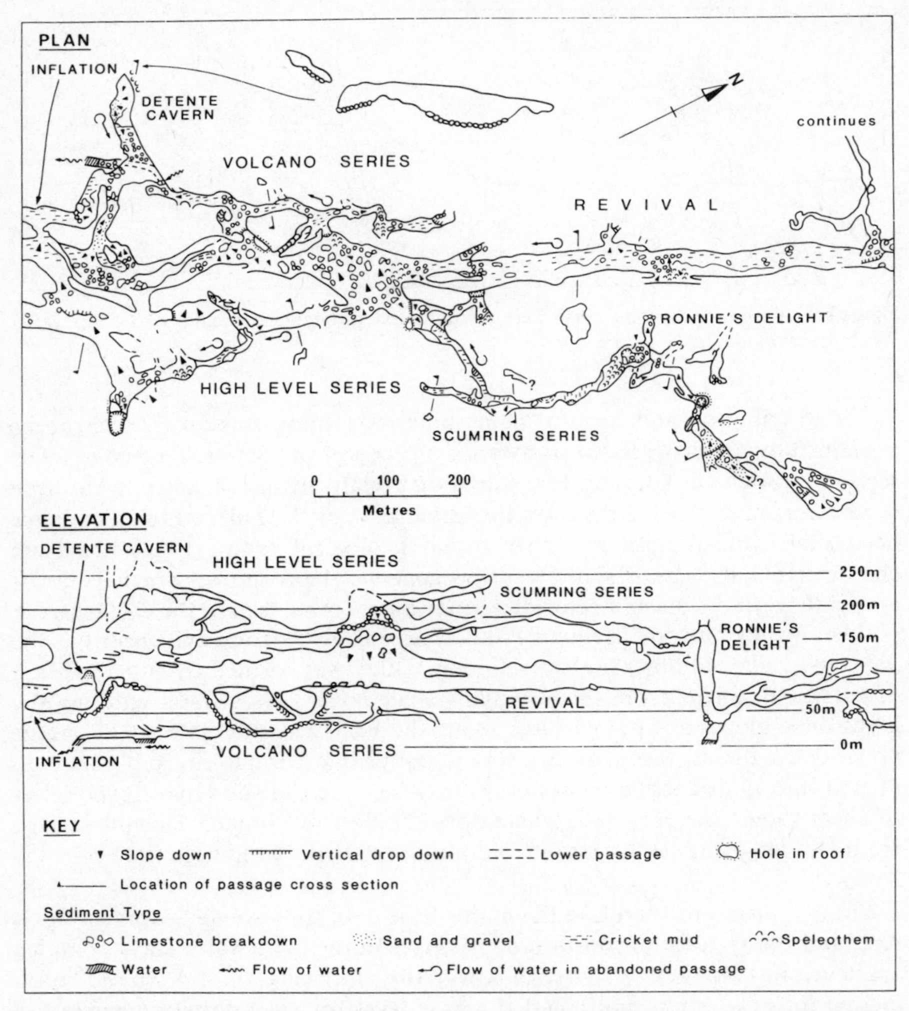

Figure 6.5 Plan and elevation projected on 030° part of Clearwater Cave, showing passage morphology and sediments.

passage developed, as for instance the 30 m high Revival passage. However, when base level fell faster than the rate of passage incision, the route was abandoned as higher hydraulic heads were developed over lower passages, which therefore developed preferentially, capturing flow from the higher levels. The passage morphology (which clearly indicates the difference between phreatic and vadose development), combined with the vertical distribution of cave passage, may therefore give an indication of the rate at which regional base level fell, and more significantly the levels at which it accelerated due to renewed uplift.

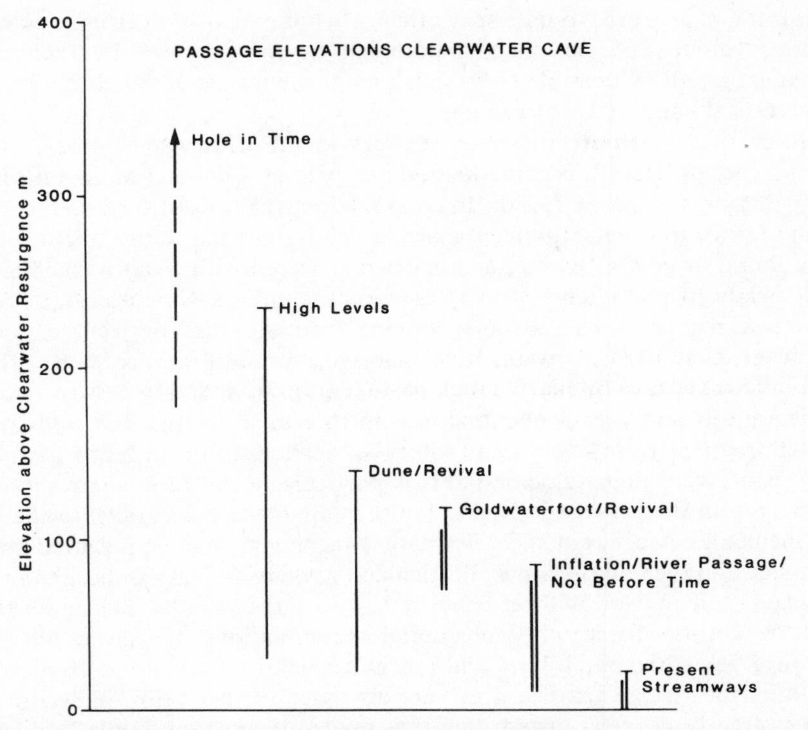

Figure 6.6 Passage elevations for the Clearwater Cave complex. The graph indicates the elevation range of active passages for a particular phase of development: phreatic passages, single line; vadose passages, double line. The Clearwater Cave resurgence elevation is 25 m a.s.l.

The three lowest levels in Clearwater Cave (below 105 m above the present entrance elevation) were formed under predominantly vadose conditions, although all have substantial phreatic precursors. Thus the rate of base-level fall compared to that of cave passage erosion has probably remained more or less constant during much of the time period for the development and fossilisation of these levels. However, the changes responsible for the abandonment of the Revival level into Inflation at *c.* 75 m may well indicate a sudden and short-lived increase in the rate of base-level lowering. The abandonment of the predominantly phreatic Dune Series (*c.* 120 m), the High Levels (*c.* 140 m) and Hole in Time (*c.* 230 m) also suggests sudden falls in base level, but, because these are part of a complex phreatic sequence, it is much more difficult to define precisely these transition levels. Furthermore, it is not impossible that changes in the input points of the underground drainage may have initiated the development of substantial new links, as is implied by the much more northerly extension and larger size of Dune Series, suggesting a source in the Melinau gorge, compared with the High Levels passages which were probably fed from Hidden Valley. By using the $^{230}Th/^{234}U$ disequilibrium method

(Harmon *et al.* 1975) to date speleothems, whose deposition from percolating waters commenced on the abandonment of the passages by their active streams, it will be possible to establish an absolute sequential chronology for the vertical suite of fossil passages.

A particular feature of some passages in the Clearwater complex is the occurrence of laterally continuous and nearly level wall notches, first described in Waltham and Brook (1980). In cross section, the roof of the notch is essentially horizontal, with the back curving gently into the floor, which may be horizontal or gently sloping, and is often veneered with clean washed gravel. The height of the notch is usually between 1 and 2 m. Notches can be found at any elevation in the passage, varying from slots in the walls of vadose trenches, as in the Clearwater River passage, to modifications of the original phreatic roof, as in Snake Track passage (Fig. 6.7), and vary between shallow indentations and spectacular features up to 6 m in width. The walls of the notch frequently show very small solutional scallops of the order of 1 cm long, suggesting very high velocities; this is particularly the case where the notch impinges on the passage roof. In plan they are often not parallel to the walls of the main passage, but show dramatic sweeping meanders, indented first on one side and then on the other. Particularly good examples can be seen in East Passage and in Not Before Time in Cave of the Winds. The notches are everywhere associated with substantial accumulations of gravel, filling the passage below the notch level and reaching thicknesses of up to 10 m in Not Before Time. These gravels, which are discussed further below in Section 6.4, have often been re-excavated such that remnants are found only in pockets, as in the Battleship, where the notch is perched 30 m above the floor of a subsequent vadose trench. Where a complete sequence is preserved, the gravel surface is concordant with or slightly higher than the floor of the notch and is generally reworked into a lag gravel.

The notches are formed by vadose (free air surface) cave streams, whose downward erosion is constrained by the build-up of a protective sediment cover on the passage floor. They, in fact, represent an end member in the

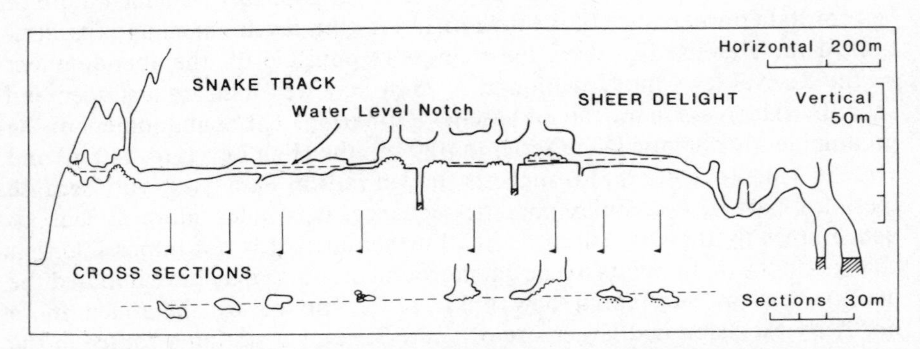

Figure 6.7 Extended elevation of Snake Track and Sheer Delight passages in Clearwater Cave (vertical exaggeration ×5). Water flow was from right to left. (After Waltham & Brook 1980, Fig. 7.)

sequence of passage shapes from very narrow vertical walled canyons whose erosive effort is almost wholly downwards, through triangular canyons which have increasing lateral erosion with depth, to notches where erosion is almost wholly lateral. This lateral concentration of the erosive power of the cave stream forms a notch whose floor is concordant with the level of the masking sediment, and whose roof is equivalent to the water surface for the dominant erosive discharge at the time of formation. The accumulation of a protective sediment cover may arise by reduction in cave stream gradients as base level is approached. However, the widespread evidence of substantial gravel infill below the base of the notch suggests that externally controlled aggradation is probably more important in Mulu.

The development of wall notches, as described above, is the vadose equivalent of paragenesis in phreatic (water-filled) passages (Ewers 1972). In fact the same equilibrium mechanism that permits paragenesis is also significant in the formation of notches where the water surface impinges on the cave roof. If further aggradation occurs in this situation, the cross-sectional area of the passage decreases and the velocity must increase to maintain the stream discharge. This allows entrainment of part of the aggradational increment. However, if the sediment removal continues, the velocity decreases as cross-sectional area increases, and erosion is halted. Eventually a lag gravel develops at the surface of the sediment, whose size is adjusted to the dominant discharge. It is this mechanism that permits the continued transport of coarse gravels through phreatic loops. This equilibrium mechanism is not, however, capable of explaining the vertical localisation of all the observed notches, which frequently are far from the passage roof, and it must be concluded that this is controlled primarily by resurgence level, and the water surface slope of the cave stream.

The levels of the notches observed in the Clearwater complex (Fig. 6.8) form a record of aggradational events occurring when the resurgence elevation was at or near the notch level, which have been preserved by the successive abandonment of the upper cave levels.

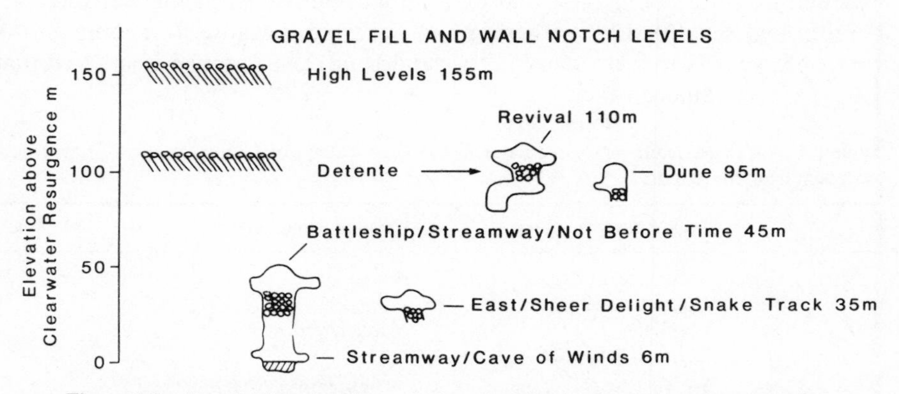

Figure 6.8 Gravel fill and wall notch elevations in the Clearwater Cave complex.

6.4 The clastic cave sediments

Because of the abandonment of cave passages by their formative streams as base level lowers, and the considerable time before surface erosion intersects these fossil passages, caves commonly preserve an ancient and detailed sequence of past sediments. The composite sedimentary sequence for the Clearwater Cave complex is presented in Table 6.1. A basal fluvial gravel deposit is overlain at a sharp boundary by laminated ponding deposits termed Cricket Muds. These Cricket Muds consist of similar upper and lower members, usually separated by a distinct band of allophane, believed to be a surface-derived weathered volcanic ash. The whole sequence is capped by flowstone and stalagmite which are rarely found as distinct horizons in the lower part of the sequence. These deposits, with the exception of the *in situ* speleothem deposition, are surface-derived materials, deposited in response to changing hydrologic conditions in the caves. Thus each unit has a different mechanism of deposition, and both the areal distribution and vertical sequence at any one locality may vary (Fig. 6.9) depending on the local conditions. The recognition of a composite sequence does not infer that all sediments of one unit were deposited throughout the cave during a single event, but merely that sedimentation tended to occur in a sequence that was repeated at different levels at different times.

The basal gravels comprise either matrix-supported or matrix-poor gravels. The former contain a sand, silt and clay matrix, rafting well rounded clasts up to 50 cm in diameter. The clasts do not generally appear to be orientated. The matrix-poor gravels in contrast have strongly imbricate fabrics, with the dip of the major plane up to 40°. The gravel unit may grade upwards into cross-bedded sands and finer gravels, but the contact with the overlying Cricket Muds is always sharp. The gravels are composed predominantly of sandstone pebbles derived from the Mulu Formation (Table 6.2) which is also the probable source for the siltstone, shale and quartz pebbles.

Limestones only account for 0.5% of the total pebbles sampled. Field examinations did not suggest that differential removal of pebbles of susceptible lithologies by *in situ* weathering had occurred, because no pebble ghosts were observed. However, many of the sandstone clasts were so weathered that

Table 6.1 Idealised sedimentary sequence in the Clearwater Cave complex, with related cave environmental conditions.

Sedimentary member	Cave environment
stalagmite/flowstone	sub-aerial
Upper Cricket Muds allophane Lower Cricket Muds	ponded, vadose
gravels	fluvial, vadose or phreatic

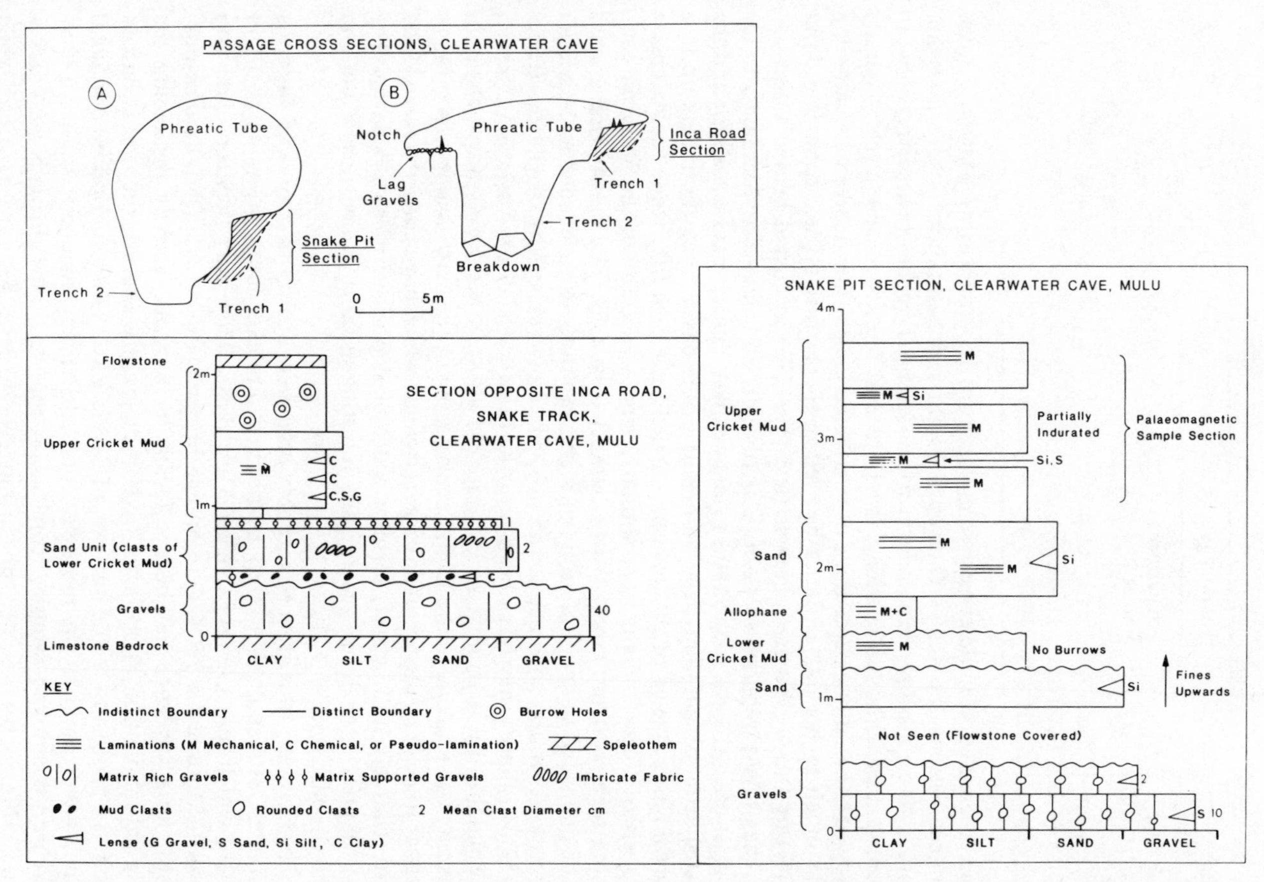

Figure 6.9 Passage cross sections and sedimentary sections at Snake Pit and Inca Road (Snake Track passage), Clearwater Cave.

Table 6.2 Relative abundance of lithologies in pebbles from the Clearwater Cave complex ($n = 1000$).

Lithology	Percentage
sandstone	51.9
shale	31.2
quartz	11.5
siltstone	5.8
limestone	0.5
speleothem	0.1

they crumbled completely on removal from the supporting matrix, and at some locations the shales showed a bright red patination. There does not appear to be a simple explanation for differential pebble weathering between sample sites, although entrance locations and the lower cave levels generally showed the smallest percentage of weathered clasts. The degree of weathering was much higher (40%) than observed in the surface terrace deposits (about 20%) or the present river gravels (less than 1%) discussed below.

The gravels were deposited both by phreatic and vadose cave streams. In the former case, there is indubitable evidence from chatter marks, scallops and the distribution of different-sized sediments in the passage that bounders up to a maximum diameter of 50 cm were carried up a 5 m lift, while the extensive gravels of Sheer Delight were similarly emplaced through the 50 m phreatic rising segment from Great Wall Chamber (Fig. 6.7). The matrix-rich gravels were deposited under conditions of high simultaneous suspended and bedload transport, similar to those found in rivers from the snouts of glaciers (Eyles 1979). In places they have been reworked to form a clean washed lag gravel, presumably by a later sediment-poor flow. At other locations, such as in Detente Cavern, and in the High Levels passages, the matrix-supported gravels appear to be deposited from invasive streams, which have carried a mixed load into relatively static water, forming a deltaic deposit with cross-bedding. The matrix-poor gravels in general are indistinguishable from those found in gravel-bed rivers.

The Cricket Muds are a remarkably homogeneous silt-to-clay sized siliceous sediment with some thin sandy partings, and thicker sandy layers up to 0.5 m thick, which may contain Cricket Mud clasts. They vary between chocolate brown (Munsell 7.5 YR 4/4) to orange-brown (Munsell 7.5 YR 7/8) in colour. The whole sequence has been affected by biogenic activity disturbing the original lamination as is shown by X-radiographs. Indeed, the unit is named from the profusion of the burrows of cave crickets (*Rhaphidophora oophaga*) which have colonised the more vesicular silts. Palaeomagnetic studies undertaken on orientated samples from a 1 m core sectioned through the Cricket Muds (Fig. 6.10) revealed no strong geomagnetic secular variation, while exhibiting normal geomagnetic polarity (Noel 1982). If the sediments were to exhibit primary depositional fabric, they would be characterised by a well

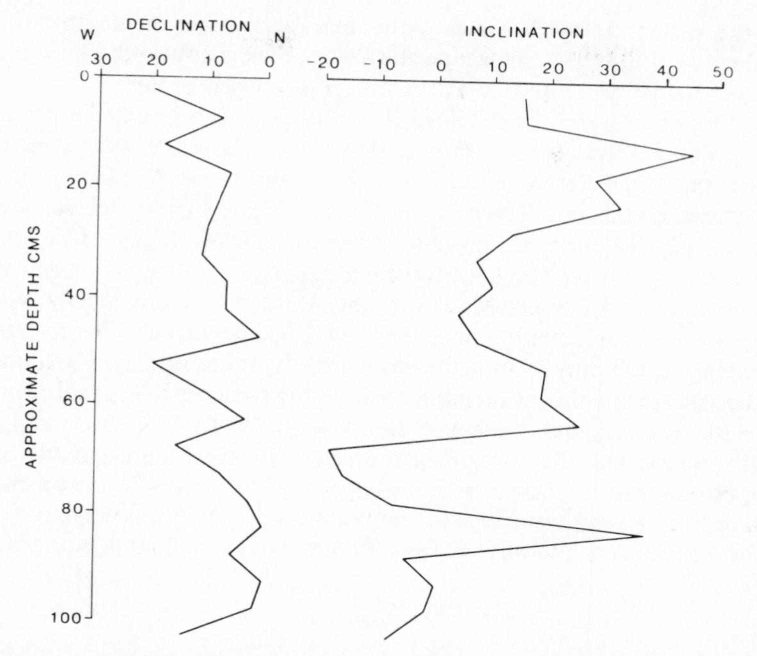

Figure 6.10 Variation of declination and inclination of remanent magnetisation with depth in Upper Cricket Muds of Snake Pit section. Samples partially demagnetised in an alternating magnetic field at 50 oersteds.

developed magnetic foliation plane close to the bedding and q values in the range 0.06–0.67 (Hamilton & Rees 1970). In contrast, these sediments exhibit no clear foliation and were found to have a scattered range of q values from 0.17 to 1.69. This supports the strong post-depositional disturbance of the sediments, discussed above. The evidence of the normal polarity must therefore be treated with caution, but the indication is that these sediments were probably deposited in the Brunhes normal-polarity epoch, which commenced 700 000 years BP (La Brecque *et al.* 1977). This is thought more likely than deposition prior to 890 000 years in the preceding normal epoch on geomorphological grounds.

The allophane layer forms a distinctive break in the monotonous Cricket Mud sediments. The basal contact is remarkably sharp, while the return to Cricket Mud sediments is often a more merging boundary. The allophane itself is a distinctive, laminated, waxy-white layer, locally up to 1 m thick. Analysis indicates that it is predominantly an X-ray amorphous aluminosilicate, with chemical properties indicative of the allophanes (Fieldes & Perrott 1966, Laverty 1982), intermixed with minor quartz and gibbsite. This unit is thought to be derived by the weathering of a volcanic ash deposit, although it is not yet certain if this occurred *in situ*. Like the Cricket Muds, it was deposited under ponded vadose conditions by an episodic translatory flow mechanism from the surface through solutionally enlarged fissures in the limestones, as described by Bull (1981), in temperate areas. This is confirmed by sedimentation to the cave roof, and in places even into closed domes.

The allophane layer forms a laterally continuous and important marker bed in the Cricket Muds, which proves that these sediments were deposited at different times in different levels of the cave. Thus while allophane is generally present as a continuous layer in the Cricket Muds of the Revival level and above, only rolled clasts are present at the Inflation level, suggesting sedimentation after the Lower Cricket Mud times. At the lowest active levels there is no allophane, and Cricket Muds are only thin, or present locally where re-erosion of the sediments of the upper levels has occurred. This re-erosion is very extensive and may include the basal gravels. It has occurred predominantly under vadose conditions forming meandering trenches in the sedimentary fill (Fig. 6.9), before abandonment of the passage levels. The Cricket Muds are thought to represent the remains of the Setap Shales which originally overlaid the limestone, but this has not yet been proven. Neither is it certain that they are all derived from the surface of the limestone, and an unknown proportion may be introduced laterally in cave streams from the same sources as the gravels.

6.5 Fluvial terraces

While the preservation of ancient sediments in fossil caves has been emphasised above, more recent alluvial deposits are also preserved at the surface in the form of terrace remnants (Fig.6.2). The terraces of the S. Melinau were first described by Leichti *et al.* (1960) and Wilford (1961), who recognised four different levels related to a regional terrace sequence found throughout northwestern Borneo and attributed to major cycles of erosion and sedimentation. Woodroffe (1980), using short levelling traverses to determine terrace heights above the contemporary river, identified terrace levels of 1–6, 8–12 and 17–24 m. He associated these with the youngest of Leichti's cycles, and placed their age in the middle and late Pleistocene. These studies have now been extended by detailed levelling (Fig. 6.11), along both discrete terrace fragments and the present floodplain (Rose 1982).

The evidence throughout the region shows that the terrace fragments in the Melinau valley comprise a single body of sediment, composed of boulders, cobbles and gravel with a matrix of alluvial clay. The upper surface ranges in elevation from 157 m above sea level (70 m above the adjacent contemporary floodplain) at the upstream end to 27 m above sea level (6 m above the adjacent contemporary floodplain) at the furthest downstream remnant. Terrace gradients decrease in a downstream direction. These patterns are interrupted only in the vicinity of the S. Lutut/Melinau junction (Fig. 6.12), where the elevation of the highest terrace fragment rises above the

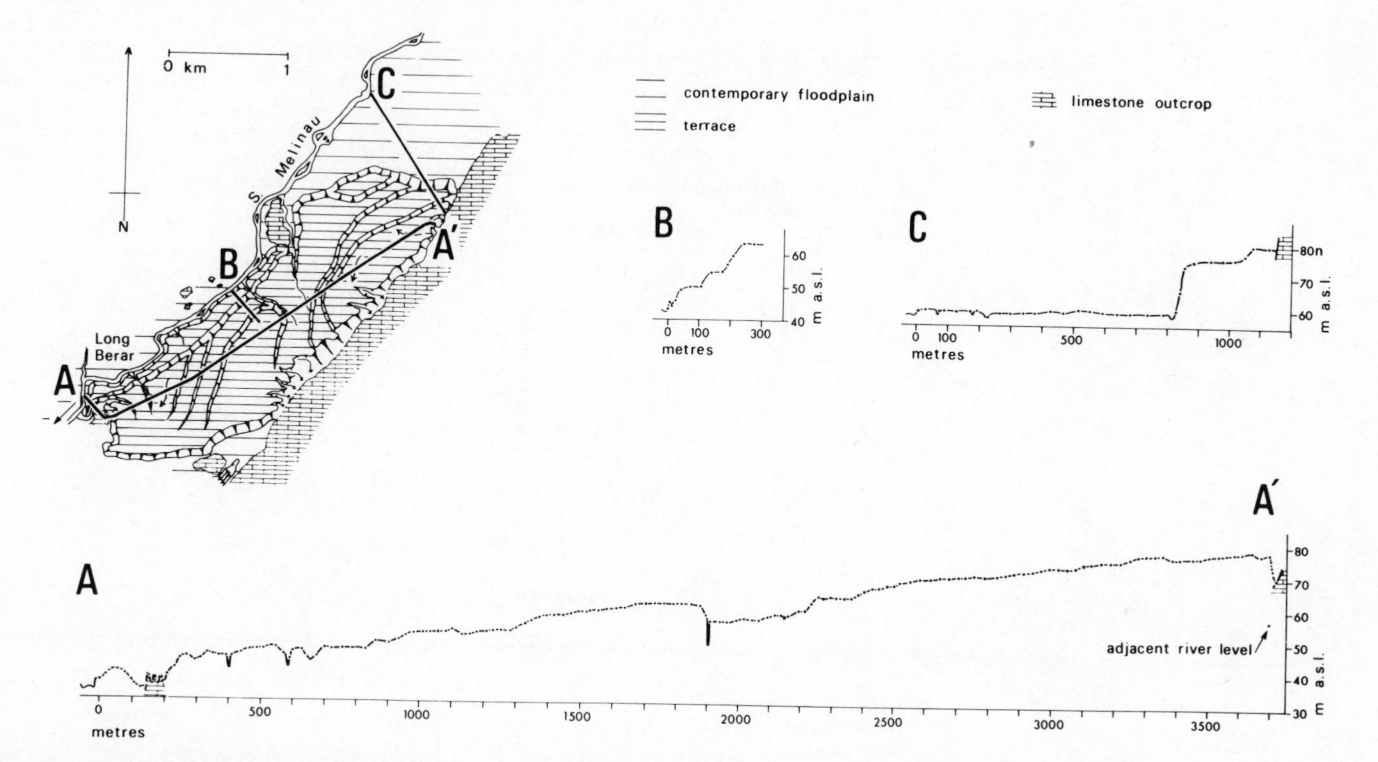

Figure 6.11 Sections of the main alluvial terrace between the S. Melinau and G. Api. AA′, parallel to present river; B and C, perpendicular to present river. (Section B after Woodroffe 1980.)

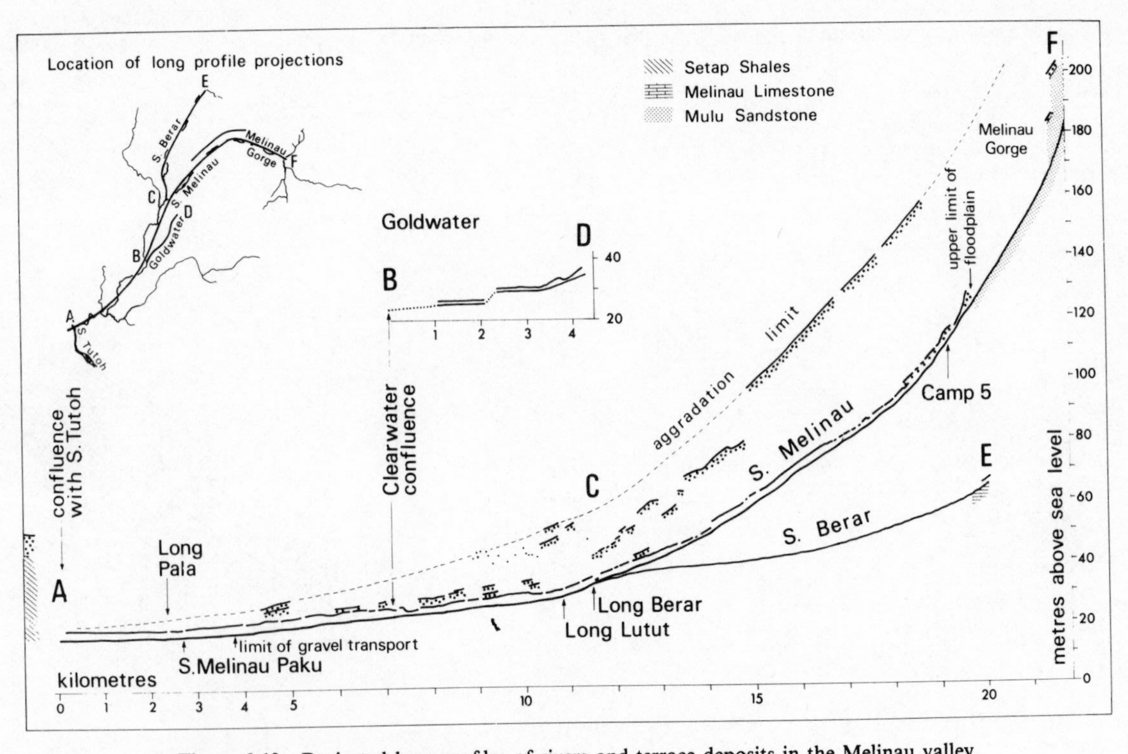

Figure 6.12 Projected long profiles of rivers and terrace deposits in the Melinau valley.

regional valley trend (Fig 6.12). Lower terrace fragments exist at all levels, and in all localities, but show a progressively lower gradient towards the present river level.

The maximum size of clasts within the terraces varies from 45.5 cm (*B* axis) at the apex of the alluvial fan to 13.5 cm (*B* axis) near the Berar/Melinau junction. Throughout, the main sediment body is composed of quartzitic sandstones with small proportions of hard shales and very small quantities of limestone, both of which become less frequent down stream, indicating preferential destruction of the relatively soft and soluble rocks during the process of terrace aggradation.

On the basis of this evidence the upper level of the terrace fragments represents the aggradation limit of a massive alluvial fan with an apex based on the Melinau gorge and slopes towards the north-east, north, west and south-west. Additionally a small subsidiary fan is located where the S. Lutut leaves the hills. The lower terrace levels represent discrete, uncorrelatable fragments with progressively lower gradients formed during the dissection of the main sediment body. Because of the fan shape, surface drainage was away from the centre of the valley towards the valley sides so that rivers such as the Berar and Goldwater are located at its margin. The S. Berar is only diverted from this position by the fan of the S. Lutut.

The landform and sedimentary evidence suggests that the fan was formed by material transported from the steep headwater channels in the mountain ranges of the Mulu Formation, deposited due to a reduction in gradient where the river entered the Melinau lowlands west of the Melinau gorge (Fig. 6.12). Sediment has been transported through the system over an adjusted bed profile, so that the coarsest boulders are deposited along the Melinau gorge and at the apex of the fan, and progressively finer material has been transported down stream in relation to the progressive decrease in channel gradient. A similar explanation is proposed to account for the fan of the S. Lutut. During the process of formation, the channel of the S. Melinau must have had a braided mode and been highly unstable, so that the drainage shifted across the fan surface from southerly routes, similar to those at the present time, to northerly routes into the Medalam catchment. The subsequent phase of incision was caused either by enhanced transport capacity in the main stream, or a decrease in sediment supply from the headwater region. An analysis of the present activity of the S. Melinau gives an insight into the processes responsible for erosion and deposition, and hence the environment in which they occurred.

The present rivers are located in trenches cut through the terrace deposits. In the headwater region incision is active, producing slope instability and extensive landslipping, which transports sediment to the stream channels. Below the Melinau gorge the river has a braided channel pattern, which changes down stream to a meandering mode with occasional braids, and finally a meandering and straight form. The maximum size of clasts transported by the Melinau decreases from 61.3 cm (*B* axis) at the bend in the river just east of the gorge, to 46.7 cm at the mouth of the gorge, and 3.6 cm just above the confluence with the Melinau Paku, where gravel ceases to be transported. Below this point, sand is the coarsest transported material and aggradation is controlled by the water level of the S. Tutoh. The clasts transported by the Melinau are predominantly quartzitic sandstones, but also include non-durable materials such as shales and limestones. All this evidence points to the fact that the Melinau is, at the present time, an actively aggrading river, which is currently building its bed, with material transported into the region from the steep headwater mountains. So great is the aggradation that the right bank of the Melinau in the region of the gorge forms the watershed between the Limbang and Tutoh drainage basins. In contrast the rivers that are not transporting sediment from the adjacent mountains, but are only draining the existing terrace deposits (e.g. S. Goldwater), are not aggrading, but continue to incise their beds as far as the zone affected by confluence with the Melinau.

The conditions operating in the Melinau at the present time are, therefore, analogous with those that must have existed when the fan was deposited. Indeed, if the present processes were to continue, the trenches cut into the fan would be infilled, and the terraces would be buried.

6.6 Discussion

It is now possible to summarise the detailed findings discussed above, before fitting them into a unified framework.

Owing to a progressive fall in base level controlled by the S. Tutoh, a vertical sequence of fossil caves exists in the limestone of G. Api, with the earliest abandoned passages the most elevated, and those still active at the base. The morphology and sequence of these caverns can give an indication of the rate of base-level lowering and by inference episodic uplift, which can be validated by the dating of speleothems. Because these caverns are protected from surface erosion, relict sediments from the active phases and any subsequent deposits are preserved preferentially compared to contemporaneous surface sediments. The basal gravels of the cave sediments are often aggradational in character, and are commonly associated with a distinctive wall notch, related genetically to this mode of deposition. Subsequent sediments are primarily fine-grained ponded deposits, but include in the higher passage levels a distinctive weathered volcanic ash, which forms a marker bed. Two aggradational and one erosive events are preserved in the terraces and alluvial deposits of the Melinau lowlands. The second aggradational event is continuing at present. In the following discussion, the relationship between external and internal aggradation is first examined, before the climatic conditions which give rise to these events are discussed.

The relationship between surface and cave aggradation

It has been suggested that the S. Melinau is actively aggrading its bed. If this is so, there must be evidence of aggradational processes currently active in the cave streams. Imperial Cave, Leopard Cave and Thargs Cave, which together form the upper part of the marginal drainage system along the west side of G. Api, are all incising into clean-washed, bedrock-floored, canyons. There is a copious supply of coarse sandy sediment derived from weathering of the sandstone gravels in the main terrace, which is still being actively eroded. Similarly the Clearwater River passage is rock-floored throughout. However, at the resurgence, the water plunges into a deep ponded sump to resurge from depth in the floor of a short sandy-floored channel, tributary to the S. Melinau. No bedrock is present in this aggraded channel, and there is ponding and deposition of sands in the Goldwater Series (the downstream segment of the marginal drainage system) and in Sump II, which is mud-floored. There is also evidence from dye tracing (Smart & Friederich 1982) that water enters the sump from lower phreatic levels, which may be sedimenting before the main streamway. Down stream, Cave of the Winds has a meandering undercut notch, floored by gravel at a similar level to the S. Melinau. The present entrances into the Melinau Paku are closed, but backflooding and ponding occur frequently during floods, depositing a fine mud on the passage floor. This is a situation analogous to that proposed below at the commencement of Cricket Muds sedimentation.

Thus is can be demonstrated that internal and external aggradation are to some extent synchronous, depending on the type of cave stream. It is now necessary to determine if the internal and external aggradation levels are accordant. At the Clearwater Resurgence, the main terrace is some 10 m above present river level, while the bed of the channel at this time would have been some 3 to 4 m lower. It is to the latter that the base of the notch should be adjusted, while the top should lie below the terrace level because the overbank and bar deposition building the terrace occur only at the higher flows, while the dominant discharges for limestone erosion are much lower. In the River Passage, the well developed notch has its roof at 6 to 7 m, which is concordant within the survey precision with the main terrace channel. Furthermore, this notch is also observed in the Goldwater Series and Cave of the Winds, both of which had separate sediment sources at the time. Thus the evidence suggests that not only do the external terrace levels and internal gravel and notch levels match, but that it is the level of the resurgence end of the cave system which controls the aggradation. The disparate elevations of the multiple sediment sources which may feed a single cave are therefore not significant. If the source is significantly above resurgence level, the resulting steep channel can transport material through the cave until external aggradation reduces the height difference and deposition can occur.

A second problem that must be considered is the relative persistence of external aggradation, compared to the time required for underground processes to adjust. The development of an alluvial fan is essentially a continuous process until the aggradation limit is reached, although oscillations may occur in response to random and short-term runs of events, to give limited incision. The aggradational limit may represent an adjusted profile under the extant conditions with the sediment supply being transported across the fan without further accumulation. The fan levels will therefore remain essentially static until conditions change, giving a considerable period for adjustment of the underground passage regime. Alternatively, the aggradation limit may represent a single point in time when the river changed from aggradational to degradational mode, and no persistent underground adjustment to this level would be expected. Osmaston (pers. comm. 1981) has measured the rate of limestone tablet erosion in the S. Melinau at about 0.5 m per 1000 years. If this rate also applies to the caves, the observed notches could develop over periods of under 10 000 years. Thus a relatively limited external steady state at the aggradation limit could produce a cave notch.

The notches preserved in the caves cannot, however, represent a complete record of all aggradational still-stands, even if subsequent erosion did not occur. If an open stream sink is not present at the upstream end of a cave streamway, but the water enters the caves through a boulder choke, as at present in Cave of the Winds, then aggradation by transported gravels cannot occur, and no notch will develop. However, fine sediment will still be transported into the caves through such a blockage, and the translatory flow sediment supply will continue uninterrupted. It is therefore possible that the Cricket Muds represent the fine-grained equivalent of the aggradational gravels. As the surface river builds up the fan, water surface slopes will be reduced in the cave streams, ponding will occur and velocities will fall. The maximum height of the resulting sedimentation will be some way below the channel bed level at the resurgence, because erosion would prevent a concordant sediment floor, as can occur with the gravels which are of comparable calibre to the sediment load of the surface river. The present situation in Cave of the Winds is one of incipient Cricket Muds sedimentation, because as yet this level differential has not been established after the cessation of gravel transport. For major aggradational events, base-level passages may be completely inundated and the previously abandoned higher levels reactivated. This may well produce complex multiphase sedimentation patterns which are difficult to detect due to similarity in the different units and lack of continuous exposures.

Reworking of the sediment sequence by an episode of external incision may also occur giving the more sandy layers commonly observed in the Cricket Muds (Fig. 6.9). In fact it is important to point out that despite the long-term fall in base level, which causes the progressive abandonment of higher-level passages, all the sediment sequences observed have been re-eroded after deposition. In some cases this is due to percolation water entering the passage, leading to sapping of the sediment in a limited area. However, more commonly the trenches are laterally continuous and can be seen to incise into bedrock in the floor of the pre-sedimentation passage, giving multiple trench phases. If the bulk of the cave sediments are deposited in response to external aggradation, as is proposed, then re-erosion of the sediments would be expected when dissection of the external fan commences, and the gradients of cave streams steepen. The periodic aggradation/degradation controlled by external conditions can therefore explain not only the depositional but also some of the erosional elements of the cave geomorphology.

The causes of aggradation

Is it possible now to relate the aggradational events to particular climatic or other conditions? It has already been suggested above that, during the recent past, rapid changes in base level due to uplift did not effect the Clearwater Complex, despite the continuing tectonic elevation of the Mulu dome (Leichti *et al.* 1960). However, because of the low river gradients and dispersion of the nick-point as it migrates upstream, such effects may be damped out before affecting Clearwater Cave. Nevertheless, sudden changes in base level have been recorded previously, as at the initiation of the Clearwater River passage. Furthermore, it is significant that the S. Tutoh, whose headwaters lie outside the Mulu area, is also actively aggrading, implying a much more widespread control than local uplift.

If tectonic factors are not responsible for terrace formation, are the terraces related to aggradation following the postglacial rise in sea level? Again the answer must be no, because deposition is proceeding down stream from the headwaters in both the Melinau and the Tutoh (Rose 1982). Aggradation of the latter has caused impoundment of the lower 3.8 km of channel in the Melinau, giving overbank deposition of fine sediments, but this is of only local significance. It is also due wholly to a lack of synchroneity in the gravel transport capacity of the two aggrading rivers, and not upstream propagation of the effects of sea-level changes. There is, however, some evidence that the initial incision of the fan is related to enhanced river gradients due to a glacio-eustatic fall in sea-level (Rose 1982).

It must therefore be concluded that the present aggradational phase either is a quasi-random episodic deposition event, such as described by Schumm (1977), or is controlled by climatic factors which affect the sediment supply and transport capacity of the rivers. As the fossil fan deposit is a major feature with sediments over a 25 km long and 4.5 km wide belt, over 70 m thick at the apex, and having an estimated volume of 2.1×10^8 m^3 prior to erosion, it seems more probable that the present aggradation is a similarly major event. Furthermore, while Schumm envisaged reworking of sediments as a major source for episodic deposition, the abundance of unweathered pebbles and the presence of less persistent lithologies such as limestone, which decreases in frequency downstream, indicate that active headwater supply is occurring. Thus the present climate may indicate the conditions giving rise to fan formation. Because of the small volume of the present aggradation deposit, it is also tempting to suggest that the climatic changes initiating fan development were associated with the end of the last glacial period. The preceding erosive phase can then be tentatively correlated with the last glacial period, and the fossil fan either with a major last glacial interstadial period or with the last interglacial period itself.

There is a growing body of evidence that the climate was generally drier in the tropics during the last glacial period, compared with the present (Jones & Ruddiman 1982). Verstappen (1975) has argued that because of the extensive shelf areas which became exposed adjacent to Borneo during glacial periods, and increased poleward migration of the Intertropical Convergence Zone, the region had both a drier and a more seasonal climate than at present. A recent climatic simulation by Manabe and Hahn (1977), based on a climate model verified under the present climatic regime, also demonstrated a general glacial decrease in both tropical precipitation and runoff. However, parts of Borneo showed up to 50% higher runoff during glacial time, although it is not possible to interpret the model results for a specific locality with any precision.

The relations between climate, sediment supply and fluvial transport capacity are complex and poorly understood. However, the build-up of a fan in the Melinau valley must indicate an excess of sediment supply, compared to transport capacity. Thus climatically induced changes in either the former, controlled by weathering rates, landslip frequency or vegetation cover, or the latter, controlled by runoff and its distribution through time, must be responsible. The available evidence suggests that equatorial rainforest has persisted in the lowlands of Borneo during glacial periods (Verstappen 1975). However, at higher altitudes (above *c.* 2000 m) there was a general lowering of the vegetation zones, as discussed for New Guinea by Flenley (1979), culminating in the glaciation of the highest peaks such as Mt Kinabalu in Sabah, where the snowline was between 3600 and 3700 m (Koopmans & Stauffer 1968). While a reduced crown cover on the slopes of G. Mulu (2377 m) in glacial times could have given an increase in understorey vegetation density, this is unlikely to have significantly affected the landsliding, which is the most significant source of sediment in the present day rivers.

It is difficult to envisage a significant increase in headwater sediment supply from that observed at present in the area. The Mulu hillslopes are steep, straight and weathering-limited, with up to 80% of their surface occupied by successive and multiple landscape scars (M. J. Day 1980). The debris supplied also shows very little evidence of extensive weathering. Thus if aggradation is active under the present climate, an increase in runoff, as suggested by the Manabe and Hahn (1977) simulation, would be sufficient to explain the cessation of fan development during glacial times.

However, if the more generally accepted decrease in glacial runoff is correct, then there must also have been a substantial decrease in the rate of sediment supply. M. J. Day (1980) has demonstrated the contemporary relations between landslide occurrence and heavy rain, but the reduction in weathering is probably at least as significant in reducing the rate of sediment supply. This reduction due to decreased runoff would also be enhanced by a glacial temperature decrease, particularly if similar to the 11°C maximum reduction suggested for the Highlands of New Guinea by Bowler *et al.* (1976) on palynological evidence. Furthermore, if the reduction in total runoff is due to increased seasonality, as suggested by Verstappen (1975), then peak fluvial transport capacity during the wet season may be very similar to that observed at present. Thus incision of the pre-existing alluvial fan could occur despite the reduced runoff, but aggradation would again commence with increase in the sediment supply on deglaciation. The latter explanation is supported by the work of Sabiels (1966) on the variation of manganese concentrations in radiocarbon-dated cave sediments from Niah Great Cave, Sarawak. He proposed that between 32 000 and 20 000 years BP the rainfall at Niah was about 2100 mm compared to 3800 mm in postglacial times. However, the methods he employed are largely unsubstantiated, and a satisfactory assessment of the glacial runoff of Sarawak must await further research.

6.7 Conclusions

In this chapter, we have examined the relations between surface and underground fluvial activity in the Gunung Mulu National Park, Sarawak. The underground links form an essential part of the fluvial network in the area, and it is therefore not surprising that underground fluvial processes are intimately related to surface river activity. Three major areas of interaction can be identified:

(a) The surface drainage network controls the points of input and output of water to the limestone and, therefore, the basic network of the caves resulting from this circulation. This network owes much of its detailed form to geological constraints, but these only mediate the essentially hydraulic factors. In Mulu, runoff from the impermeable Mulu Formation into the Melinau Limestone and discharge to the south where the S. Tutoh cuts the limestone outcrop have controlled the predominantly southward drainage. This has been accentuated by the presence of a Setap Shale cover above the limestone, and laterally continuous open bedding planes favouring strike development.

(b) The resurgence point where cave waters rejoin the surface rivers forms the base level for cave development. Above this level vadose air-filled passages may form, while below it water-filled phreatic passages develop. The progressive fall in base level that results from incision of the surface rivers gives rise to a progressive modification of the underground passage form. Where base level falls more rapidly than the rate of cave passage enlargement, then phreatic passages are abandoned with little vadose development. Where the converse is true, extensive vadose canyons develop underground. An examination of cave passage morphology can therefore give an indication of the rate and continuity of surface fluvial incision. In Mulu, the evidence is for a relatively steady and continuous fall in base level, whose rate has been matched by cave incision to give extensive vadose canyons.

(c) Aggradation of the surface river systems gives rise to parallel sedimentation underground, whose level is again controlled by the resurgence elevation. Where active gravel transport occurs through open cave entrances, then underground sedimentation parallels that at the surface. However, where cave entrances are closed, then underground ponding occurs and fine mud sediments make up the aggradational unit. Only in the former case is there any modification of the passage morphology, with the development of wall notches at the level of the top of the aggradational gravel. Wall notches preserved in the higher levels of the cave systems indicate earlier aggradational events for which no surface evidence remains.

The most significant feature of past surface fluvial activity is a massive alluvial fan derived from the Melinau gorge. This corresponds with the lowest of five such aggradational events recorded by the distinctive notches in the walls of the Clearwater Cave complex. The fan was constructed by downstream aggradation, a process that can be shown to be occurring in the contemporary Melinau. Based on this observation, on the similarity in maximum sediment size in the fossil fan and present river and on the relative magnitude of the fossil and present fans, it is suggested that fan development is an interglacial process. Radiometric dating of cave speleothems is in progress to confirm this suggestion. By inference, dissection of the fan occured during glacial periods, partially in response to lower sea levels increasing river gradients and partially due to climatic change. At present, it is not known if these changes resulted in increased runoff enhancing transport capacity, or in drier, cooler conditions which reduced sediment supply by decreasing weathering rates on the Mulu Formation.

Part C

ENVIRONMENTAL CHANGE

Evidence for environmental change

Editorial comment

New dating techniques and more precise interpretations of the best records of vegetation and landform history provided by continental, lacustrine and near-shore sedimentary deposits have demonstrated the remarkable series of changes affecting the tropics in the Quaternary. An increasingly consistent, global picture of expanding and contracting rainforest and desert areas, of changing lake and sea levels, and of metamorphosing rivers has emerged, particularly for the last glacial/interglacial cycle. Although the detailed evidence comes from scarce, widely separated sites with long, continuous depositional sequences, such as Lynch's Crater (Fig. 2.5), new remote-sensing technology, including LANDSAT imagery, sideways-looking airborne radar (SLAR) (Tricart 1974a), METEOSAT (Mainguet *et al.* 1980, Mainguet 1983) and combinations of data from various sources (Haxby *et al.* 1983) have facilitated regional reconstructions of past environmental conditions. The variety, quality and growing quantity of evidence for Quaternary climatic change is well summarised by Williams in Chapter 11 for Africa, Asia and Australia, while in Chapter 10 Tricart relates his largely personal story of the unravelling of the Quaternary of tropical South America.

Chapters 7, 8 and 9 all explore extensions of the basic theme of empirical environmental reconstruction. Thus Flenley is interested in explaining individual pollen curves in terms of the population biology of the taxa that produced them. Ideally, the ultimate goal of such an approach might well be to reconstruct the spatial and temporal mosaic of the forest growth cycle (Spencer & Douglas Ch. 2). However, in order to quantify palaeoclimates from pollen spectra, there must be a relationship between contemporary climate and modern pollen rain which can be applied, by a multivariate transfer function, down core (e.g. Birks & Birks 1980; N. America: Webb 1980; Chile: Heusser & Streeter 1980). It must also be assumed that regional vegetation patterns not only reflect climate but also are in equilibrium with the climates in which they occur. Many factors complicate such potentially direct relationships. Combinations of species with no parallel in the present vegetation found in cores may represent either a set of environmental conditions that no longer exist or former aclimatic interactions between the species themselves

(Davis 1981). Furthermore, there are the problems of a migrational lag between climatic change and vegetation colonisation (e.g. Birks 1981). As Street-Perrott *et al.* point out in Chapter 8, this possibility would not be so serious if climatic change could be shown to be gradual. However, both the sea-level (e.g. Chappell 1983a) and lake-level (Ch. 8) evidence indicates the abruptness of climatic change and the likelihood of a period of disequilibrium between vegetation and climate. The potential geomorphological consequences of such disequilibria, or periods of rhexistasy, have been realised (Erhart 1955, Knox 1972, Douglas & Spencer Ch. 3). Thus although pollen analysis may give a good idea of the status of vegetation, it is a less satisfactory indicator of climate (Grove 1982) and geomorphic process.

The detailed study of lake sediments offers one solution to these problems, as Oldfield and his colleagues show in Chapter 9. Erosion rates may be linked to different ecological conditions by detailed examination of the physical, chemical and biological properties of lake core sediments and the identification of their sources. Both the Oldfield team in New Guinea and Flenley on Easter Island (Flenley & King 1984) found that, as elsewhere, early human migrants had profound effects on ecosystems and denudation systems.

With close attention to the nature of thresholds and lags in geomorphic systems, detailed knowledge of the response of landforms to climatic change now makes it possible in some cases to use geomorphic evidence to speculate on the manner of climatic change, as in the case of coastal and former desert dune evolution (Chappell 1983b). As this ability to examine climatic change from geomorphic evidence expands, the exciting and challenging prospects for modelling changes in the global circulation and in atmospheric characteristics and their ecological impacts become much closer to reality.

7

Relevance of Quaternary palynology to geomorphology in the tropics and subtropics

J. R. Flenley

Quaternary palynology originated in temperate regions (von Post 1916) and was initially a technique for geological correlation and relative dating. More recently, as techniques of absolute dating have become available, the emphasis has switched to vegetational, and hence environmental, reconstruction. Vegetational reconstruction has usually depended on modern pollen rain data and the application of the uniformitarian principle (Rymer 1978). Environmental reconstruction has similarly depended, overtly or implicitly, on some kind of transform function, and on the same principle.

As the major development of Quaternary palynology in the tropics since about 1960 occurred after the start of radiocarbon dating (Libby 1952), and was therefore free from the straitjacket of geological correlation, it has been able to flower as a technique for palaeoenvironmental reconstruction. As it can only be applied where a suitable depositional environment, usually a lake or bog, exists, palynology has provided long continuous sequences in some areas, while in others, such as the arid tropics, there are few satisfactory records.

Of the palaeoenvironmental reconstructions possible from palynology, three might be the most relevant to geomorphology: vegetational, thermal and hydrologic. Although just outside the tropics, Easter Island at 27°S provides an example of what is, up to now at any rate, a purely vegetational reconstruction. It is an entirely volcanic island, whose oldest portion has been dated at about 15 million years. Numerous craters are clearly much younger and the caldera, Rano Kao (alt. *c.* 110 m), has been estimated at 150 000 years (Baker *et al.* 1974, Clark & Dymond 1974). The present vegetation of the island is grass-dominated, with few non-introduced shrubs or trees surviving, the last native tree having disappeared about 1960. Geomorphologists would be concerned whether the weathering and erosion of the caldera had proceeded under the present short-grass vegetation, or under scrub or forest woody vegetation with a more equable ground-level microclimate which would have affected soil and weathering profile development. An outline pollen diagram (Fig. 7.1) from a core taken near the edge of the swamp suggests that the

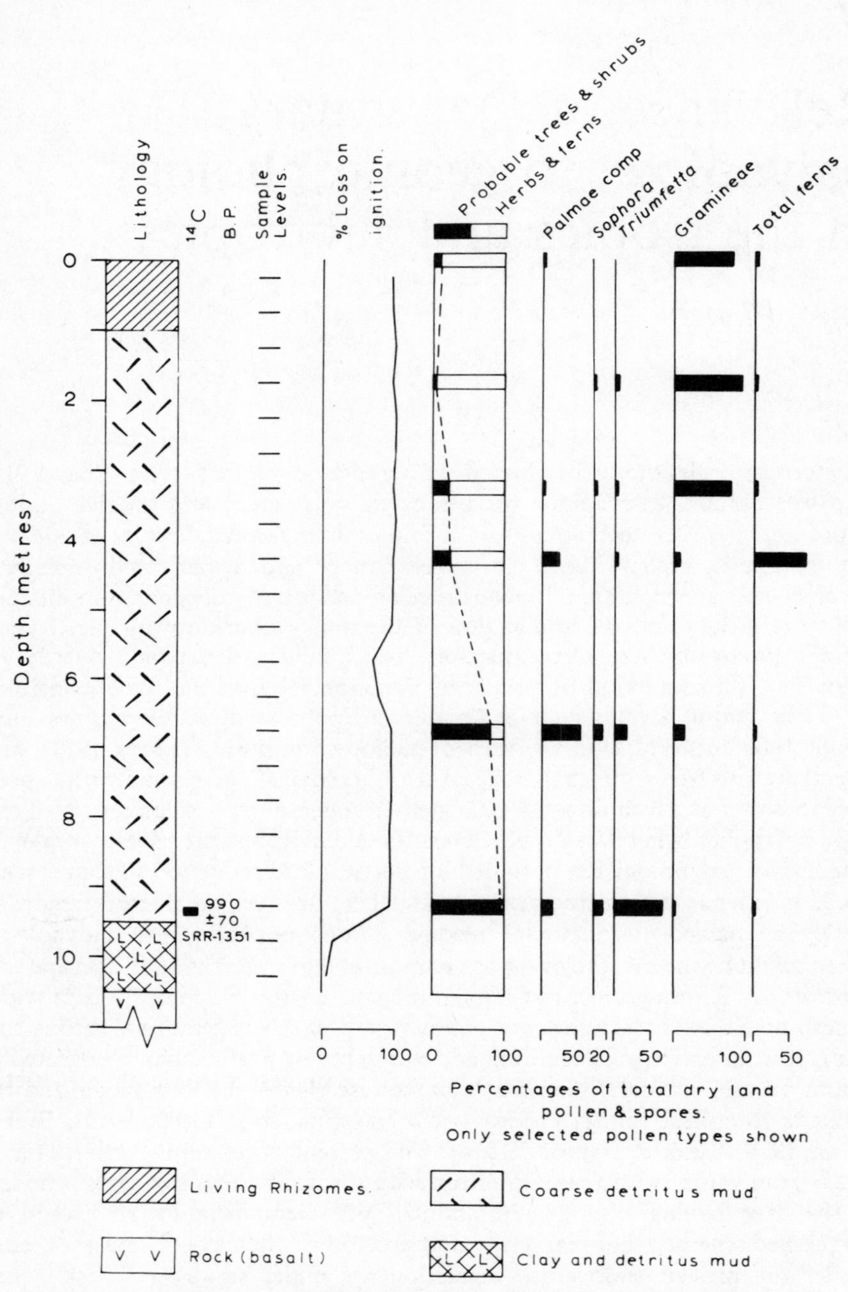

Figure 7.1 Preliminary outline pollen diagram from Rano Kao, Easter Island.

present dominance of grasses in the pollen rain is a recent phenomenon. About 1000 years ago the dominant taxa were the shrub *Triumfetta*, the small tree *Sophora* and a species of palm. The change is so dramatic that, even in the absence of modern pollen rain, it is difficult to draw any conclusion other than that a woody vegetation has been replaced by a grassy one. It is also clear that this change has taken place since the appearance of man on the island (Ayres 1971, Heyerdahl & Ferdon 1961, Flenley & King 1984).

A special type of vegetational reconstruction is that where the former vegetation provides clues to former sea level. In the tropics, such studies are frequently related to mangroves. In South America for example, the outer mangrove is usually dominated by *Rhizophora*, while in the inner mangrove (i.e. at a higher stratigraphic level) *Avicennia* grows. This zonation may vary somewhat, but appears to be the common one. Borings near accreting coasts of Guyana and Surinam have yielded pollen diagrams, published by Van der Hammen and coworkers, which can be interpreted in terms of fluctuating sea level. Interestingly, low sea-level phases seem to coincide with times when savanna was the predominant vegetation on dry land, as summarised by Van der Hammen (1974) and Flenley (1979).

For an example where the reconstruction is not only vegetational but also climatic, and in particular suggests thermal changes, we may turn to the New Guinea highlands (Fig. 7.2). The highest peaks there are still ice-covered, and the levels of cirque floors suggest a former lowering of the firn line by about 1000 m. But this, taken alone, is not very clear evidence of thermal change, since the depression could have been due to change in precipitation rather than to thermal change. The present vegetation of the highlands is mostly of various types of rainforest, giving way to tropic-alpine vegetation at a forest limit around 3800 m, which has a mean annual temperature of about 6°C (Walker & Flenley 1979). Modern pollen rain from the region shows that the tropic-alpine vegetation yields characteristic pollen spectra. Some tree pollen may be carried up into this area, but certain characteristic pollen taxa are usually present as indicators of the tropic-alpine origin of these spectra.

A pollen diagram from Lake Inim, at 2550 m (Fig. 7.3), shows that the Holocene vegetation is dominated by rainforest. But in the late Pleistocene, before *c.* 12 000 BP, although the deposit here is somewhat disturbed and dates are inverted (the disturbance itself being a possible indication of environmental fluctuation), the pollen spectra suggest a tropic-alpine vegetation, similar to that now found about 1500 m higher up. Other pollen diagrams from this area include the one from Sirunki which is a pollen influx diagram, i.e. the results are expressed in pollen grains $cm^2 yr^{-1}$. This has the advantage that the results are independent values not subject to the artefactual errors of the percentage technique. Such diagrams can only be made when a sufficient number of absolute dates has been obtained. These and other New Guinea diagrams hang together in a remarkably consistent way, to suggest movements of the forest limit summarised in Figure 7.4. The late Pleistocene lowering of the forest limit appears to have been around 1500 m. When

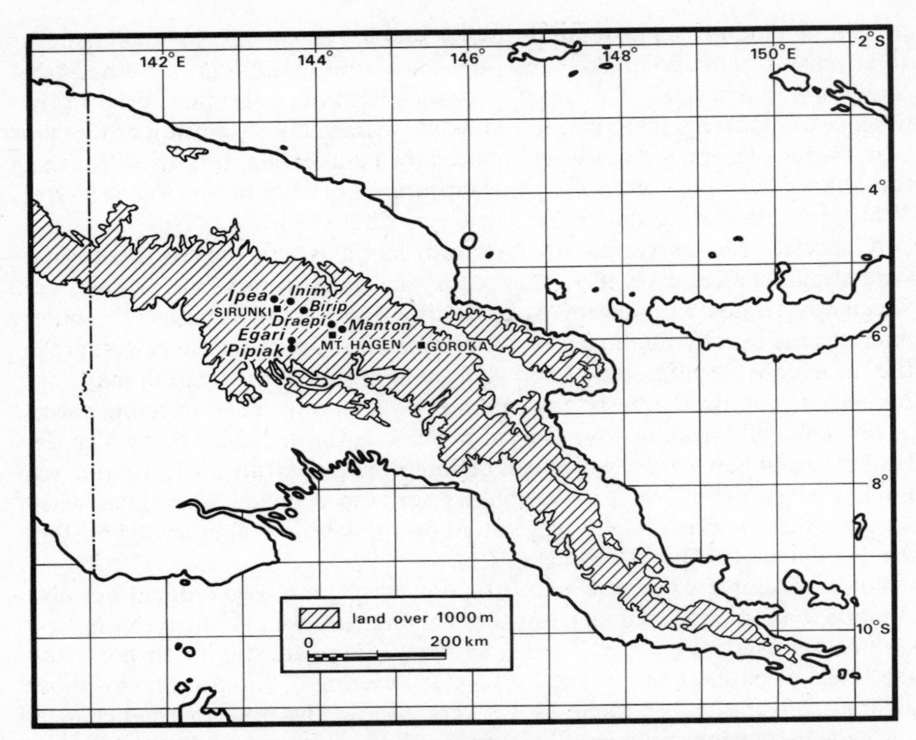

Figure 7.2 Papua New Guinea showing the highlands and localities mentioned in this book.

compared with the 18 000 BP sea surface temperature reconstructed by the CLIMAP Project Members (1976), these results seem to suggest a steeper Pleistocene lapse rate (Fig. 7.5). The lowering of the snowline by only 1000 m can be explained as the result of less precipitation, which is also consistent with a steeper lapse rate (Walker & Flenley 1979).

A Hull University team working in Sumatra has been able to extend these findings to a lower altitude. Inevitably this research was concerned not with the clear biogeographical boundary of the forest limit but with more subtle changes within the forest. There are gradual changes in forest composition with altitude on a Sumatran mountain (Table 7.1) (Morley 1981). The highly diverse lowland rainforest occurs up to 1000 m. Submontane forest, between *c.* 1000 m and *c.* 1400 m, also has a diverse flora of canopy trees, including *Aglaia* spp., *Aphamixis grandiflora*, *Ardisia lanceolata*, *Celtis* sp., *Elaeocarpus* spp., *Eugenia* spp., etc. Above 1400 m we are in the Montane Orographic Zone of Van Steenis (1972). The forest here can be divided into two types; a montane forest I from 1400 to 1800 m and a montane forest II from 1800 to 2400 m. Montane forest I is a mixed forest characterised by the abundance of oak trees (*Quercus* and *Lithocarpus* spp.) in the canopy. Montane forest II is much less diverse and is characterised by the abundance of the gymnosperm

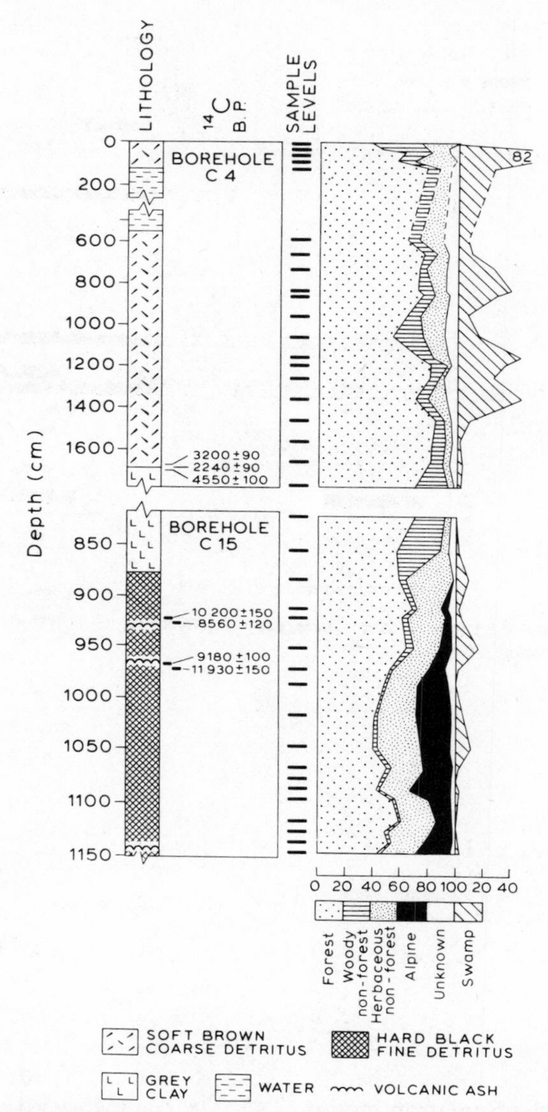

Figure 7.3 Summary pollen diagram from Lake Inim, New Guinea highlands. (After Flenley 1972, Walker & Flenley 1979.)

Podocarpus (*Dacrycarpus*) *imbricatus*, along with angiosperm trees such as *Weinmannia blumei* and *Symingtonia populnea*. Above 2400 m the forest gives away to the Sub-Alpine Zone of Van Steenis. Of course, it must be clearly understood that the boundaries between these forest types are statistical ones based on the relative 'abundance' of species. Naturally, it would be possible to find individuals of most or all the montane forest II species below

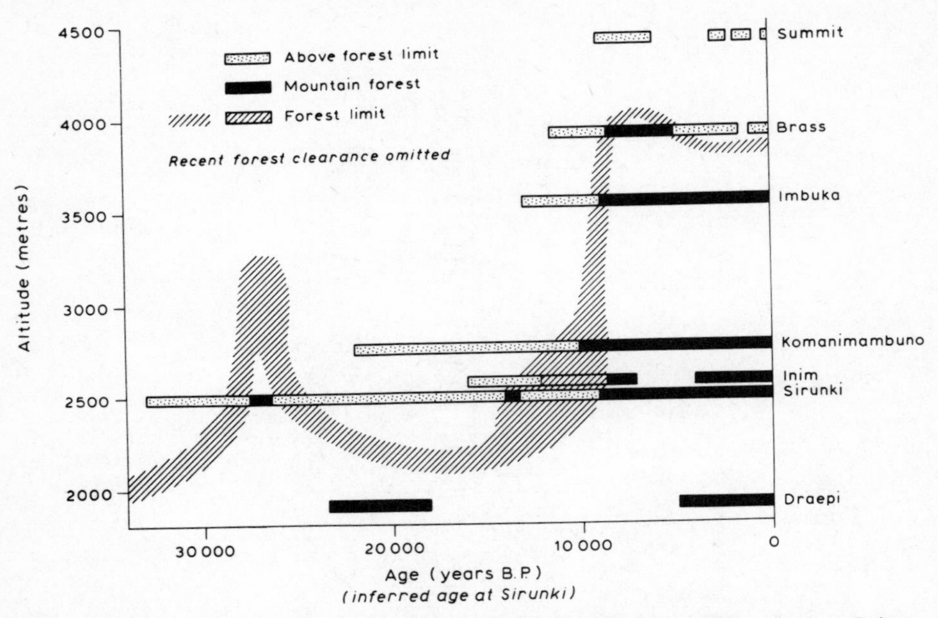

Figure 7.4 Summary diagram of late Quaternary vegetational changes in the New Guinea highlands. (After Flenley 1979.)

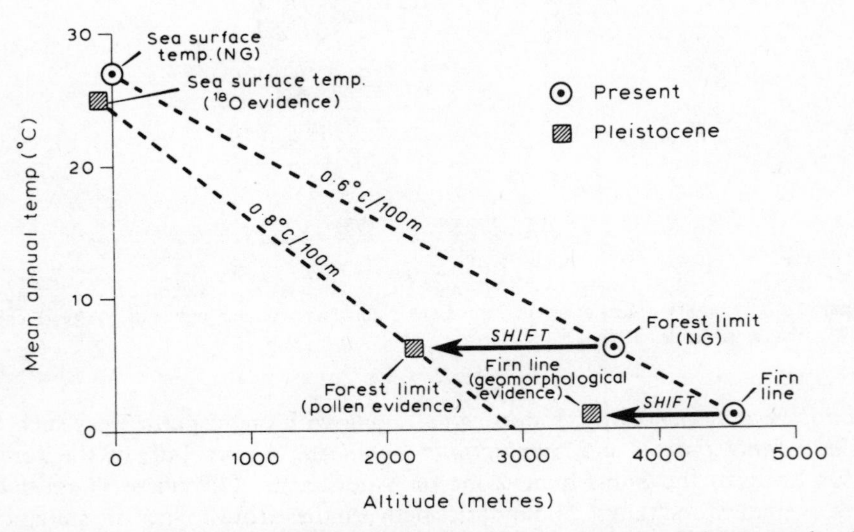

Figure 7.5 A possible model for lapse rates in New Guinea during the late Pleistocene and at the present day. (Data from Walker & Flenley 1979.)

Table 7.1 Altitudinal zonation of vegetation in Sumatra. (After Morley 1982.)

Vegetation type	Structural characteristics	Altitudinal range (m)	Orographic Zone (van Steenis 1973)	Rainforest Formation (Grubb 1974, Whitmore 1975)
ericaceous scrub	low microphyll shrubs with herbs	3000–3600	Sub-Alpine Zone	Upper Montane Rainforest Formation
ericoid forest	closed, low mossy microphyll forest	c. 2800–3000		
Gleichenia scrub	Gleicheniaceae with low microphyll trees and shrubs	c. 2400–2800		
montane forest II	closed, high-stemmed, floristically little diverse, mossy mesophyll forest, lianes rare, ground flora rich	c.1800–2400	Montane Zone	Lower Montane Rainforest Formation
montane forest I	closed, high-stemmed, floristically diverse, mesophyll forest, mosses and lianes common, buttresses and emergents rare, ground flora rich	c.1400–1800		
submontane forest	closed, high-stemmed, floristically diverse, mesophyll forest, emergents and lianes common, buttresses present, little moss, poor ground flora	c.1000–1400	Sub-Montane Zone	

1800 m. For instance, isolated individuals of *P. imbricatus* are fairly frequent down to *c.* 1600 m and it has been collected even much lower. But it would be exceptional to find all of these species together over a considerable area of land below 1800 m, whereas this becomes the rule between 1800 m and 2400 m.

The boundary at 1800 m, termed for convenience the montane I–II boundary, and that at 1400 m, the submontane–montane boundary, are readily detectable in modern pollen rain samples, and their movements can be traced in pollen diagrams.

From Lake Di-Atas at 1535 m, a pollen diagram covering at least the last 30 000 years suggests that the montane I–II boundary was below the lake until approximately 12 000 BP (Whitehead 1984). From an even lower altitude, 950 m, comes the first pollen diagram to cover the complete Holocene within the lowland zone of Malesia (Morley 1982). At this site, Lake Padang, there is evidence that the submontane-montane boundary lay below the lake at the end of the Pleistocene. Pollen of *P. imbricatus* is also recorded at this time in fair quantity, suggesting that it may have grown on the hills surrounding the site at *c*. 1250 m. These results combine to suggest a late Pleistocene depression of zone boundaries by about 350 m or more (Fig. 7.6). It is difficult to explain these results without invoking some climatic cooling at that time affecting not only montane zone but also, at least marginally, the lowland zone.

Results from similar studies in the equatorial mountains of South America and East Africa are shown in Figures 7.7 and 7.8 in summary form. As might be expected, these results differ somewhat from those obtained in New Guinea and Sumatra. In Africa, for instance, the mountain climates tend to be drier, so we should probably not expect the vegetational sequences to be similar. Nevertheless, there is a surprising measure of agreement between the diagrams, suggesting that worldwide climatic fluctuations may have been the primary controlling mechanism.

Kershaw's work on the Atherton Tableland of Queensland, Australia (Kershaw 1976, 1978), illustrates how palynology can lead to conclusions about palaeohydrology. There is today a sharp rainfall gradient across the Tableland. The wetter east bears mixed rainforest and the drier west sclerophyll woodland dominated by *Eucalyptus* species. A series of pollen diagrams from volcanic craters along the rainfall gradient shows evidence for the absence of rainforest during the late Pleistocene, and for its migration across the Tableland during the Holocene. The longest record (Fig. 2.5) from

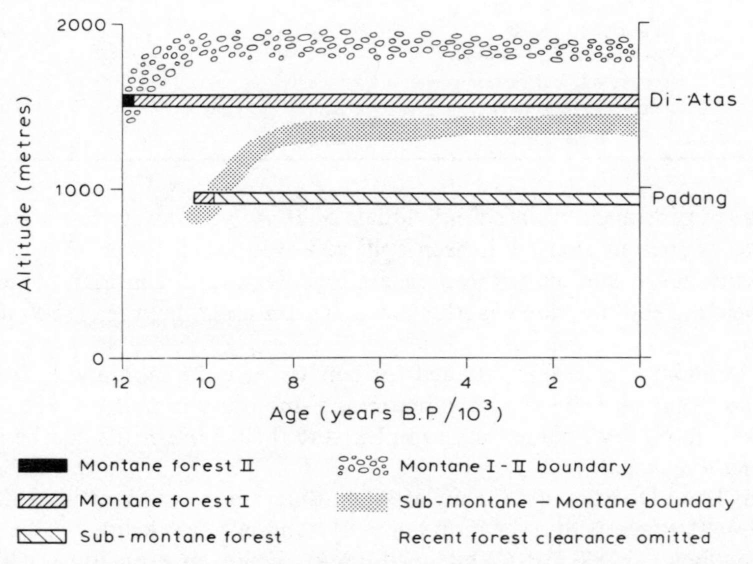

Figure 7.6 Summary diagram of late Quaternary vegetational changes in Sumatra. (After Morley 1982, Whitehead 1984.)

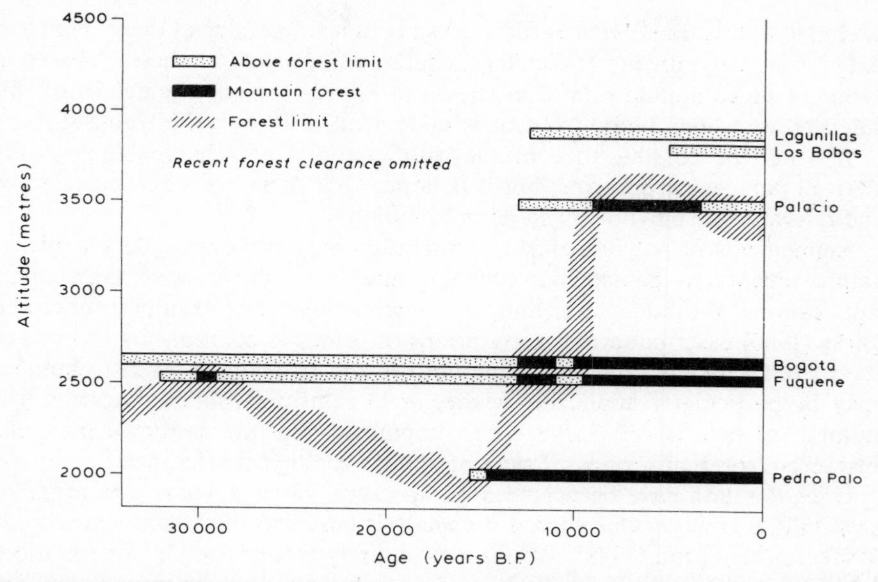

Figure 7.7 Summary diagram of late Quaternary vegetational changes in the mountains of Colombia, South America. (After Flenley 1979.)

Lynch's crater (Kershaw 1978) shows an earlier vegetation still, a rainforest rich in *Araucaria*, which flourished from before the range of radiocarbon dating to *c*. 38 000 BP. The demise of this was probably partially at least the result of desiccation, but the possibility of firing by the aborigines cannot be

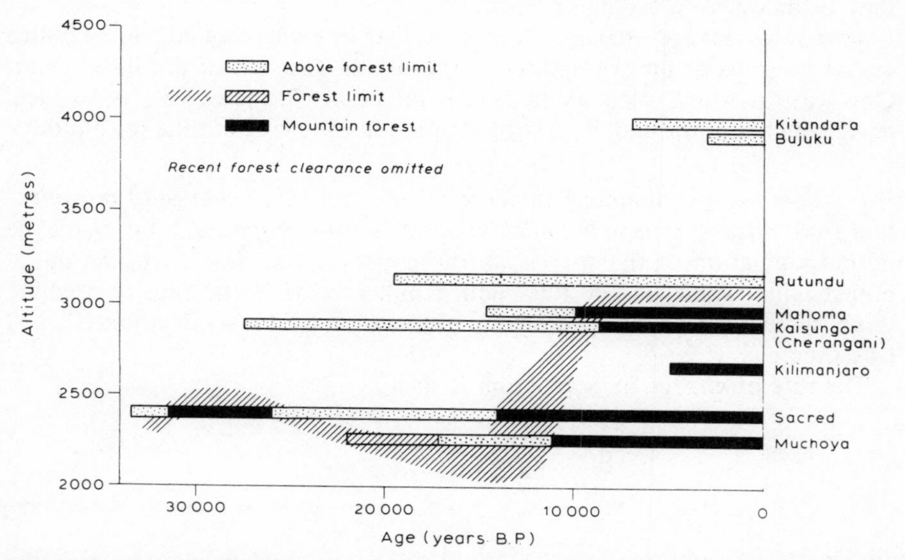

Figure 7.8 Summary diagram of late Quaternary vegetational changes in the mountains of East Africa. (After Flenley 1979.)

excluded at this time. Even earlier, possibly in last interglacial time, there was again a mixed rainforest. Kershaw's interpretation of all these changes in terms of mean annual rainfall is shown in Fig. 2.5. While the details of this interpretation may be doubted, it is difficult to disbelieve the whole story.

It would be possible to continue with examples of other palynologically derived palaeoenvironments, but it is hoped that these are sufficient to show the relevance of palynology to geomorphology.

Refinements of palynological interpretation may make possible yet further improvements in palaeoenvironmental inference. There have been many applications of numerical techniques to palynology. For example, Birks and Birks (1980) describe several applications of principal-components analysis to data from temperate regions, using both *R* and *Q* analyses. Such techniques may be particularly applicable where, as in rainforest pollen diagrams, the number of taxa is very large. The components that are extracted may, for instance, represent groups of taxa of similar ecological tolerances.

Once the data have been placed on an influx basis, a whole new range of possibilities emerge. One such is the analysis based on likelihood statistics by Walker and Wilson (1978). We are now entirely free of the idea of zonation, which was a hangover from geology. For each taxon, a division of the record is first made into sections that are essentially invariant, and sections where there is variation. Some attempt at curve fitting may then be made to the latter (Walker & Pittelkow 1981). A related idea, used by Green (1981), is time-series analysis. Palynologists have long wanted to use this, but were unable to do so while their data were mere percentages. Using influx values, Green has applied a frequency analysis to the record for individual taxa, producing results which may be analysed in ecological terms.

These analyses are working towards the idea of explaining individual pollen curves in terms of the population biology of the taxa which produced them. One way in which such an approach might be furthered, so as to yield environmental information, is suggested in the following outline for a model.

An outline for a preliminary model for deriving rates of change of reproduction from rates of change in pollen influx data for one species. Let N_0 be the initial population of the species, N the population at time t, A_0 the initial pollen influx value, A the final pollen influx value, r the rate of seedling establishment ('birth rate'), s the rate of loss of individuals ('death rate'), and t be time.

The rate of change of population is then

$$\frac{dN}{dt} = \text{birth rate} - \text{death rate}$$

$$= rN - sN$$

$$= N(r-s)$$

Therefore, we can write

$$\frac{1}{N}\frac{\mathrm{d}N}{\mathrm{d}t} = (r-s)$$

Integrating with respect to t, we obtain

$$\log_e N = (r-s)t + c$$

When $t = 0$, $N = N_0$, and therefore

$$c = \log_e N_0$$

Thus the equation governing population is

$$\frac{N}{N_0} = \mathrm{e}^{(r-s)t}$$

If we make the assumption that pollen influx is a direct indicator of population (i.e. we ignore physiological factors, sedimentological factors, etc.), then

$$\frac{N}{N_0} = \frac{A}{A_0}$$

and so

$$\frac{A}{A_0} = \mathrm{e}^{(r-s)t}$$

$$\log_e\left(\frac{A}{A_0}\right) = rt - st$$

$$r = \frac{1}{t}\log_e\left(\frac{A}{A_0}\right) + s$$

Example of application. Consider a tree species, known from present-day studies to have a death rate of 1/500. Suppose that over a 1000-year period the pollen influx value for this species doubles, i.e. $A/A_0 = 2$. The seedling survival rate is

$$r = \frac{\log_e 2}{1000} + \frac{1}{500} = 0.002\ 69$$

Compare this with the previous millenium in which the rate had been stable, i.e.

$$A/A_0 = 1:$$

$$r_0 = \frac{\log_e 1}{1000} + \frac{1}{500} = 0.002$$

Therefore

$$r/r_0 = 1.35 \text{ or } 135\%$$

In other words, during the 1000 years of increase, the chances of survival of this species were increased by 35% on average. Now we can find, by experiment and field observation, the environment which is associated with particular establishment rates, and thus we could estimate the degree of climatic amelioration which would result in the measured increase.

This basic outline would need modification before use. As there is a delay between seedling establishment and the production of pollen, there must be a lag factor. Also, the death rate may not necessarily be a constant. In fact, changes in population could result as easily from changes in the death rate as from changes in the 'birth rate'. The model can, however, be used equally easily to give a measure of r (assuming s constant), or of s (assuming r constant), or of $r - s$.

Of course, such a model could only be applied once the population biology of the taxa were known. In the tropics this all too rarely so, although studies of some tropical species in which the ages of individuals can be determined by their annual rings are now under way (Ogden 1981). This will make it possible to determine the age structure of populations under known environments.

Palynologically-based palaeoecology in the tropics, and especially in the lowland tropics, is still underused. Inevitably, therefore, what I have written has been overgeneralised and in places speculative. I hope, however, that I have shown the great potential for this technique in the study of tropical environments – a potential which is still virtually untapped.

8

Geomorphic implications of late Quaternary hydrological and climatic changes in the Northern Hemisphere tropics

F. A. Street-Perrott, N. Roberts & S. Metcalfe

8.1 Introduction

For more than 150 years, geographers and geologists have speculated about the nature and causes of climatic change in the tropics (von Humboldt 1811). Fluctuations of water level in closed lakes have been central to this debate, on account of the direct relationship between the area and depth of any non-outlet lake and the net water balance over its catchment and water surface (Halley 1715, Jackson 1833, Drew 1875). In this chapter we present lake-level evidence for climatic changes in the Northern Hemisphere tropics (0–23.5°N) for the period since the last glacial maximum *c.* 18 000 years BP. We first analyse variations in the distribution and relative extent of lakes in Africa and western Eurasia (20°W–60°E), for which there are abundant data. We then compare these results with the more limited amount of evidence for moisture variations in Central America, to determine whether the pattern of changes there was comparable. Finally, we suggest some implications of our findings for the geomorphic evolution of the tropics. Our detailed conclusions regarding late Quaternary climates in low latitudes are published elsewhere (Street-Perrott & Roberts 1983, Roberts *et al.* 1981).

8.2 The water balance of closed lakes

The water balance of any lake is given by

$$P + R + G_{\mathrm{I}} = E + O + G_{\mathrm{O}} \pm \Delta S \qquad (8.1)$$

where P is precipitation onto the lake surface, R is runoff from its catchment, G_{I} is groundwater inflow, E is evaporation from the lake, O is surface outflow,

G_O is groundwater outflow and ΔS is change in storage (Street 1980). For a hydrologically closed lake, $O = G_O =$ zero. Separating the lake surface and catchment components of equation (8.1), the water balance of a closed lake in equilibrium can be given by

$$A_L(E_L - P_L) = A_B(P_B - E_B) \tag{8.2}$$

where P is precipitation, E is actual evaporation or evapotranspiration, A is area, and the subscripts L and B refer to the lake and its drainage basin respectively (Street 1979b). The ratio z of lake area to catchment area is dependent on the net water balance of the whole basin (Snyder & Langbein 1962), as indicated by the equation:

$$z = \frac{A_L}{A_B} = \frac{P_B - E_B}{E_L - P_L} \tag{8.3}$$

which has recently been verified empirically for Australian salt lakes by Bowler (1981). Mifflin and Wheat (1979) found that 75% of the variance of the estimated z values for late Pleistocene palaeolakes in the Great Basin, USA, could be explained by lake altitude and latitude, which they used as climatic surrogates. Even at the global scale, there is a clear relationship between the percentage land area in each 5° latitude belt occupied by lakes and the corresponding latitudinal value of $P - E$ (Street-Perrott & Roberts 1983).

Attempts to use past variations in lake area to calculate former precipitation using water-balance or combined water- and energy-balance models have proved controversial (Galloway 1965a, Dury 1973, Brakenridge 1978, Churchill *et al.* 1978, Street 1979b, Kutzbach 1980). These methods furthermore require very precise information on lake configuration, catchment characteristics and either palaeotemperature or albedo. In this chapter, we reconstruct past variations in net water balance using a simpler and more direct approach which does not require the relative contributions of precipitation, evaporation and runoff to be evaluated. This approach has the additional advantage of enabling us to use the large amount of reliable data from sites which do not meet the stringent requirements of existing numerical models.

8.3 Methods used in this study

The data set discussed in this chapter consists of a large amount of published information on the changes of water level of closed-basin lakes. Relative water depth is used here as a surrogate for the hydrologically significant variable lake area. Lake levels are evaluated at 1000-year time intervals over the last 30 000 years, and this information is stored in a computer databank at Oxford, along with lists of relevant ^{14}C dates and dated materials. We include only lake-level records dated by ^{14}C, or in exceptional cases by tephrochronology or historical sources. As far as possible, however, lake levels have been assessed on the basis of ^{14}C-dated stratigraphic *sequences* and not from individual ^{14}C dates.

In order to standardise information derived from different techniques of investigation, lake levels within each basin have been divided into three classes, each of which has a broadly similar frequency of occurrence in the data set:

low	0–15% of total altitudinal range of fluctuation, including dry lakes
intermediate	15–70%
high	70–100%, including overflowing lakes

This classification, established by Street and Grove (1979), has proved to be straightforward and widely applicable.

The 1981 version of the databank (Street-Perrott & Roberts 1983) has been modified and improved in several respects compared with the 1979 version used by Street and Grove (1979), Street and Gasse (1981) and Street (1981). The stimulus for this change came from the US Department of Energy, which requires high-resolution palaeoclimatic data covering the Holocene 'thermal optimum' in order to predict the effects of increasing atmospheric CO_2 on future global climate.

First, many more lake basins have been included. Secondly, the basis of the compilation has been altered to give the analysis greater precision. In the 1981 version, 'lake status' was defined as the class (low, intermediate or high) in which the water level of any lake stood at a specific time horizon (e.g. 1000 BP, 2000 BP, etc.). Previously, lake status had been based on the highest water level recorded within each millennium – a method that underemphasised phases of low lake level but made best use of single ^{14}C dates. Where isolated dates do occur, they are now allotted to the closest time horizon. Thirdly, a new variable referred to as 'lake-level trend' has been incorporated into the data set. As with lake status, this is assessed for specific time horizons, and is classified as (1) up, (2) stable, (3) down and (4) unclassifiable/no data. In general, classes (1) and (3) are assigned when there is a change in lake status between one time interval and another.

In the discussion that follows we make the assumption that the main control on lake levels since 18 000 BP has been climate. The effect of climatic variations upon runoff has undoubtedly been modified by the associated changes in vegetation and soils (see 'Geomorphic implications' below for further discussion) and it is also likely that human impact on vegetation has been a cause of lake-level fluctuations during recent decades (Makin *et al.* 1975). On the other hand, there is no evidence that man has been a significant agent in lake-level change over the timescale under consideration except in areas with a long history of irrigation (Harding 1965, Palerm 1973, Hollis 1978, Kes 1979). This conclusion is reinforced by the close correspondence in the timing of major shifts of lake-level regime between Africa, the Middle East, south Asia and the Americas.

8.4 Fluctuations in lake level in Africa and Arabia

General character of the record

The most comprehensive data on late Quaternary lake-level changes in the Northern Intertropical Zone come from the Afro–Arabian bloc (20°W–60°E), which contains some 60 basins with published lake-level records. Four of the best-dated sequences in Africa, those from Bosumtwi, Chad, Ziway-Shala and Abhé (Fig. 8.1) are illustrated in Figure 8.2, and demonstrate such remarkable similarities as to suggest strongly that the major factors controlling lake hydrology were of regional or subcontinental rather than of local scale. A similar degree of consistency is found within the whole Afro–Arabian data set, as is apparent from Fig. 8.3a, which illustrates the temporal sequence of variations in lake status during the last 30 000 years. Evidently, the late Quaternary record in tropical Africa and Arabia comprises two prolonged lacustral phases (from 27 000 to 21 000 BP, and from 12 500 to 5000 BP) separated by an arid interval which spans the last global ice-volume maximum (*c.* 18 000 BP as identified by CLIMAP (CLIMAP Project Members 1976, Denton & Hughes 1981). Arid conditions returned around 5000 BP and have continued, with only minor interruptions, to the present day. These prolonged periods of predominantly high or low lake levels, lasting for 5000–9000 years, were separated by rapid changes in hydrological regime.

The abruptness of these major transitions from dry conditions to wet is confirmed by a contingency table analysis, which shows that status and trend are very strongly correlated ($p < 0.001$). The combination intermediate/stable is significantly less common than its expected frequency in the data set, whereas intermediate/down and low/stable are overrepresented (S. P. Harrison, pers.

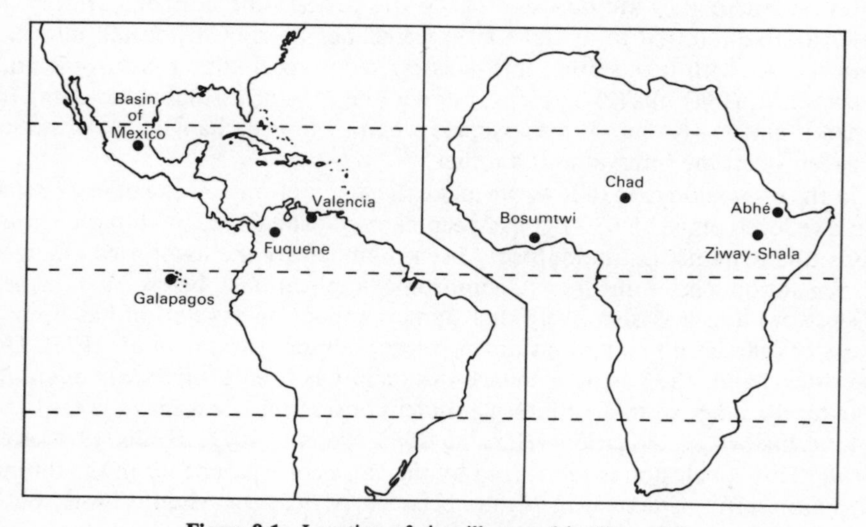

Figure 8.1 Location of sites illustrated in Figure 8.2.

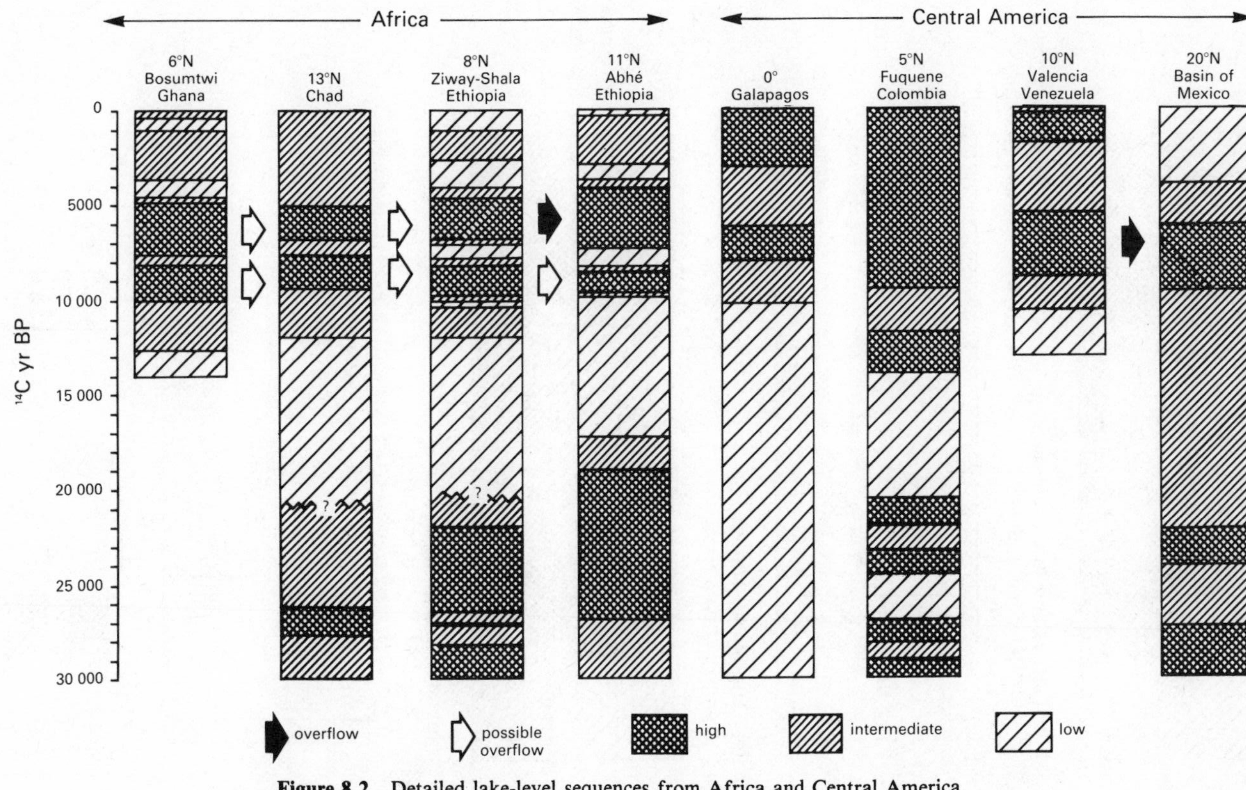

Figure 8.2 Detailed lake-level sequences from Africa and Central America.

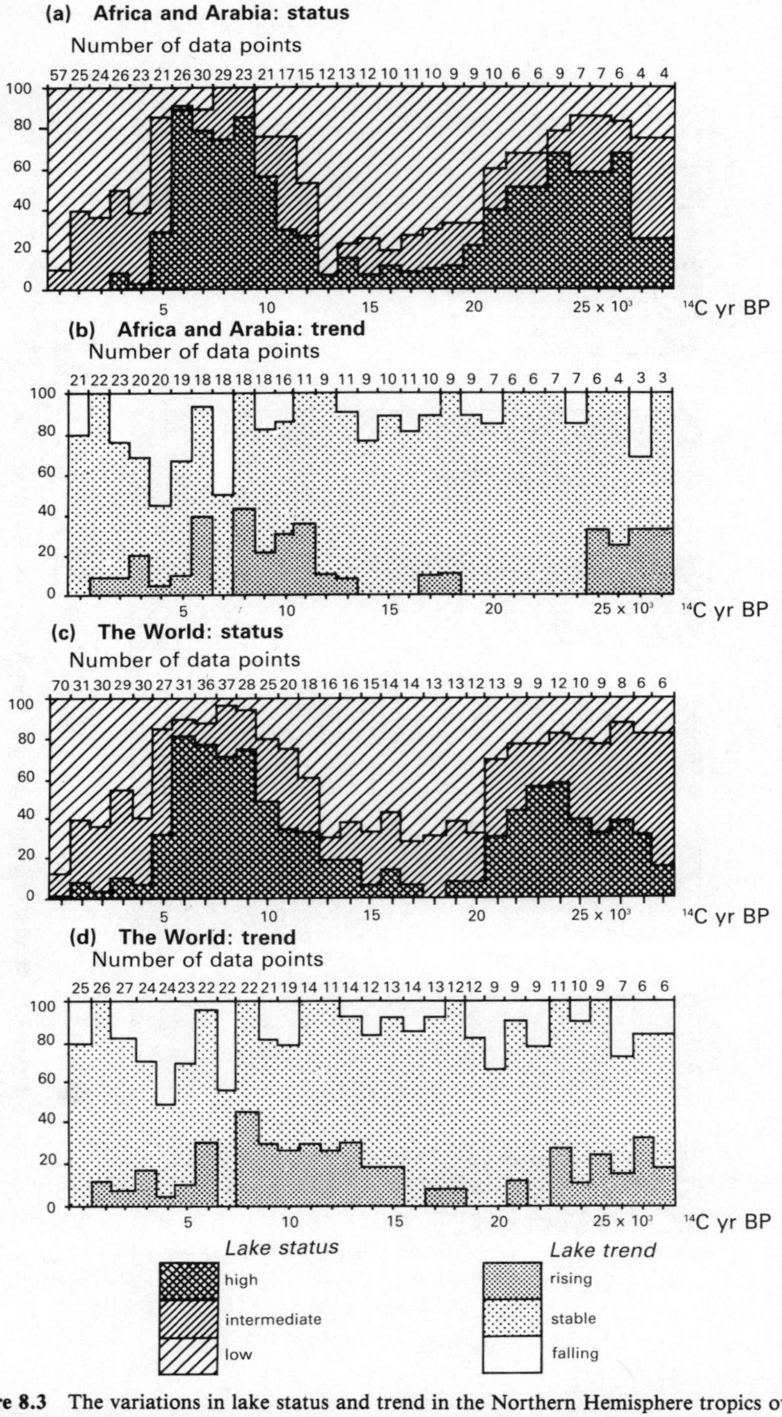

(a) **Africa and Arabia: status**

Number of data points

57 25 24 26 23 21 26 30 29 23 21 17 15 12 13 12 10 11 10 9 9 10 6 6 9 7 7 6 4 4

100 80 60 40 20 0

5 10 15 20 25 × 10³ ¹⁴C yr BP

(b) **Africa and Arabia: trend**

Number of data points

21 22 23 20 20 19 18 18 18 18 16 11 9 11 9 10 11 10 9 9 7 6 6 7 7 6 4 3 3

100 80 60 40 20 0

5 10 15 20 25 × 10³ ¹⁴C yr BP

(c) **The World: status**

Number of data points

70 31 30 29 30 27 31 36 37 28 25 20 18 16 16 15 14 14 13 13 12 13 9 9 12 10 9 8 6 6

100 80 60 40 20 0

5 10 15 20 25 × 10³ ¹⁴C yr BP

(d) **The World: trend**

Number of data points

25 26 27 24 24 23 22 22 22 21 19 14 11 14 12 13 14 13 12 12 9 9 9 11 10 9 7 6 6

100 80 60 40 20 0

5 10 15 20 25 × 10³ ¹⁴C yr BP

Lake status: high, intermediate, low

Lake trend: rising, stable, falling

Figure 8.3 The variations in lake status and trend in the Northern Hemisphere tropics over the last 30 000 years, for the Afro–Arabian (a, b) and global (c, d) data sets.

comm.). It therefore appears that most tropical African and Arabian lakes have tended to stabilise at low levels and to expand markedly only when a major climatic and/or hydrological threshold is crossed. In the absence of an outlet, the maximum level reached during any high stand seems to be controlled as much by the duration as by the intensity of humid conditions. Hence intermediate lake-level maxima, which are relatively rare, may reflect short-lived, rather than less intense, humid episodes.

The temporal variations in lake-level trend corresponding to those in lake status are summarised in Figure 8.3b. They confirm the first-order pattern of lake-level change outlined above, and emphasise the importance of the major shift in regime at *c.* 5000 BP. Figure 8.3b also highlights some second-order oscillations in lake level on a timescale of 1000–3000 years. Events with a duration less than 1000 years are obviously not resolved. Two particularly widespread regressions took place during the intervals 10 800–10 200 BP and 8000–7400 BP. These dramatic oscillations, which are described briefly below, have been discussed by us in detail elsewhere (Street-Perrott & Roberts 1983, Roberts *et al.* 1981).

Detailed evaluation

The lake-level record for tropical Africa and Arabia can be placed within a standard sequence established for the whole of Africa and western Eurasia (Fig. 8.4). The main stages now recognised for the last 25 000 years are as follows:

E pre-17 000 BP
D 17 000–12 500 BP
C 12 500–10 000 BP
B 10 000–5000 BP
A 5000–0 BP

In northern intertropical Africa, phase B can be subdivided into phases B_1 (10 000–7500 BP) and B_2 (7500–5000 BP). The pattern of water-balance anomalies during each of these phases is illustrated by selected maps which show the changing distribution of lake status over Africa and western Eurasia (Fig. 8.5–8.11).

Phase E (pre-17 000 BP). At 18 000 BP (Fig. 8.5), water levels in northern intertropical Africa and Arabia were low or intermediate and falling, except at Lake Mobutu Sese Seko (Albert) near the equator (Harvey 1976), the Jebel Marra massif (Williams et al. 1980) and Mundafan in Saudi Arabia, where a late Pleistocene phase of high lake levels was drawing to a close (McClure 1976). Although the scarcity of data for 18 000 BP undoubtedly reflects severe desiccation and deflation in many basins, it is nevertheless clear that the glacial maximum was significantly less arid, taking Africa as a whole, than the succeeding phase D. Figure 8.5 implies an equatorward shift in the

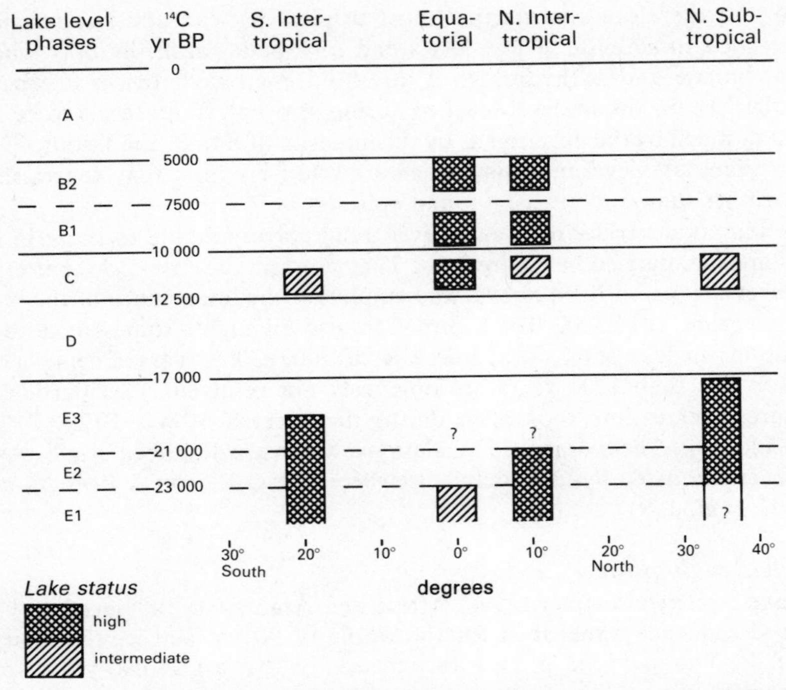

Figure 8.4 Major time divisions of the late Quaternary lake-level record in tropical and subtropical Africa and western Eurasia (schematic).

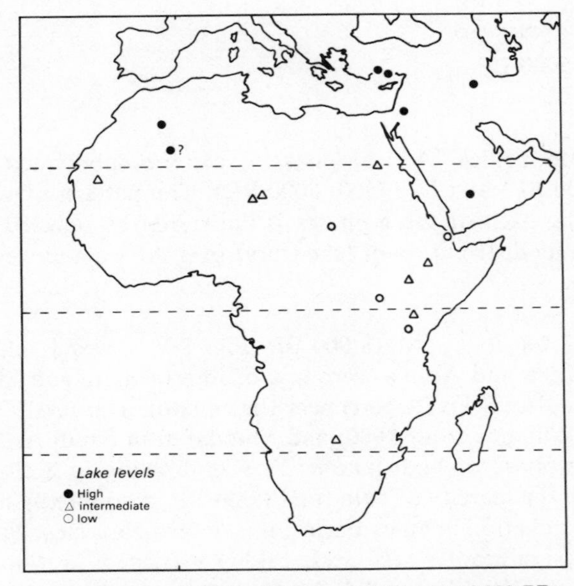

Figure 8.5 Lake-level status in Africa at 18 000 BP.

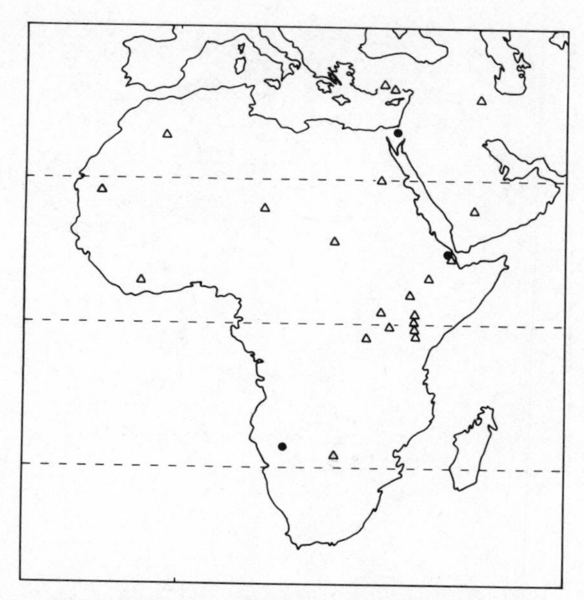

Figure 8.6 Lake-level status in Africa at 13 000 BP.

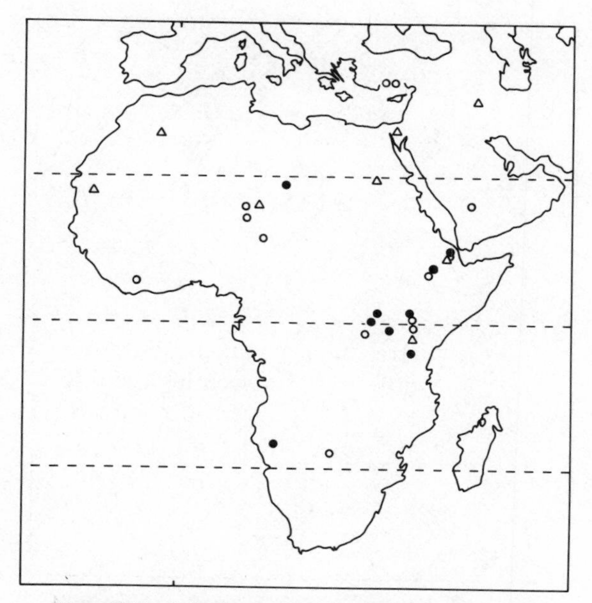

Figure 8.7 Lake-level status in Africa at 11 000 BP.

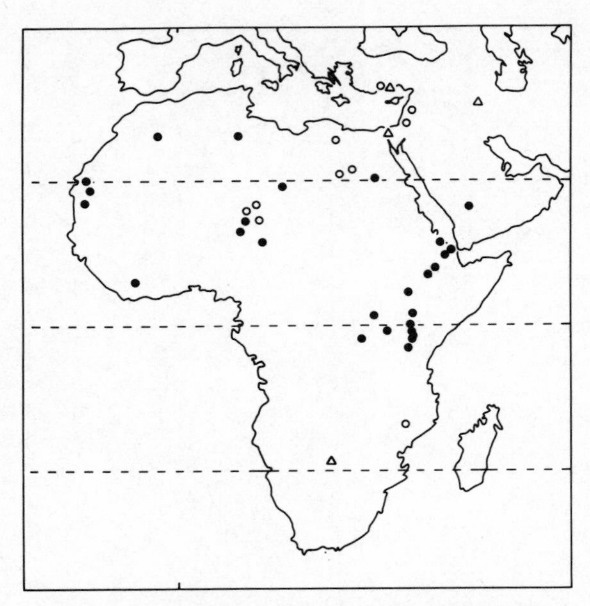

Figure 8.8 Lake-level status in Africa at 7000 BP.

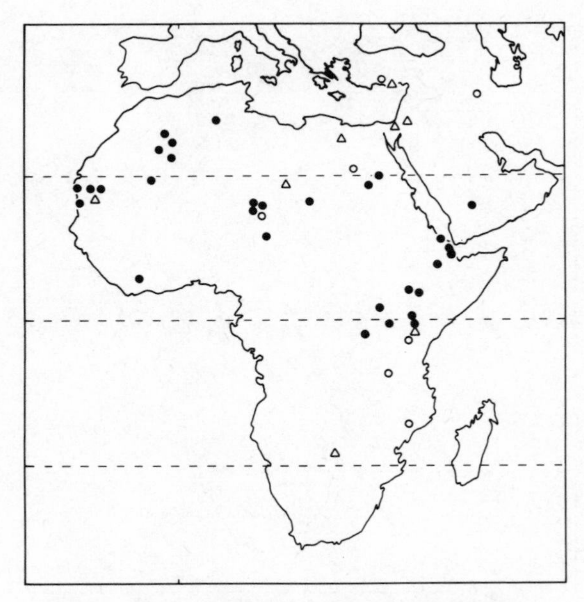

Figure 8.9 Lake-level status in Africa at 6000 BP.

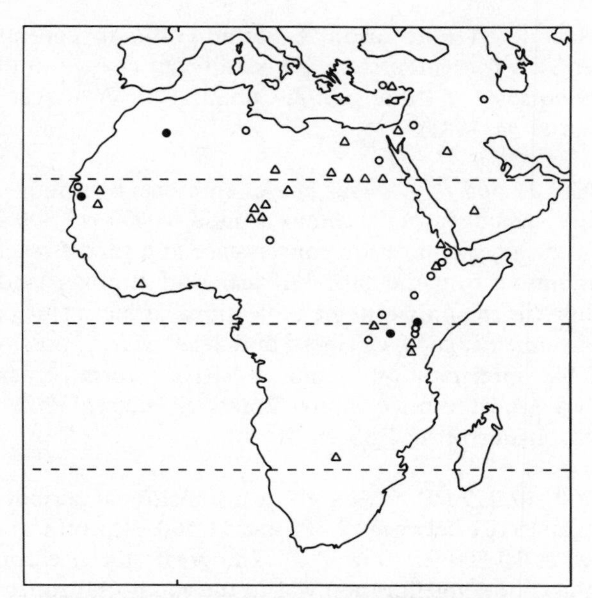

Figure 8.10 Lake-level status in Africa at 4000 BP.

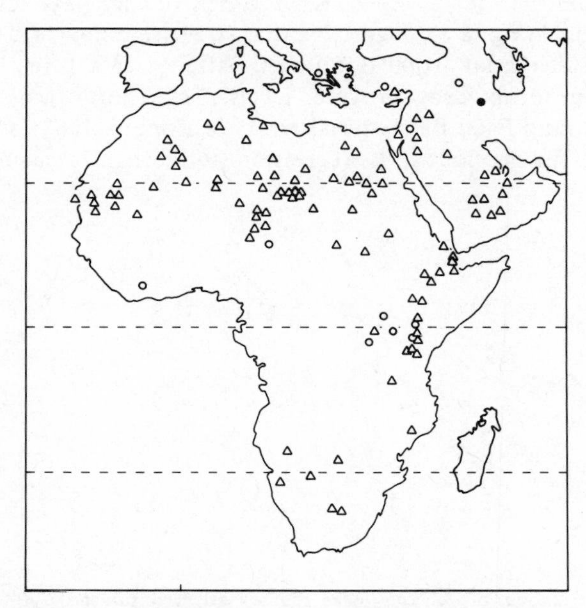

Figure 8.11 Lake-level status in Africa at the present day.

Saharo–Arabian desert (Nicholson & Flohn 1980), in contrast to aeolian evidence which fails to demonstrate any significant change in the latitude of the Saharan wind systems (Mainguet & Canon 1976, Fryberger 1980, Talbot 1980, Sarnthein *et al.* 1981).

Phase D (17 000–12 500 BP). This phase represents a prolonged and intense period of aridity, which reached a climax around 14 000–13 000 BP (Fig. 8.6). At this time, atmospheric moisture convergence and runoff reached their late Quaternary minimum over the entire African and Arabian land area. Apart from Lake Mobutu, the only reliable indications of lacustrine conditions are in upland environments, such as Tibesti and Jebel Marra, possibly associated with orographic precipitation from westerly storm tracks displaced southwards by the northern ice sheets (Butzer & Hansen 1968, Nicholson & Flohn 1980, Street-Perrott & Roberts 1983).

Phase C (12 500–10 000 BP). Phase C is a transitional period, for although many lakes began to rise between 12 500 and 11 500 BP, some of them receded dramatically after 10 800 BP (Fig. 8.7). The greatest concentration of lakes experiencing this initial amelioration was in the equatorial zone, but basins as far north as Tibesti (25°N) were affected (Roberts *et al.* 1981).

Phase B (10 000–5000 BP). An important shift in hydrological regime occurred around 10 000 BP. Data on lake-level trend indicate that water levels responded first near the equator and rose progressively later northwards towards the central Sahara. By 9000 BP a belt of high lake levels extended from 4°S to 33°N (Fig. 8.8), suggesting that large areas now arid were regularly receiving substantial tropical rainfall. After *c.* 8000 BP, many basins experienced an abrupt drop in water levels. This short B_1/B_2 arid episode culminated around 7400 BP (Street-Perrott & Roberts 1983) and affected a similar area to the previous C/B_1 regression (Fig. 8.12). It was not, however,

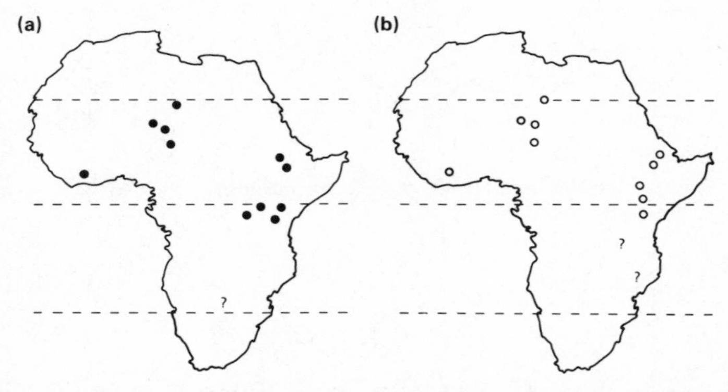

Figure 8.12 Distribution of African lakes that experienced (a) the C/B_1 and (b) B_1/B_2 regressions.

registered by lakes which receive a substantial input of ground water (Gasse & Street 1978, Busche *et al.* 1979, Street 1980). Those tropical lakes that had fallen at the end of phase B_1 recovered between 7300 and 6800 BP, and then generally remained high until around 5000 BP. During phase B_2, a broad belt of high lake levels covered Africa and Arabia from 2°S to 32°N, as is well illustrated by the pattern for 6000 BP (Fig. 8.9).

Phase A (5000–0 BP). The belt of high lake levels disintegrated rapidly after 5000 BP (Fig. 8.11), and intermediate-level lakes were soon restricted to the narrow latitudinal range (2°S–13°N) that they occupy today (Fig. 8.12). Severe desiccation transformed the Sahara in only a few centuries from a land of lakes to the arid desert we know today. A few basins are known to have experienced significant, if short-lived, water-level changes during phase A, but these minor oscillations cannot readily be correlated between regions.

8.5 Variations in moisture conditions in Central America

Existing lake-level data from the American tropics are relatively sparse (Table 8.1). Figure 8.2 illustrates four of the most complete and best-dated sequences from Central America. These show certain broad similarities with the record from northern intertropical Africa: notably, humid episodes at some sites between 30 000 and 20 000 BP, relatively dry conditions during the last glacial (although the period between 22 000 and 9500 BP seems to have been less arid than today in the basin of Mexico), and a general wet phase in the early Holocene. The late Holocene appears to have been moister than in Africa, at least in the zone 0–10°N. So far there is little evidence for the dramatic regressions experienced in Africa around 10 200 and 7400 BP, except for a study of Lake Valencia recently published by Lewis and Weibezahn (1981). This discrepancy could be attributed to the climatic buffering effect of the large ocean area surrounding Central America, but it is in fact more likely to result from the generally lower stratigraphic resolution and inferior dating control of the sequences from northern intertropical America.

The scarcity of good lake-level data points in Central America makes it impossible to compile maps similar to Figures 8.5–8.11 and to test the

Table 8.1 Basins in the Northern Hemisphere tropics included in the Oxford databank on late Quaternary lake-level fluctuations.

	1979 version	1981 version	Increase (%)
Africa	43	53	23.3
Arabia	6	7	16.7
Americas	2	5	250.0
Total	51	65	27.5

Afro–Arabian model of water-balance fluctuations. It is, however, feasible to reconstruct the temporal and spatial pattern of moisture availability, at least in outline, by combining lake-level data with other evidence for palaeohydrological conditions. Figures 8.13–8.15 cover the region with most data (i.e. Mexico and south-west USA), and have been constructed from pollen diagrams (Peterson *et al.* 1979), lake-level information from the Oxford databank, analyses of macrofossils from packrat middens (Wells 1976, Van Devender & Spaulding 1979), zoogeographic data and analyses of plant and animal remains from archaeological sites (e.g. Byers 1967). They show the broad patterns of available surface moisture at the last glacial maximum, during the glacial/interglacial transition and at 5000 BP.

At 18 000 BP, there was a wide, coherent zone of increased moisture in northern Mexico and south-west USA between about 35 and 25°N. South of 25°N and east of 100°W, conditions were generally drier than today, although not to the same extent as in Africa and Arabia between 5 and 25°N. By 12 500–10 000 BP, the belt of effectively wetter conditions had weakened and retreated north of the US border, while a change to a more humid climate was taking place in central Mexico and Yucatan. At 5000 BP (Fig. 8.15) the pattern of moisture anomalies was almost a mirror image of the glacial situation.

Figures 8.13–8.15 therefore confirm the broad similarity between the glacial/interglacial fluctuations in moisture conditions in Central America and in the Afro–Arabian tropics. The palaeohydrological evidence is, unfortunately, still inadequate for comparison of second-order fluctuations in the

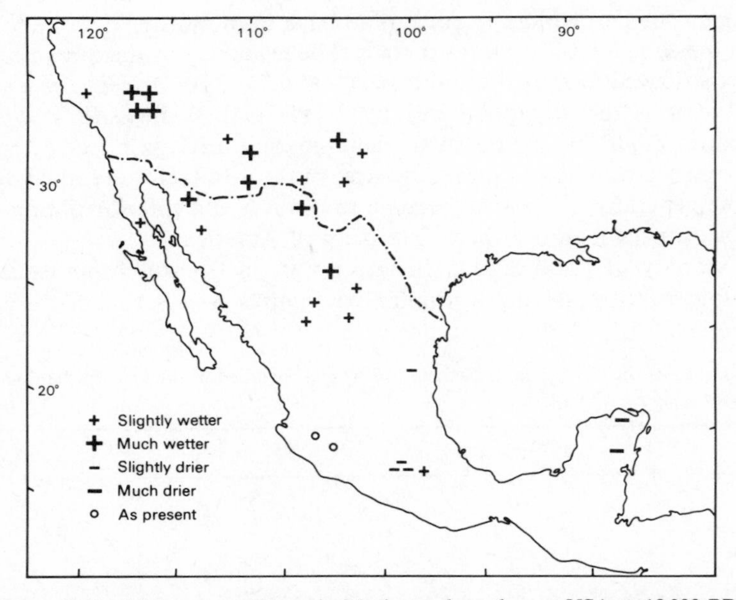

Figure 8.13 Moisture conditions in Mexico and south-west USA at 18 000 BP.

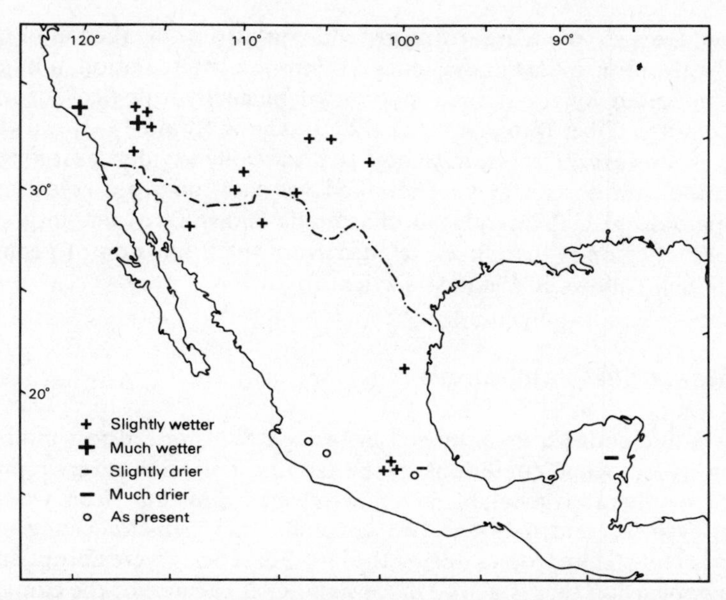

Figure 8.14 Moisture conditions in Mexico and south-west USA at 12 500–10 000 BP.

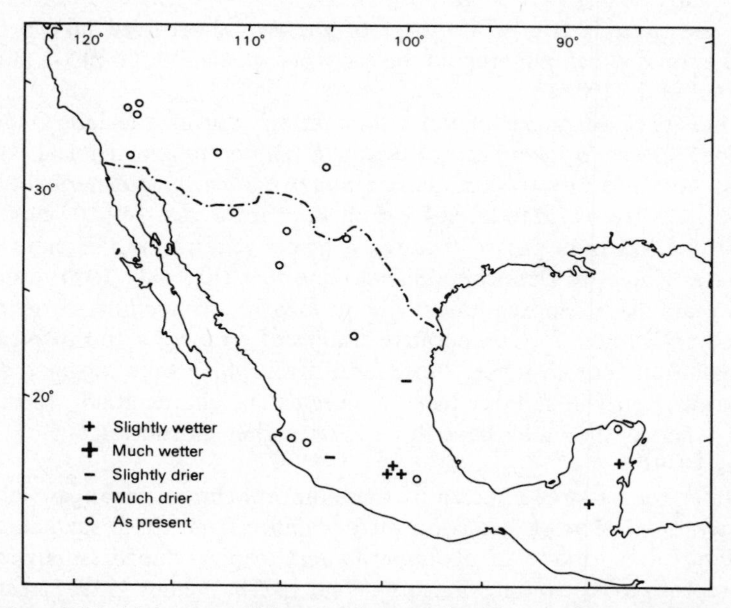

Figure 8.15 Moisture conditions in Mexico and south-west USA at 5000 BP.

two areas. Even so, there are strong grounds for attributing the major changes to large-scale adjustments in the general atmospheric circulation, a hypothesis that is supported by the limited amount of palaeohydrological information from India and Tibet (Singh *et al.* 1972, Bryson & Swain 1981, Shi & Wang 1981). It is, however, important to note that the pollen evidence from Malesia, a region that has not so far yielded lake-level data, does not reveal any sign of comparable glacial/interglacial changes in moisture availability (Flenley Ch. 7). This region is therefore excluded from the discussion of geomorphic impact which follows in the next section.

8.6 Geomorphic implications

Lake-level fluctuations, as discussed in Section 8.2, are a direct indicator of changing hydrological conditions in the tropics, and, moreover, represent a data source only rarely available in temperate environments. From the analysis of lake levels presented above, we conclude that climatic changes in the Northern Hemisphere tropics during the late Quaternary were abrupt and step-like – a conclusion that is all the more significant because of the climatic sensitivity of our data set. In essence, the two intervals, 21 000 to 12 500 BP, and 5000 BP to the present, were marked by low lake levels, indicating reduced runoff and a relatively arid climate; while the intervening interval (12 500–5000 BP) witnessed high levels of lakes and of runoff, and humid climatic conditions (Fig. 8.3c & d). In Africa, at least, the transitions between these climatic phases were so abrupt that they cannot easily be resolved by ^{14}C dating; the same is true in the case of the two short-lived lake recessions (= arid events) which punctuated the early postglacial humid phase at around 10 200 BP and 7400 BP.

The incidence of major climatic fluctuations during the late Quaternary therefore appears to have been different in temperate and tropical latitudes. Whereas northern intertropical Africa and America have experienced a full climatic cycle, from arid to humid and back, over the past 14 000 years, north-west Europe has only passed through a hemi-cycle during the same period, from (late) glacial to (late) interglacial. The fact that only 5000 rather than 10 000 years have elapsed under the present morphoclimatic regime has obvious implications for comparative studies of temperate and tropical land-form evolution. For instance, those landforms which take longer than 5000 years to develop will not yet have achieved their characteristic form in the tropics, whereas they may have done so at higher latitudes (cf. Brunsden & Thornes 1979).

The intertropical area affected by these morphoclimatic changes was extensive. Palynological work has confirmed evidence from fossil sand dunes and fluvial landforms for a relatively impermanent tropical rainforest environment during recent geological time (Tricart 1974a, 1977b, Flenley 1979, Street 1981). In a comparable way, lake-level and other evidence shows that the present zone of active sand-dune movement in the Saharo–Arabian desert may hardly have existed at all between 10 000 and 5000 BP (Figs 8.8 and 8.9; see also Sarnthein 1978). There would appear to be very few parts of the tropics, except possibly Malesia (Flenley Ch. 7), where the established morphoclimatic regime has operated unchanged over the last 20 000 years. The spatial and temporal variability of geomorphological processes in low latitudes must therefore have been at least as great as in the temperate zone.

Potentially even more significant for overall process rates has been the alternation of different geomorphological regimes through time, whose combined effect may have been responsible for some characteristic landform assemblages, especially in climatically marginal zones such as the Sahel. For example, stratigraphic studies of Lake Turkana (Butzer 1980a), the Nubian Desert (Wendorf *et al.* 1977, Haynes *et al.* 1979) and southern Libya (Pachur & Braun 1980, Jäkel 1979, Busche *et al.* 1979) show that dry phases in arid areas tend to be represented by erosional intervals of aeolian origin, whereas wet phases are associated with the development of extensive fine-grained alluvial, lacustrine and marsh deposits. The early Holocene humid episode was therefore responsible for the deposition of much of the sediment which is presently being reworked and transported by aeolian processes in the Sahara (Cox *et al.* 1982). As a result, the sediment supply now available for deflation may be controlled less by current rates of rock weathering than by former geomorphic processes. The implication of this is that any attempt to establish long-term denudation rates for the arid zone is doomed to failure if only current processes are measured.

Another powerful combination of processes has operated in the humid tropical lowlands, notably in Amazonia, where chemical weathering and landscape stabilisation during interglacials appear to have alternated with episodes of fluvial dissection during the glacials (Tricart 1975, 1977a,c, Damuth 1977, Baker 1978). This cyclic production and stripping of a deep-weathered mantle must have significantly increased long-term erosion rates in the Amazon lowlands (Damuth & Fairbridge 1970).

On the other hand, evidence from many parts of the tropics suggests that episodes of geomorphic instability and temporarily increased erosion rates were associated with times of climatic transition rather than with the longer intervals of relatively stable arid or humid climate. Theoretical and empirical arguments suggest that vegetation is a critical variable controlling temporal changes in surface runoff and sediment yield, and a plausible sequence of adjustments in relative vegetation cover and erosion rates following abrupt changes in the level of effective precipitation has been proposed by Knox (1972). According to Knox's model, maximum geomorphic work in semi-arid or sub-humid fluvial systems is carried out immediately after the shift from a drier to a relatively wetter climate. In the transitional stage, vegetation is not fully established and hillslope erodibility temporarily remains high, a situation which can be exploited by the increased erosivity of precipitation and runoff. Note, however, that only an abrupt climatic change would produce this geomorphic response, since with a gradual increase in precipitation there would be no disequilibrium between vegetation and climate.

Lake-level evidence strongly suggests that moisture fluctuations in northern intertropical Africa were abrupt and step-like and these data may therefore be appropriate for testing Knox's (1972) biogeomorphic response model. We may identify the period 13 000–9000 BP in particular as a phase during which episodes of accelerated erosion could be predicted. Two abrupt shifts to a wetter climate occurred within that time interval, the more important (according to the lake-level record) being that between 10 200 and 9500 BP. The earlier transition, around 12 500–10 800 BP, was a less marked hydrological event, but might be expected to have had a greater geomorphic impact because it was preceded by a longer period of low runoff and aridity.

A difference in the rate of response of the hydrological and vegetational systems to the same two, apparently abrupt, climatic events is suggested by the divergence between the African pollen and lake-level records. The small number of well dated pollen diagrams from sites in forest or savanna areas such as Lake Victoria (Kendall 1969) and Lake Abiyata, Ethiopia (Lézine 1981) suggests a lagged vegetational response in lowland environments, in contrast to high-altitude sites where vegetation responded more or less in phase with lake levels (Flenley 1979, R. A. Perrott pers. comm.). Palynological evidence therefore confirms that in areas of low relief there was a temporary imbalance between climate and vegetation cover during the terminal Pleistocene and early Holocene. This was due in part to biogeographic controls such as the proximity of refugia and varying rates of plant migration (Hamilton 1976).

Phases of temporary geomorphic instability are identifiable stratigraphically from changing rates and patterns of sedimentation in lacustrine, alluvial and submarine deltaic sequences. Though often limited by insufficiently close dating control, there is a measure of empirical support for the predicted terminal Pleistocene variations in sediment yield – for instance, from the Niger River basin in West Africa. Much of the Niger catchment is in the climatically marginal Sahelian zone, with both fossil dune fields of late Pleistocene age and relict watercourses that were active during the early Holocene (Talbot 1980): clear evidence of major spatial and temporal changes in morphoclimatic regime. The Niger basin fluviatile sequence includes extensive coarse-grained, braided stream deposits immediately overlain by finely laminated early Holocene silts and clays; the former are consequently likely to be terminal Pleistocene in age. Talbot (1980 p. 48) attributes this phase of high sediment yield to the initial transition from lower to higher rainfall, when large volumes of sediment would have been available for fluvial erosion and transport.

His suggestion would appear to be confirmed by the study by Pastouret *et al.* (1978) of a long, well dated core taken on the outer Niger Delta, covering approximately the last 30 000 years. Sedimentation rates changed dramatically within this core and increased by almost 20 times during the interval 11 500–10 900 BP, presumably because of accelerated erosion in the upper and middle basin at the onset of the early postglacial wet phase. Notwithstanding the dangers of extrapolation from a single core, this sequence suggests that more sediment may have been removed by the Niger River during two or three millennia in the terminal Pleistocene than during the entire Holocene. Similar evidence for exceptionally high discharges and sediment yields during the transition between glacial aridity and early Holocene moist conditions comes from the fluvial and deltaic sequences of the Nile (Rossignol-Strick *et al.* 1982, Butzer 1980b, Adamson *et al.* 1980) and other rivers in West Africa and the Saharan uplands (Thomas & Thorp 1980, Servant & Servant-Vildary 1980). A number of lake basins, such as Lake Kivu (Degens & Hecky 1974) and Lake Abiyata (Street-Perrott unpubl. data), also experienced enhanced rates of sediment accumulation at the onset of lake-level phases C and/or B_1.

Low-latitude environments appear to have been subject to a major climatic perturbation at least once in every 5000–10 000 years which often led to a massive increase in sediment removal before the geomorphological system was able to attain dynamic equilibrium. Consequently it may be difficult to explain tropical landform evolution in terms of the stochastically based concept of 'magnitude and frequency'. Perhaps more appropriate is a deterministic model of the type proposed by Knox (1972) which involves biogeomorphic response to a known pattern of climatic variation. By the same token, the current phase of man-induced accelerated geomorphic activity may be far from unique in recent geological history. If, as Fairbridge (1976) has suggested, the present-day situation is analogous to previous, climatically controlled periods of enhanced erosion – such as the terminal Pleistocene in northern intertropical Africa – then at least we can hold out some hope that current geomorphological trends are not irreversible.

9

Evidence from lake sediments for recent erosion rates in the highlands of Papua New Guinea

F. Oldfield, A. T. Worsley & P. G. Appleby

9.1 Introduction

Where rates of sedimentation in lakes can be measured accurately for well dated time intervals, estimates of rates of erosion from the catchment can sometimes be derived. The reliability of the exercise will depend on a wide range of variables, some of which may be intrinsic to the field situation and lie outside the control of the investigator. The present study uses evidence from small lakes in the highlands of Papua New Guinea. The many environmental and logistic limitations placed on the work mean that the reliability and degree of detail of the estimated rates necessarily vary between sites as well as between time intervals for any single site. Consequently, methodological problems are treated in some detail and discussed both in general terms and in relation to the specific case studies considered here. The timescale studied is that of the last 1200–1600 years, during which the area has undergone major changes in land use and human impact (O'Garra unpubl.). All estimates should be regarded as tentative and subject to large and poorly quantified errors.

The area studied lies between 1800 and 2500 m in the New Guinea highlands (Fig. 7.2). Throughout this mountainous region there are broad high-altitude valleys which are densely settled by subsistence horticulturalists. In many areas, gardens are surrounded by extensive tracts of now unproductive grassland thought to be the result of human activity. In the western and southern highlands considered here, direct, effective, European cultural, technological and medical impact began after the end of World War II, and non-commercial socio-economic patterns persist through to the present day. Although archaeological evidence shows that man has practised some form of subsistence in the highlands for over 5000 years (Golson 1977), the main subsistence staple, sweet potato (*Ipomoea batatas*), is not a native plant. As an introduction from the New World its arrival in the region certainly lies within the last millenium and is quite likely to post-date AD 1500. Its impact on the intensity of gardening, the growth of population and the altitude at which subsistence could be carried out may well have been not only relatively recent but very dramatic.

The combination of circumstances presented by the human ecology of the region thus provides scope for widely divergent views about timescales of anthropogenic transformation of the landscape. At the same time, there is both an extreme proverty of conventional data on the subject and a need for some informed assessment of recent ecological impact in the light of possible future demographic, cultural, socio-economic and technological trends in the region. Reconstructing the history of erosion in presently or previously cultivated areas is one important facet of this assessment.

The three lakes referred to in the present study, though altitudinally in the zone of lower mountain forest, lie within intensively gardened regions with extensive anthropogenic grasslands nearby. They are, however, sharply contrasted in terms of the within-watershed terrestrial vegetation.

Lake Ipea at 2500 m is surrounded by a large catchment predominantly of grassland and garden at the present day. Lake Egari, a small volcanic crater lake at 1800 m, has garden and forest in its small catchment but little or no established grassland. Lake Pipiak, a small crater lake close to Egari, was, at the time of sampling in 1973, surrounded by forest with very few small, cleared patches of recently established gardens.

9.2 Aims and requirements

The primary aim of this aspect of our work in the New Guinea highland lakes has been to reconstruct changes through time in total sediment input to each lake. Only when this can be done for the whole lake can meaningful figures be derived for total output from the whole catchment. Accurate whole-lake/whole-catchment sediment budgeting places demands on the study which are rarely satisfied either by the environmental context or by the repertoire of techniques available (cf. Dearing 1982). These demands are discussed below. In addition, we have attempted to address the problem of sediment sources in view of the possibility that both surface soil and material from road construction (cf. Dunne 1979) may have contributed in recent times.

Chronology

Precise and accurate timescales of sedimentation are very difficult to establish. In the present study several techniques have been used in combination. For the period before AD 1600 we have been entirely dependent on ^{14}C dates for estimates of age. However, some confirmation of the dates obtained in connection with the present study (Oldfield *et al.* 1981) has come from correlation of the tephra sequence in these lakes with dated tephra sequences further east (Blong 1982). Problems of dating within the period AD 1600 to AD 1800 have centred on the age of the most recent volcanic ash, the Tibito tephra of Blong

(1982), here termed ash A. There is by now an extensive literature on this ash fall, ranging from oral history to geochemistry. Earlier estimates have fallen between AD 1600 and AD 1890 depending on the evidence used. There is, however, much more agreement between the most recent dates. Recorded western contact with Long Island, the source of the tephra, from AD 1824 onwards precludes a post AD 1800 date and Dampier's sketch of the island from the sea in AD 1700 portrays a topography so comparable to the present outline that an 18th century date for a major tephra explosion seems unlikely (Blong 1982). Comparison of the geomagnetic secular variation record in the Lake Ipea sediments for the period from the ash A to 1973 with the calculated variation from observatory records from AD 1650 to AD 1975 also indicates a pre AD 1700 rather than a post AD 1700 age for the ash (Thompson & Oldfield 1980). Interpolation between ^{210}Pb and ^{14}C age – depth profiles for Lake Egari and Lake Ipea gives an estimated age of AD 1680–90 for the ash (Oldfield *et al.* 1980) as does the most recent recalibration of the best ^{14}C dates on wood and charcoal directly associated with the ash elsewhere (Polach 1984). Thus documentary, palaeomagnetic and radiometric evidence converges on a 17th century date. Although ~AD 1685, the date used here, is the best estimate from the assemblage of radiometric evidence available, the true age is likely to be somewhat earlier in view of the noted luxuriance of the vegetation of Long Island at the time of Dampier's passage. Between this date and the age of the earlier Olgaboli tephra (ash B), dated both in this study and in Blong's work to 1200 BP, dates and rates have been determined by interpolation using the methods summarised in Oldfield *et al.* (1980).

Two techniques, ^{137}Cs and ^{210}Pb assay (Oldfield 1981), have been used for the period after AD 1800. The fall-out radioisotope ^{137}Cs has been used to give some indication of the levels in the Lake Ipea sediments which represent 1955 and 1964, the years of onset and peak (Southern Hemisphere) fall-out as a result of atomic testing. Much more crucial has been the use of ^{210}Pb dating to obtain both age–depth curves and estimates of dry-mass sedimentation rates for the last 150–200 years. The assumptions and mathematical techniques used in these calculations are fully documented in Appleby and Oldfield (1978, 1984), Oldfield *et al.* (1980, 1981) and Oldfield (1981). Before ^{210}Pb results can be used in this way, the data must satisfy the requirements of the dating model chosen. This in turn implies that carefully designed strategies of sample selection and treatment, and of radiometric assay, must be adopted and tests of internal consistency and external validation carried out (Appleby & Oldfield 1984, Oldfield & Appleby 1984).

Total influx estimates

Since patterns and rates of sedimentation vary within lakes, it is seldom, if ever, possible to use the results from one or two cores to develop estimates of total net sediment input to the whole lake for a given time interval, no matter how accurately the chronology of sedimentation has been established for the core(s) used. There have been three types of approach to this problem. Lerman

(1975) has sought ways of extrapolating from single central cores using mathematical models of sedimentation for geometrically simple basis. Davis (1976) has used multiple cores correlated by exhaustive pollen analytical studies to build up a firm empirical base for total sedimentation estimates. Bloemendal *et al.* (1979) and Dearing *et al.* (1981) have used rapid non-destructive magnetic measurements, especially whole-core volume susceptibility logging, to establish core correlations and provide an empirical base for total influx estimation much more rapidly than is possible by other means; though even here, marginal shallow-water deposition can seldom be quantified effectively because of discontinuities in sedimentation. These latter empirical studies suggest that rarely, if ever, do lakes satisfy the simple formulations of Lerman's models. Moreover, shifts in sediment grain size, type and source generate materials of different terminal velocity, which in turn may lead to changes in the spatial distribution of sedimentation rates and patterns (cf. Dearing 1979) that cannot be modelled by data from one or two 'central' cores. Unfortunately, for logistic reasons the present study falls far short of the density of cores required to provide a basis for fully quantitative estimation. Only in the case of L. Egari, which has a simple shape, has this been attempted and even then on minimal data (see below). The few coring sites in the other lakes, although unsuitable for extrapolating input to the whole lake, do however provide a basis for comparing rates in relative terms (see Oldfield *et al.* 1980). Nevertheless any inferences must be qualified by some recognition of the way in which differences within and between cores can be affected by the positions of the coring sites in relation to the lake shore and to specific sediment sources. Fortunately, in the present studies significant river inputs are absent and there are no deltaic features within the lakes themselves.

Sediment origin

Ideally, the fraction of the total net sediment input which is catchment-derived (allochthonous) should be distinguished from the fraction derived from primary production within the lake waters (autochthonous); only the allochthonous fraction is of interest here. In addition, the allochthonous component should be subdivided in terms of its likely sources within the catchment.

The problem of quantifying the allochthonous component varies very much according to the type of lake and sediment studied. Several authors have used only the inorganic fraction, often estimated from the loss-on-ignition 'residue', as the basis for influx estimation (e.g. Dearing *et al.* 1981). In the present study the whole sediment is used as a basis for estimation. Mackereth (1966) showed that the bulk of the organic input to the sediments of upland lakes in the English Lake District came from stable organic residues derived from the terrestrial soils and vegetation of the catchments, rather than from the degradation of organic matter derived from the water column. In common with those of Mackereth, the present lakes have low pH and low biological productivity, no lack of hypolimnetic oxygen and, for the most recent periods, abundant evidence for the input of stable organic residues from terrestrial soils (O'Garra unpubl.). Bearing these factors in mind, we have chosen to regard the organic matter in the sediments as largely allochthonous. The resulting overestimation will be more serious for the early periods of minimal erosion than for the later episodes, and thus one consequence will be an *under*estimation of any inferred acceleration. This is compounded by inability to distinguish autochthonous from allochthonous silica in the sediment, since proportionally, the autochthonous contribution from diatom frustules will be greater during episodes of low erosive input.

In the present study, new evidence is included on the differentiation of potential sediment sources within the catchments and work in progress suggests that mineral magnetic parameters will permit characterisation of material from different parts of the regolith in suitable lithological contexts (cf. Oldfield *et al.* 1979, Walling *et al.* 1979).

Storage, output and changing water levels

In order to use sedimentation rates as a direct basis for reconstructing past erosion rates, it is necessary to assume that detached sediment is not stored within the catchment and that particles move from source to lake fairly rapidly in terms of the timescales on which measured changes are to be resolved. In the present study these conditions appear to be most nearly met in the small volcanic crater lakes of which Lake Egari is an example. At Lake Ipea, there is some sediment retention in two major swamps to the south-east of the lake, and in several of the smaller lakes studied, inwashed material is retained close to the lake margin as a result of bringing reedswamp development. Sediment retention in marginal peaty environments of high autochthonous organic productivity makes both quantitative estimates and correlation with true limnic deposits virtually impossible. Estimates of total input based on underwater sedimentation may also be distorted by the results of marginal reworking associated with changing lake levels. Although the pre 1600 BP sediments from Lake Ipea contain clear evidence of stratigraphic discontinuities arising from lowered lake levels, the more recent sediments from all the lakes used here appear to have accumulated without any such disturbance.

Loss of particulate matter from the lake via its outflow will clearly require consideration in many cases. In the present study however, the lakes are either within closed basins (Egari and Pipiak) or else drained by a minor stream carrying little particulate matter.

9.3 Previous results

Work completed in the three lakes studied so far (Table 9.1) has established a chronology of sedimentation for 1 m Mackereth minicores (Mackereth 1969) taken from each lake using the methods outlined above and in Oldfield *et al.* (1980). The results confirm that over the last 350 years, in Ipea and Egari, the

Table 9.1 Characteristics of the lakes and catchments in Papua New Guinea.

Lake[a]	Elevation (m)	Watershed	Watershed land use and vegetation (1973)	Regional land use and vegetation
Ipea	2400	>20×	grassland, garden and secondary forest	
Egari	1800	~2×	garden and secondary forest	grassland, secondary forest and garden
Pipiak	1800	~2×	forest and a little new garden	

[a]Egari and Pipiak are small crater lakes < 2 km apart.

two lakes with subsistence gardening and anthropogenic grassland in their catchments, rates of dry-mass sedimentation have accelerated by 4–32 times depending on site and core location (Table 9.2). Generally speaking, the greatest acceleration has taken place in marginal rather than central cores. At Pipiak, the catchment of which was almost entirely covered in forest in 1973, the year of sampling, the acceleration in sedimentation over the same time interval was negligible. The following sections are concerned with the problems of converting the previously published evidence for sedimentation rates at specific core location into quantitative estimates of changing erosion rates for the whole watershed, and of identifying the dominant source of allochthonous sediments within the lakes.

Table 9.2 Summary of sedimentation rates for selected periods in cores from each lake (see Oldfield *et al.* 1980). For Ipea and Egari, dry-mass accumulation is recorded as 10^{-2} g cm^{-2} yr^{-1}; for Pipiak, wet-volume sedimentation is recorded as 10^{-2} cm yr^{-1}.

	(a) 500 BP	(b) 200 BP	(c) 100 BP	(d) 40 BP	(e) 1973	(e)/(a)	(e)/(b)
Ipea							
max.	0.21	0.26	0.77	1.92	5.26	31.6	25
min.	0.13	0.18	0.57	1.42	3.88	19.2	14.9
Egari							
1	~0.10	~0.30	~0.80	~1.10	~2.35	~23.5	~7.8
2	0.20	0.18	0.53	0.78	1.58	7.9	8.8
4	0.32	0.20	0.59	0.87	1.75	5.5	8.8
5	0.44	0.21	0.61	0.90	1.81	4.1	8.6
Pipiak							
A	1.91	2.01	2.15	2.24	2.35	1.23	1.17
B	1.74	1.98	2.18	2.30	2.43	1.40	1.23

9.4 Erosion rates

Table 9.3 summarises requirements for establishing sediment budgets and erosion rates on a whole-lake/whole-catchment basis (see Section 9.2 above). Of the three lakes studied, only Egari comes close to satisfying the requirements. Although only five cores were taken, they are representative of the main depositional environments in the lake and ^{210}Pb profiles have been obtained on three of them. Figure 9.1 shows the location of each core and the portion of the lake bed of which it is taken to be representative. Using the dry-mass sedimentation rates derived from the previously published chronology (Oldfield *et al.* 1980), it is possible to estimate total sediment input per year for each sector and thus for the whole lake. This has provided the basis for estimating the changing rates of erosional loss from the whole catchment and from the potentially cultivable part of the catchment (Fig. 9.2) as t km^{-2} yr^{-1} for the last 1200 years. The letters A–F identify the following sedimentation/erosion episodes:

- A peak AD 1973 rate
- B AD 1950–73 period of effective Australian administrative control
- C AD 1800–1950 pre contact forest clearance and subsistence gardening linked probably with the introduction of sweet potato to the area
- D AD 970–1800 period of predominantly closed forest cover and only sporadic gardening in the catchment
- E AD 1680–1790 period of minimum erosion rates within D
- F AD 770–970 earlier prehistoric forest clearance and gardening episode

The reconstructed rates lie within the range of those calculated for a roughly comparable range of land-use types within the Potomac system at the present

Table 9.3 Some criteria for selecting lakes suitable for calculating whole-lake/whole-catchment sediment budgets.

Requirements	Egari	Pipiak	Ipea
well defined catchment	√	√	–
no long-term sediment storage in catchment	√	√	–
no major deltaic deposition	√	√	?
minimal outfall loss	√	√	√
no major reedswamp development	√	√	*
low autochthonous organic yield	√	√	√
stable lake levels	√	√	√
detailed chronology of sedimentation	√	–	√
surveyed grid of cores	?√	–	–
rapid detailed core correlation	√	–	√

*Sirunki swamp.

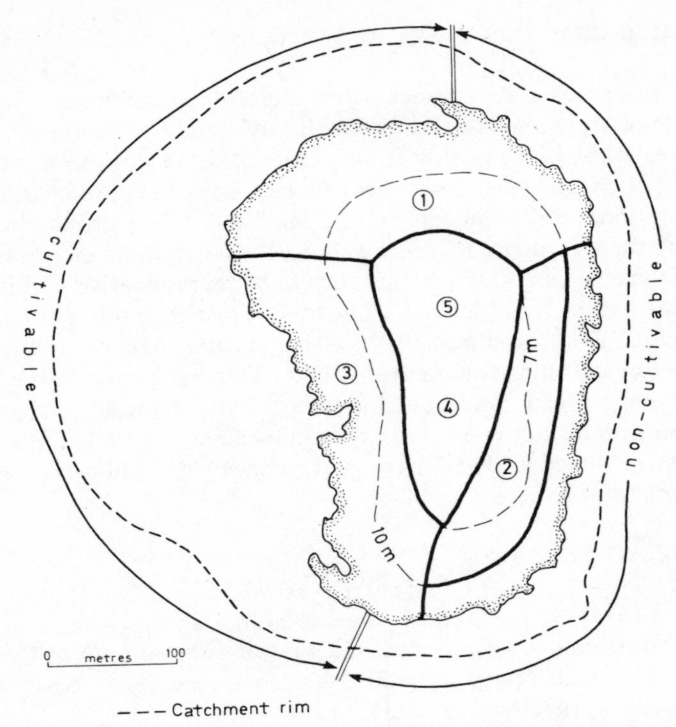

Figure 9.1 Plan of Lake Egari and catchment drawn from 1973 air photograph. The numbered cores taken in 1973 are located in sectors of lake bed which have been drawn to delimit the zones for which each core is taken to be representative of the main depositional environments.

day (Wark & Keller 1963). Both periods of subsistence agriculture in the Egari drainage basin gave rise to comparable rates of 50–100 t $km^{-2} yr^{-1}$, whereas the peak 1973 rate is some 2–4 times higher.

The only nearby road lies mostly just outside the catchment along the eastern edge of the uncultivable part of the rim, in the form of an ungraded vehicle track to Egari village. The wooded uncultivable part of the catchment lies between this road and the water's edge. The core closest to this edge of the lake shows the least acceleration whereas those on the opposite side, close to the gardened area, show the maximum acceleration. In consequence, we infer from the pattern of sedimentation and the lack of roads within the catchment that the main source of sediment in the recent past has been the cultivated area.

9.5 Sediment sources

In the case of the much larger Ipea catchment, the problem of identifying the dominant source of the recent inwashed sediments is more complex. Several

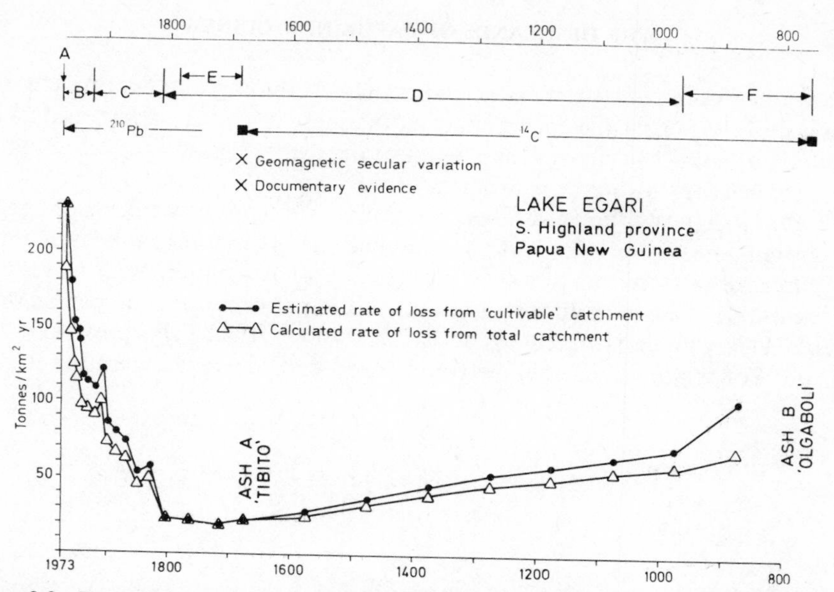

Figure 9.2 Rates of dry-mass sedimentation for Lake Egari converted to erosional loss from the catchment. The main bases for the timescale are noted above (see Oldfield *et al.* 1980). The figures refer to total sediment and are not adjusted for organic content or biogenic silica (see text). No allowance is made for a possible progressive enlargement of the area of lake bed over which deposition has occurred.

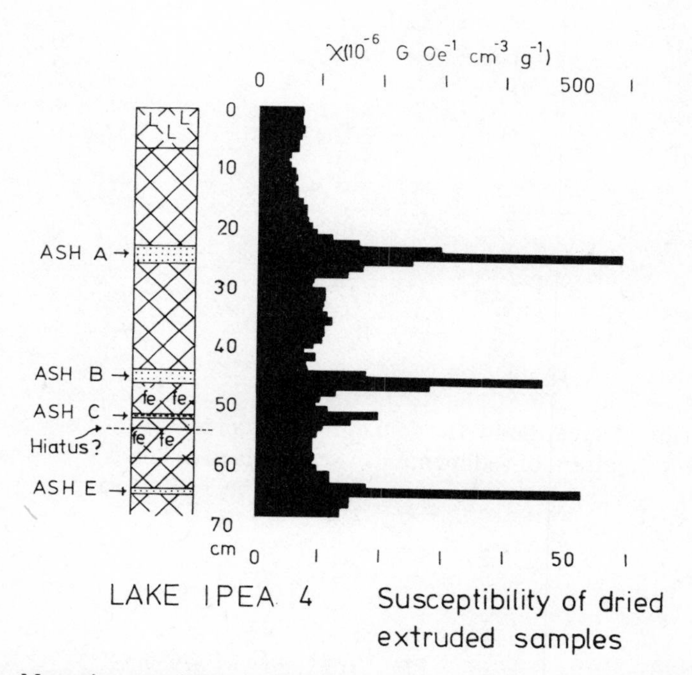

Figure 9.3 Magnetic susceptiblity of the Lake Ipea sediments (core 4). Each volcanic ash gives a distinctive peak in this and all other cores.

roads have been constructed and maintained, including one coming close to the water's edge near the coring sites. At the same time, the pattern and intensity of gardening has changed as a result of the demographic and technological changes induced by western contact.

Figure 9.3 shows the variations in magnetic susceptibility in one of the six measured cores from Lake Ipea. This and all the other cores show an increase in susceptibility in the top 6–8 cm (Fig. 9.4) associated with a visually recognisable change in lithostratigraphy to a paler more minerogenic sediment. This change is dated by both ^{137}Cs and ^{210}Pb to the post AD 1950 period. For the marginal shallow-water cores studied, the acceleration in dry-

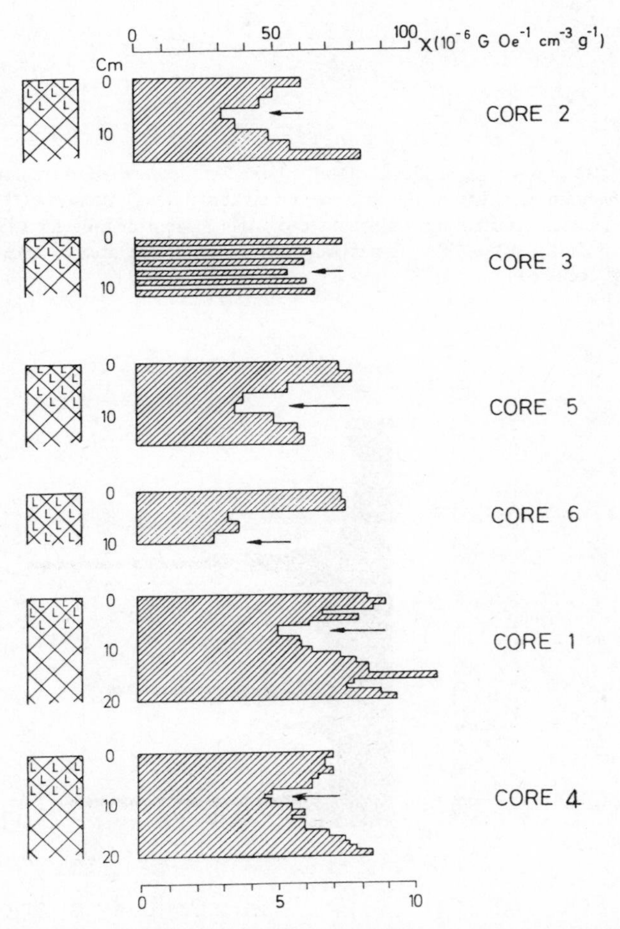

Figure 9.4 Magnetic susceptibility of the near-surface sediments in all the 973 Lake Ipea cores. Common to each one is a decline from peak values reflecting gradually diminished tephra inwash after the 17th century ash-fall. This is followed by an increase associated with paler clayey sediment (cf. Fig. 9.3).

mass sedimentation over the last 350 years ranges from 15 to 32. The origin of this most recent sediment is of considerable interest.

It is clear from magnetic measurements (Fig. 9.5) that the uppermost sediment can be distinguished from the underlying material. The decline in SIRM/χ ratio (the ratio of saturation isothermal remnant magnetisation

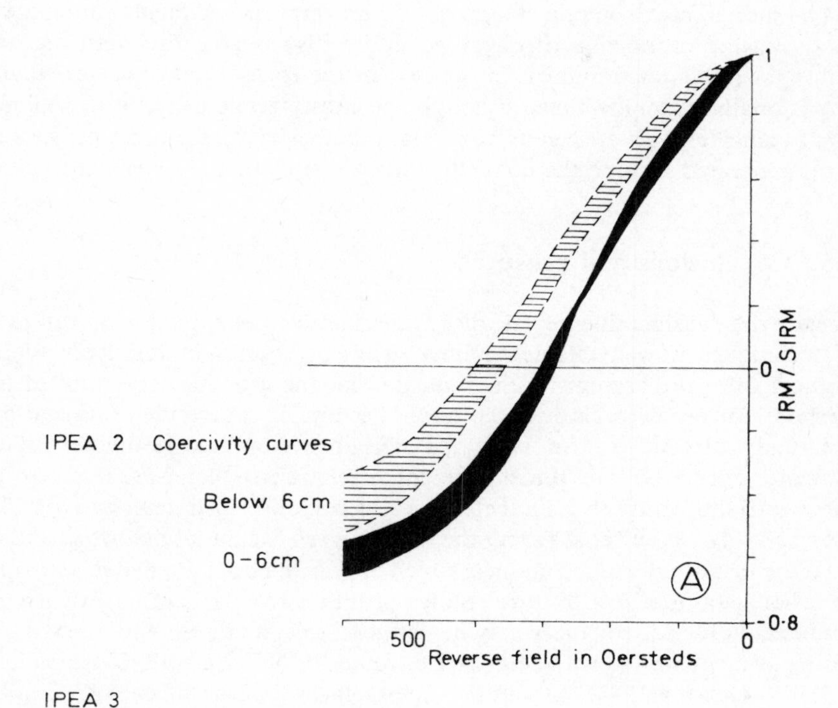

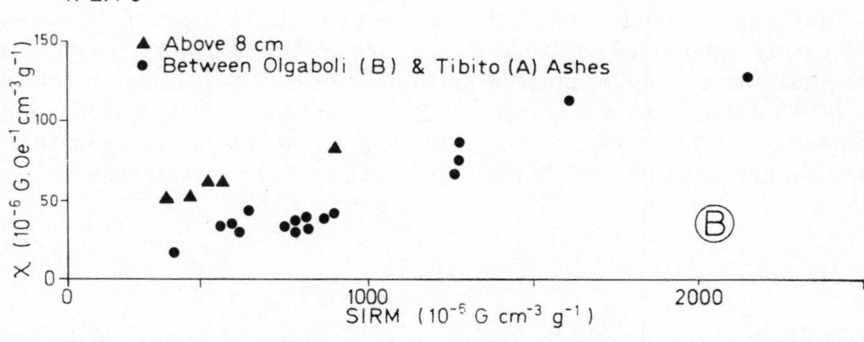

Figure 9.5 Some additional magnetic parameters for the upper sediments at Ipea. (A) Partial coercivity of SIRM curves for groups of samples from core 2 above and below the recent increase in susceptibility recorded in Figure 9.4. The two envelopes of values diverge, with the surface samples showing lower coercivity values. (B) Susceptibility (χ) versus SIRM values for samples above and below the recent increases for core 3. χ/SIRM is consistently higher for the uppermost samples.

(SIRM) to specific magnetic susceptibility (χ): see Oldfield *et al.* 1979) is open to several alternative interpretations, one of which is that recent SIRM values have been depressed by a proportionally higher contribution to the bulk susceptibility (X) from very fine-grained viscous and superparamagnetic crystals derived from secondary magnetic minerals formed in surface soils (cf. Mullins 1977).

The sharp increase in pollen influx in the most recent sediments coupled with an increasing proportion of degraded grains also points to a major surface soil input (O'Garra unpubl.). In the case of the Ipea drainage basin, we may provisionally conclude that, although the most recent increase in sediment input coincides with a surge in road-making activity, the sources of the sediment generated include the cultivated surface soils of the catchment.

9.6 Conclusions and prospects

Despite the considerable logistic difficulties involved in applying the full range of techniques now available to lake sediment studies in relatively remote tropical sites, the present results indicate that the growing repertoire of new methods for core correlation, dating and sediment source identification may eventually provide a firm basis for deriving from lake sediment studies estimates of erosion rates under different vegetation and land-use regimes, and for identifying shifts in sediment source in response to human activity. The present study has suffered from time lags between the initial fieldwork and the development of crucial techniques, which have often been applied in retrospect to residual sample sets. Future studies profiting from recent experience and from technological advances may be designed to systematise and intensify the coring strategy in line with recent case studies in NW Europe (Bloemendal *et al.* 1979, Dearing 1984). In addition, close co-ordination of chemical, pollen analytical and mineral magnetic measurements from the onset of the project will greatly enhance the ecological significance of the results. In the context of the present case study, samples taken from Lake Ipea and several other sites in 1979 suggest that the accelerating trend in erosion rates indicated in this chapter and earlier (Oldfield 1977, Oldfield *et al.* 1980) is not universal in the New Guinea highlands and has not continued over the past decade.

10
Evidence of Upper Pleistocene dry climates in northern South America

J. Tricart

10.1 Introduction

When I published with A. Cailleux, some 20 years ago, a paper about biogeographic regionalisation of Atlantic Brazil, I considered that Amazonia had not been affected by any significant climatological change. I was wrong. Since then, one of the geographic advances made about tropical South America has been the gathering of evidence of important climatic changes in the area (Tricart 1977b). Central Amazonia has suffered the effects of dry periods when it ceased to be covered by the present type of rainforest. On its margins, fossil dune fields indicate the former prevalence of truly semi-arid conditions. These important climatic changes resulted in the drastic modification of both geomorphic processes and the distribution of vegetation communities. They have given to the geographical environment some specific features affecting resource use which must be taken into account in any development plans.

I will begin with the problems of central Amazonia and examine later its northern and southern margins.

10.2 Evidence of climatic change in central Amazonia

In Brazil, the area called central Amazonia is the strip of land that stretches on each bank of the Amazon River from the Atlantic Ocean to the confluence of the Rio Negro with the Solimões (Marañon in Peru). Up stream of this confluence, the same main river is no longer called the Amazon, but rather the Solimões. Nevertheless, the term 'central Amazonia' is extended here to include the margins of the lower Solimões, as is frequently done in the Brazilian literature (Fig. 10.1).

Central Amazonia is a tectonic trough, perhaps an old rift valley. Between the two tectonic units of the Brazilian and Guyana Shields, which were folded and metamorphosed about 1500 million years ago, the central Amazonia trough has suffered intensive movements of subsidence since the Lower

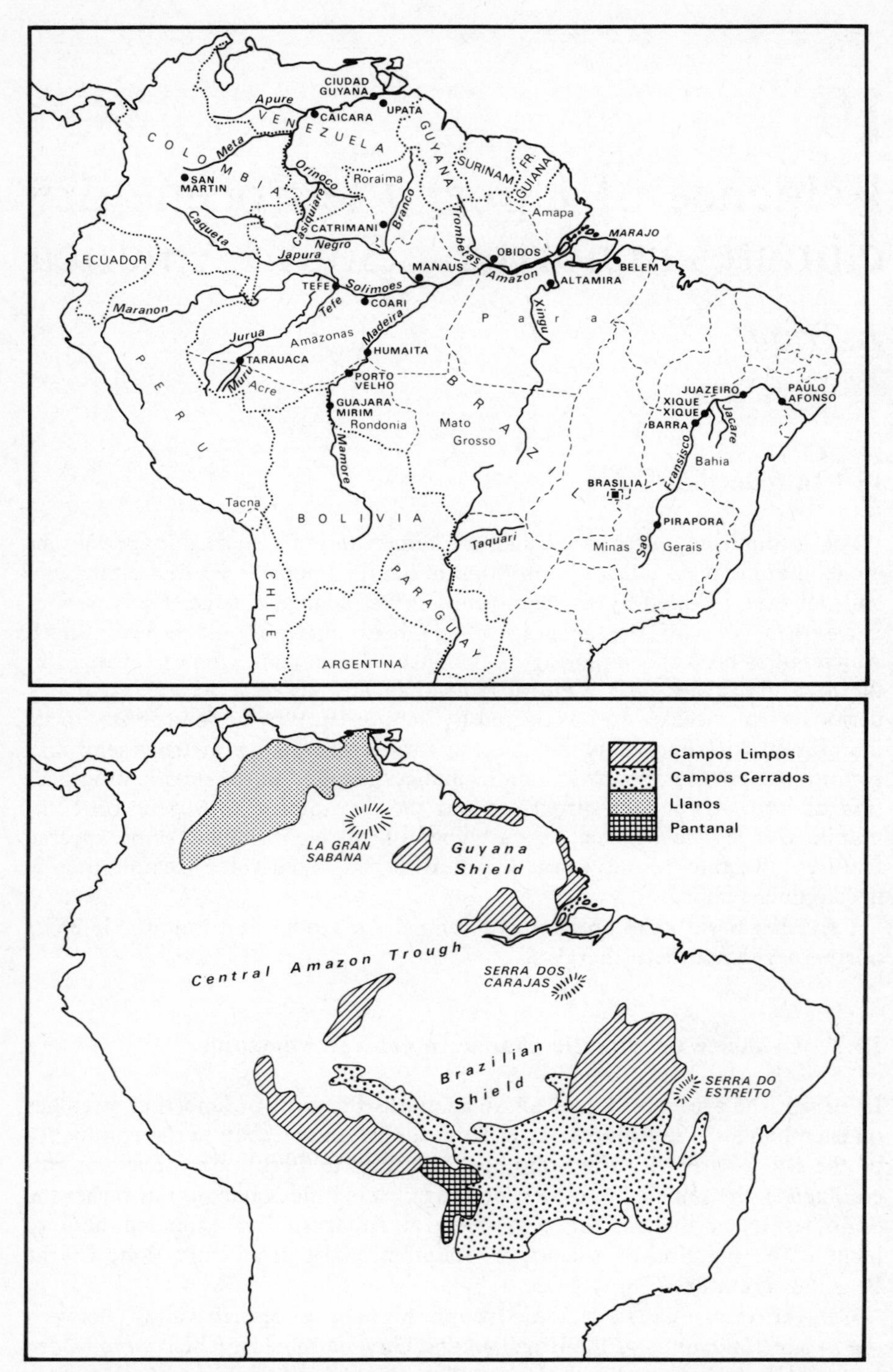

Figure 10.1 Amazon basin showing places and rivers, and vegetation zones and structural units mentioned in the text.

Palaeozoic. Sedimentary rocks (limestones, sandstones, siltstones and slates) of Upper Cambrian to Carboniferous age have been deposited and affected by cover-folds. During Permian times, this sedimentary basin was transformed into a quasi-rift-valley, with the deposition of a thick formation of evaporites (salt, gypsum), which have been studied near Manaus for oil exploration. Later, Cretaceous sandstones and slates were deposited. Finally, the Barreiras Formation, of Neogene age, unconformably overlies the older deposits, covering the margin of the Brazilian Shield south of the Amazon River, near Altamira and the Serra dos Carajás. The Barreiras Formation consists of the products of intense weathering which have been stripped and redeposited by flood outwash under semi-arid conditions. It is a mixture of quartz sands and silts and kaolinitic clays, with some quartz and siliceous gravels. This type of material may be relatively easily eroded by rivers, but is able to support steep slopes.

The Barreiras Formation is capped, WNW of Obidos (Pará State), by remnants of a bauxitic duricrust. The largest outcrop is now worked as the Trombetas bauxite mine, named after the Trombetas River, the waterway used for access to the mine whose proven reserves exceed 2 billion tons. This duricrust forms plateaux, of mesa type, at a higher altitude than their intensively dissected surroundings. The mesas are flat-topped and limited by a steep margin, indented by coalescing embayments, which seem to have been shaped by landslides originating in the underlying clays. In some places, such as near the airfield of Obidos, some remnants of iron duricrusts (laterite) have been observed. They form benches at an intermediate level between the valley bottoms and the mesas. Where extracted near Obidos airfield, the material consists of siliceous pebbles and some duricrust pebbles cemented by iron oxide. It has not been possible to correlate such remnants with a specific stage of the geomorphic evolution of the area.

The Barreiras Formation has been intensively dissected during a recent period. Its topography consists of numerous ridges between narrow valleys with steep sides. The modal slope is about 20°. In spite of its altitude, under 100 m above sea level, central Amazonia is anything but flat. The ridges have rounded tops, radar mosaics giving a false impression of watershed sharpness (Tricart 1977c). At the present time, the forested slopes are not being eroded by runoff and no alluvial deposits are being laid down in the local valleys that dissect the Barreiras Formation. Most of the landforms are stable, being inherited from an earlier period when the vegetation communities were different from those of today. Unfortunately, this paleoenvironment cannot be reconstructed because there has been no systematic study of the pollen types of Amazonian plants and no radiocarbon dates are available.

Thus the only way to date this major palaeoclimatic fluctuation is to use geomorphic evidence.

As far up stream as between Tefé and the confluence of the Japurá River (Caquetá in Colombia) with the Solimões, the main Amazonian trunk stream is incised deep into the late Tertiary unconsolidated material and has been subsequently drowned. The same sequence has occurred along the lower course of all the river's tributaries. There is no floodplain along the valley bottoms, but rather swamps and lakes. Radar and LANDSAT imagery show some elongated clusters of trees which have the size and shape of alluvial levées. However, the trees do not grow upon dry land (*'terra firme'* in Brazilian) but in fact constitute an amphibious forest. This forest is able to colonise these shallow stretches of water where depths are less than 3 m for a long enough period of the year to enable plant reproduction successfully to take place; elsewhere, deeper water does not allow amphibious forest growth to occur. Thus for these ecological reasons, drowned levées are enhanced by vegetation and are quite recognisable on suitable remote-sensing imagery. It is clear, however, that the sedimentary balance was characterised by an important deficit during the recent past, for otherwise alluvial levées should not be drowned. A generally dry floodplain would have been built in the valley bottom. Between these drowned levées, deeper basins are colonised by other types of endemic vegetation including floating meadows composed of plants that are bottom-rooted during low-water stages but which float under flood conditions. Most of the species of these different vegetation communities are endemic; their specific adaptations to a very distinctive environment suggests a long period characterised by similar conditions, perhaps most of the Quaternary. However, these environments may have suffered distributional changes during this long period as aquatic plants can migrate easily along river courses.

Further evidence of a negative sedimentary balance during a recent past period is given by the extraordinary depths of the Amazon River bed. At Obidos, where it narrows to only 2 km in width and the strong current is probably too rapid to allow any settling of suspended matter, a depth of 87 m has been measured. The bottom of the river bed is thus at an altitude of 83–4 m below sea level. As the distance to the Amazon mouth is about 700 km (the mouth cannot be strictly defined), this altitude is compatible with the sea level of −100 m which is currently accepted as corresponding to the maximum regression of the last glaciation at *c.* 18 000 years BP. It must be taken into account that an excavating river would have a steeper slope than a drowned valley and that some deposition may have taken place since the maximum of the last glacio-eustatic regression at Obidos. Along the lower stretch of Amazon River, it is probable that during glacio-eustatic low sea levels, this powerful stream (the mean annual discharge at Obidos is estimated to be about 175 000 $m^3 s^{-1}$) incised its bed sufficiently to reach its equilibrium profile. The Amazon River, in its turn, has functioned as a base level for its tributaries. In the weak rocks of central Amazonia, all of them have deeply incised their beds, which has resulted in the very intense dissection to which reference has already been made. These numerous V-shaped valleys with steep sides were most probably cut during periods of low sea level and, principally, during the most recent regression.

Evidence supporting this view is provided by the geomorphology of the lower parts of these valleys. There, sedimentary balance has been even more strongly negative than along the Amazon River itself. The upper V-shaped valleys end in 'finger-lakes', the valley floors being drowned many kilometres up stream of their ends in the main valley. It is clear that these valleys were cut in a period when the Amazon River itself was flowing at a lower altitude during the last glacio-eustatic regression. In view of the similarity between these finger-lakes and the drowned valleys of many coasts, they have been called 'fluvial rias'. The upper ends of these fluvial rias are triangular in shape, with very acute angles. No sedimentation has modified them and no deltas can be observed, evidence that practically no alluvium has been deposited during the last stages of the Flandrian transgression. The phytostabilisation of the area has been very efficient. The dense and sharp dissection of the ridges is an inherited feature, developed during the last period of low sea level, i.e. the last stage of the Würm (Wisconsin) glaciation. During this period, it is highly probable that central Amazonia was not covered by the present phytostabilising rainforest, but by drier types of vegetation under which runoff was able to produce slope dissection.

Some biogeographical arguments favour this view: Dias de Avila-Pires (1974) has pointed out that, as early as 1945, Sampaio revealed the existence of many species of plants and of *Crotalus durissus* in both the 'cerrados' of central Brazil and the 'tabuleiros' of western Amazonia. The 'cerrados' are a xeromorphic plant association, with an open stratum of trees, a denser shrub stratum and an open lower stratum. Unconcentrated runoff is active under the 'cerrados'. The 'tabuleiros' of Amazonia are considered to be a relict formation of similar physiognomic aspect found on plateaux remnants with poor drainage during heavy rains and with an extremely low soil-moisture retention capacity, so that they have an extreme regime like that of the 'cerrados'. Usually, 'tabuleiro' vegetation grows on structureless, nutrient-deficient, white quartz sands. Splash erosion and subsequent runoff are important. The same type of vegetation occurs in eastern central Amazonia, between Obidos and Oriximiná (Pará state). This tabuleiro vegetation and its specific morphodynamics can be used to explain the type of dissection observed in central Amazonia, where the Barreiras Formation outcrops. For Dias de Avila-Pires, tabuleiro vegetation was widespread on coarse detrital rocks in central Amazonia during a recent drier period. As the climate became more humid, its area of distribution became more and more confined and discontinuous, as is the rule for relict biocoenoses. Presently, it survives on soils that offer the worst edaphic conditions for rainforest, because, under such conditions, the competition from the pioneer community is weak. The waters of the catchments occupied by tabuleiro vegetation are 'black waters' (aguas negras). Sioli (1975) has specified their characteristics:

clear water	pH: 5.2,	Al: none.	Fe: $0.03\ mg\ l^{-1}$,	NH_3: none
Black water	pH: <4.2,	Al: trace,	Fe: $0.15\ mg\ l^{-1}$,	NH_3: a little

Black waters are nearly abiotic, with a very low primary productivity. Their turbidity is not the result of a suspension of mineral particles, but of polymerised organic matter, coming from podsolic soils and the very acid environment. Presently, these tabuleiros are not suffering further dissection; this is apparent from the lack of fan or delta deposits at their upper margins.

Other biogeographical features support the hypothesis that the biocoenosis of the Amazonian rainforest is of pioneer type. Fittkau (1973–4) has shown that the number of species of plants and animals is higher on the foothills of the Andes, and on the Guyana and Brazilian Shields. Downstream, the total number of species declines progressively. Some species, like bamboo and *Cecropia*, grow on the divides in the foothills, but become exclusively riparian downstream. One can see along the Juruá how the *Cecropia* colonise sand-banks recently abandoned by the river, such as the convex banks of meander bends. Obviously, their seeds are transported by the river, thereby enabling them to colonise those regions downstream. This view is supported by Fittkau (1973–4), who has demonstrated that the waters coming from the higher regions on the margins of Amazonia have a significantly higher nutrient content than the black waters of central Amazonia. This is the result of ferrallitic weathering of basement rocks. These nutrients generate an 'oasis effect' along these allochthonous streams, since they produce ecological conditions that are much more favourable for the *Cecropia* than the environment of the dissected landforms. Dias de Avila-Pires (1974) has observed the same type of migration for various animal species: the Madeira River forms the western limit to the range of the birds of the Upper Amazon. The *Saguinus* sub-genus is formed by two groups; tamarin occupy the region at the foot of the Andes as far as the Japurá and Madeira rivers, while nigricollis are distributed from the Caribbean coast of Guyana to the Rio Negro and the Amazon River, with a small extension to the south around Belém. Cabuella species are found from the Andes to the Japurá River and to the town of Tefé. The distribution of different forest birds is similar. For Dias de Avila-Pires, such distributions support the hypothesis of a contraction in the area of rainforest during recent drier periods. Rainforest would have survived only in isolated refugia, where it received sufficient rains as a result of orographic effects: Guyana and Brazilian Shields, eastern slopes of the Andean cordillera.

It is now possible to give a general view of the morphogenesis of Central Amazonia and of the role played by climatic changes.

(a) Central Amazonia was characterised by tectonic subsidence which was still very active at the end of the Tertiary. Near Manaus, at the confluence of the Madeira and Amazon rivers (Ilha da Trindade), drilling has shown the Barreiras Formation to be composed of 400 m of terrestrial facies. During the Quaternary, important tectonic movements were still taking place. Near Manaus (Ilha do Careiro), the base of Quaternary alluvium has been encountered at – 120 m. Lower R. Negro lake, reaching about – 140 m, is an assymetric trough, probably still undergoing tectonic deformation. At the mouth of the Amazon River, at Cururu, boreholes have revealed 500 m of marine sediments of Miocene age, with below them 3300 m of sediments from the Miocene to the Cretaceous (Mendes 1957). The partially drowned delta of Marajo Island is the result of the complicated interaction of sea-level changes, climatic fluctuations and tectonic movements. The same applies to many fluvial landforms. The lakes of the lower Rio Negro and the Tapajós can be explained by a combination of recent and probably present sinking, the Flandrian transgression and a sediment deficit as a result of the phytostabilisation of landforms by the pioneer rainforest.

(b) During the Quaternary, important climatic changes occurred. The last episode, which cannot yet be dated with accuracy, took place during the last stage of low sea level (Grimaldian regression). Most outcrops of the Barreiras Formation, which have not been affected by subsidence since the middle of the Quaternary, were intensely dissected as a consequence of the base-level lowering resulting from the regression. The huge discharge of the Amazon River enabled it to cut rapidly into the weak sediments and to attain an equilibrium profile. The same process occurred along the lower courses of its tributaries. This dissection was favoured by the occurrence of a drier climate which eliminated the rainforest from the lowlands along the Amazon River. Rainforest refugia were restricted to mountainous slopes receiving orographic rainfall on the eastern side of the Andes and on the Guyana and Brazilian Shields.

(c) During the Holocene, there occurred, simultaneously, a change to a wetter climate and a rapid rise of sea level, the Flandrian transgression. Abundant alluvium would have been necessary for the streams to have built floodplains during the sea-level rise. However, in the same period, climatic change has caused a progressive phytostabilisation of the dissected landforms, with a consequent decrease in sediment yield. This yield steadily diminished as the lowlands were colonised by the pioneer rainforest. Conditions became increasingly unfavourable for the accumulation of alluvium, thereby explaining the absence of solid land (terra firme) along the completely inundated valley floors. Along the Amazon River, although some alluvial levées were built when the base level was some metres lower than at present, they are no longer developing and are thus covered by some metres of water. Along the local tributaries, no sedimentation has taken place, with the result that their lower courses have been transformed into rias (fluvial rias).

There is a general rule, too frequently forgotten by geomorphologists, that landform development is the result of the interaction of many factors. Climatic influence is not exclusive of tectonic influence, of structure or of base-level changes. We need a systems approach in order to reach a better understanding of these interactions. Central Amazonia provides one example of how this may be done.

10.3 Climatic changes on and around the margins of Amazonia

In western Amazonia, valley landforms are different although, between valleys, intensive dissection still prevails. In some regions, other evidence of semi-arid climates has been observed. In view of the extension of the area and the consequent diversity of present and past climatic conditions, this account will be presented on a regional basis.

Western Amazonia

Alluvial forms are quite different in western Amazonia from those described above for central Amazonia. This results from the relative proximity of the Andes, which means that this region can be considered as a piedmont of the Andes.

During the Holocene, detrital material was always sufficiently abundant to allow for the building of typical alluvial plains. The last great fluvial ria, up stream from the Solimões, is the Tefé Lake, at the mouth of the Tefé River. Further up stream, the confluence of Solimões and Japurá (Caquetá) is a kind of ballet of meanders. In radar imagery, ox-bow lakes and levées are enhanced by the different heights and types of vegetation. Free meandering, characteristic throughout the recent period, has produced frequent changes in the two stream courses. In this region, the sediment yield was sufficient to allow the streams to keep pace with the base-level rise during the Flandrian transgression. This stems from the relative proximity of the Andes, which yield a large amount of detrital material. Nevertheless, the junction of the Solimões with the Japurá is a straight-line distance of about 1500 km from the foot of the Andes. Such a distance means that most of the alluvial material consists of suspended matter, with a small amount of sand from the undercutting of the foot of the valley slopes consisting of sandy rocks. The relative abundance of fine material explains the great instability of the river beds, and the very rapid cutting of meanders. Data on the load of the Solimões in the area are lacking, but this stream, just before joining the Rio Negro, near Manaus (Gibbs 1967, Irion 1976), has a suspended sediment concentration of 40–300 mg l^{-1}. Obviously, its suspended load is more abundant at the junction with the Japurá, about 1000 km up stream. Another factor that enhances stream shifting is the important variation in the water level, the annual mean difference between high and low water levels being as much as 20 m.

The influence of tectonics is clear: the enormous, intensely dissected, chain of mountains of the Andes has always been able during the Quaternary, whatever the climate, to yield sufficient detrital material for the building of an alluvial plain. The climatic factor, in landform development, is manifest just through the influence of vegetation. The dense forest of the floodplain results in the deposition, up stream, of the coarsest fraction of the alluvium, so that the floodplain is constructed mainly from suspended material. On the other hand, the same vegetation enhances the deposition of this fine material, by acting as a filter. Sioli (1957) has measured a decrease in concentration from

128.4 mg l^{-1} at a distance of 0.4 km from the main channel of the Amazon River, to only 20.9 mg l^{-1} at 4 km from it.

Terraces in the Juruá basin

Although in SW Amazonia, the Juruá basin has no Andean mountainous areas. It consists only of sedimentary rocks of Tertiary and Cretaceous age, mainly clays, silts and sands, accumulated in an external trough along the Andes geosyncline. Lithological characteristics are somewhat similar to those of central Amazonia, which makes for easier comparisons. In the upper catchment, the Acre Clays ('Argilas do Acre'), which cover a wide area, are impervious and contribute to a torrential stream characterised by increases in discharge during rainy periods.

The writer has studied the terraces along Muru and Tarauacá rivers, up stream of the town of Tarauacá (Acre State), which show the following succession:

(a) In the valley bottom, the present floodplain consists of alternating layers, at most 0.3 m thick, of clay and sand, with the clay being predominant. These beds are usually flat, although in some places they appear lenticular and have filled former channels. No soil formation has been identified.

(b) A low terrace reaches an altitude of 5–6 m above low water level. Its surface is gently undulating. In some places, it falls abruptly to the floodplain, where it is undercut in concave curves. The higher parts of this terrace are no longer flooded and soil formation has taken place. The soil has a brown-reddish colour, slightly orange, but always pale and somewhat greyish. Its depth is irregular, with tongue-like pockets in its lower part. These reach a depth of 0.2–0.8 m below the bottom of the soil, which is about 0.6 m deep. On the lowest parts of the terrace, thin humic grey soils are covered by accumulations of fresh sand, at most 0.3 m thick. This sand has a pale grey-yellowish colour and is obviously deposited by the highest floods of the present period. The alluvium of this low terrace has no obvious stratification, which probably results from bioturbation. It consists of a mixture of sand, silt and clay.

(c) A higher terrace has been preserved, mainly on the left side of the rivers. Its surface is at about 12 m above low water level. The terrace scarp is concave-convex and cut by small valleys. Its stage of geomorphic evolution is clearly more advanced than that of the low terrace. The soil, too, shows a more important evolution. Its depth is about 3 m and its colour is brown-red, being slightly browner than a brick. It looks to be ferrallitic, but no analytical data are available. The alluvial material is clayey–sandy in texture and without visible stratification.

The terraces of the Muru and Tarauacá rivers are built from the same type of material as a result of the homogeneity of their catchments. Both their geomorphological pattern and the differences in the soils formed upon them show that the present is a period of river bed incision. River aggradation has occurred during two former periods, separated by a phase of channel incision, similar to that of the present. A study of the regional geology suggests that this area is being slowly uplifted and this would explain the staircase pattern of the terraces. The alternation of channel aggradation and incision would seem to be related to climatic changes as the area is too far from the Atlantic Ocean to be affected by changes in sea level.

The same sequences, from aggradation to downcutting, can be seen on the Solimões River between Tefé and Coari. The present phase is one of weak aggradation. Alluvium from the Andes is leading to levée construction: these levées, in turn, dam numerous fluvial rias occupied by lakes. Erosional features along these rias appear to be recent in origin. They have been carved into a former aggradational terrace, whose surface is some metres higher than the stream low waters. It seems likely that the lower terrace of the Muru and Tarauacá Rivers are flooded from time to time. Some abandoned channels are still identifiable on the terrace surface; differences in vegetation are associated with changes in water regime resulting from small topographic differences. Remnants of fluvial rias, clearly obliterated, are dammed by this terrace. Most of the rias have undergone infilling. They appear to have been cut into a higher terrace whose microtopography is very difficult to decipher. Nevertheless, in some places, the vegetation pattern suggests meanders.

Along the Solimões River, regressive incision has been generated by low sea levels of glacio-eustatic origin. Very probably, aggradational phases, like the present one, coincided with higher sea levels and were made possible by the permanent yield of detrital material from the Andes. These terraces would be eustatic in origin, rather than of climatic causation. The drier climates of the Quaternary simply resulted in intensive dissection phases. Biogeographical arguments also suggest the former existence of ecosystems associated with drier climates in this part of Amazonia. The present humid tropical ecosystems are pioneer, as has been emphasised.

Southern margins of Amazonia

Other evidence around Porto Velho (Rondônia Territory) deserves attention. In the valley of the Katira River, 120 km north of the town, Van der Hammen (1972) has established a pollen diagram for 25 m of dominantly clay sediments, resting on a duricrust. At 13 m depth, he found the Pliocene/Quaternary boundary, with a flora characterised by 90% tree pollen, characteristic of swampy forest, and 10% herbaceous plant pollen, mainly Gramineae. Between 6 and 12 m in a layer of grey clays, only 5% of the pollen was from trees, the remaining 95% being typical of a dry savanna, with small Gramineae and *Cuphea*, a species now common in the Colombian savanna lakes. From 2 to 6 m depth, a colluvial red clay contained 10% tree pollen, only of *Mauritia* type, and 90% pollen from 'clean' savanna herbaceous plants ('campo limpo'). As a matter of comparison, near Porto Velho, experimental plantations of cocoa trees have given very good results and the Brazilian

Government has planned a large-scale production of cocoa. The cocoa tree has very specific climatic requirements, being a plant of typical rainforest. Van der Hammen's analyses demonstrate a drastic climatic change in this area. The present rainforest conditions are recent and follow a single long period, or a succession of shorter periods, of dry climate.

A geomorphological survey of the area, for Radambrasil, in 1975, (Radambrasil 1976a,b) clearly demonstrated the effects of these dry climates on landform evolution. At about 7–8°S, near Humaitá, on the Madeira River, the following succession of Quaternary forms may be seen:

(a) The present depositional features on the valley bottom, with notable levées.
(b) A higher deposit, slightly higher than the valley bottom, where the levées have already suffered from incipient obliteration. In part they reveal former meanders as opposed to the present braided channels. Remnants of ox-bow lakes not yet infilled suggest a relatively recent age. Some fluvial rias are dammed by the levées of this depositional unit.
(c) Still higher is found another accumulation, with large meanders associated with the levées of braided channels. As usual, these fluvial forms are detectable by vegetation differences. This terrace has been cut by some fluvial rias, which are dammed by the preceding accumulation. This deposit could date from before the pre-Flandrian regression.
(d) A still higher deposit is nearly at the same altitude as the Neogene plateau. Enhanced by vegetation, numerous former river beds can be identified. Their width is variable, as is the radius of such meanders as can be found. The general direction of these old streams trends towards ENE, somewhat different from the present NE course of the Madeira River. This extensive sedimentary unit looks like a very low fan, possibly dating from the early Quaternary. The present course of the Madeira River is guided by a tectonic lineation which begins in the vicinity of Tacna (Peru), on the Pacific coast, and runs as far as Manaus, where it ends in a field of faults.

This geomorphic sequence is similar to that already described for the Solimões down stream of Tefé, but clearer for the older Quaternary events. Here, one clearly sees three periods of accumulation and two intermediate periods of river cutting, the last of these being contemporaneous with the last low sea level. Also in this region one can assume that this last low sea-level period was characterised by a dry climate. Large, relict clean savannas, completely surrounded by rainforest, effectively only persist to the west of Humaitá. They are probably the last remnants, subject to extreme edaphic conditions, of the clean savannas identified by the palynological work of Van der Hammen.

South of Porto Velho, the major landforms consists of inselbergs with pediments. The slopes of the inselbergs, in metasediments, are steep and rocky outcrops are frequent. In spite of the humid climate, they are mainly covered by savannas as a consequence of the edaphic conditions (shallow soils with low moisture-retention capacity). The low slope of the pediments means that drainage is poor: shrub savanna ('campo sujo') grows on their surfaces. Near Guajará Mirim along the Bolivian border, the right bank of the Mamoré River is flanked by an old alluvial deposit, whose altitude is not noticeably higher than the sandy alluvial levées of the present stream. The initial topography of channels and levées in this accumulation has been completely obliterated. In its turn, this old deposit grades into the lower part of the pediments.

Porto Velho is situated on the border of the Precambrian basement which has not been cut by the streams. Just up stream of the town, the Madeira is no longer navigable, Porto Velho being the limit of navigation. Between this town and Guajará Mirim, there is a succession of rapids which have blocked all the Quaternary retrogressive incision. This is the reason why, above the rapids at Guajará Mirim, the Mamoré River flows on the surface of the pediment. The lack of Quaternary incision has favoured the development of this pediment. The pediment is polygenetic but has been sculpted during the successive dry periods of the Pliocene and the Quaternary. There has been enough time to transform it in a pediplain. Effectively at the end of the Tertiary, the Barreiras Formation was deposited under a dry climate. Near Porto Velho, the pediment cuts the contact between the Precambrian basement and the Upper Tertiary sediments which are the stratigraphic equivalent of the Barreiras Formation. Van der Hammen's pollen studies have demonstrated the existence of a succession of periods of savanna vegetation. However, it seems likely that in the Katira valley, as along the Madeira valley, only the dry periods of the Quaternary favoured alluviation. This is why all the pollen in the profile represents savanna vegetation. During wet periods, accumulation was interrupted and no pollen was preserved. The Katira River has probably not been able to incise into its valley bottom like the Madeira River near Humaitá, but we have no information on the geomorphology of the area studied by Van der Hammen. Nevertheless, the shallow depth at which he encountered Pliocene sediments demonstrates that the accumulation rate was very slow, probably because it was discontinuous and frequently interrupted.

Further south, in the Upper Paraguay basin, the writer has recently identified features demonstrating the occurrence of semi-arid conditions in a recent period:

(a) Enormous fans have been constructed in the sinking area of the Pantanal. On the biggest of them, the Taquari fan, deflation has excavated numerous basins, usually small but sometimes up to 3 km in diameter. Presently, the basins are flooded or colonised by swamp vegetation. Some of them have typical sebkha features and alkaline soils have been observed around their margins.

(b) Fluvial geomorphology indicates that the Paraguay River, during the same period, was unable to flow out of the Pantanal.
(c) South of the Pantanal, on both sides of its valley near the border of Brazil with Paraguay, some remnants of temporary saline lakes have been identified on LANDSAT and radar imagery. Some endoreic ephemeral streams end in these lakes.

The southern part of the Pantanal and the surrounding area to the south, in Brazil and Paraguay, have been affected, during the recent Quaternary, by semi-arid climates. The type of evidence outlined above has been found as far south as 21°S.

Southeastern Amazonia: São Francisco valley

The interior of NE Brazil is currently a dry area. It remained semi-arid throughout the Quaternary. However, some features provide evidence for the existence of even drier climates during certain periods of the Quaternary. The clearest example is given by the region of the great bend of the S. Francisco River, up stream of the town of Juázeiro.

The middle of S. Francisco valley, in the southern part of Bahia State, lies in a depression carved into the sedimentary cover of the Precambrian basement. This consists of convergent pediments at the base of inselbergs. Most of the pediments are of lithological origin. Many are formed by the rocks of the sedimentary cover, consisting mainly of sandstones and quartzitic sandstones of the Palaeozoic and, in the west, of the Cretaceous cover rocks. Some pediments are composed of very hard limestones, like the Morro de Bom Jesus da Lapa, an example of tropical karst, while others are formed in Precambrian folded metasediments. This latter type consists of appalachian ridges which have very probably been exhumed from a weaker sedimentary cover, possibly of Cretaceous age, like the Serra do Estreito. As the Brazilian Shield dips gently towards the SSW, its cover has persisted extensively up stream and, in contrast, has been eroded extensively in the north, around and up stream of Juázeiro.

The main geomorphological features of the area are semi-arid in character: pediments and inselbergs, which result from evolution after a long period of time. Good arguments suggest that these landforms had already been shaped during the Neogene, in the period when the Barreiras Formation was being deposited along the Atlantic coast. As in eastern central Amazonia, it consists of detrital material which has been subsequently redistributed under semi-arid climates.

A closer examination reveals alluvial sedimentary units along the middle S. Francisco River. Deposited on the lower part of the converging pediments, they consist mostly of sand with some finer material and gravels. The floodplain lies at a slightly lower level, so that the deposits form a low terrace although, in general, no clear scarp has been excavated along this contact. A detailed study would be needed to establish the existence of various units within this fill. Up stream, near Pirapora (Minas Gerais), different terraces have been observed, but they converge down stream as a result of the lack of regressive incision up stream of the waterfalls at which the Paulo Afonso hydroelectric plant has been constructed.

Likewise along the Madeira and Mamoré rivers, this lack of regressive incision results from a combination of climatic and lithological factors: the Precambrian basement is very resistant, and the Paulo Afonso cascades are located where the S. Francisco River literally 'falls' into a gorge excavated relatively easily along a strip of densely fractured gneiss. On the other hand, like most tropical streams, the S. Francisco River carries mostly fine material and sand, which is all thrown into suspension where the flow is rapid, so that little abrasion of hard rock occurs. This lack of regressive incision results in the good preservation of the pediment and inselberg topography.

However, in the middle S. Francisco valley, another process has played an important role: the interruption of river course during two, or perhaps more, arid periods. During them, the S. Francisco River became endoreic and its waters disappeared in a flood outwash area near the small towns of Barra and Casanova. Rapid observations were made there before the flooding of the S. Francisco valley by the Sobradinho dam.

Between the river and the S. do Estreito, on the left bank of the S. Francisco, lies an extensive dune field. The river makes a large bend northeastwards around the margin of this dune field. This is evidently an adaptation, achieved when the climate became more humid, allowing the S. Francisco to become exoreic again. The dune field can be described as follows (Tricart 1974a):

(a) Along the S. Francisco valley, still partly active, there are reticulate dunes which require an abundant sand supply. Such sand cannot come from the eastern margin of the S. Francisco, most of which consists of inselbergs.
(b) Gradually, towards the west, these dunes become less and less reticulated and give way to longitudinal dunes, which reach the very footslope of the S. do Estreito. This transition and the orientation of the dunes denote winds coming from east to south-east.
(c) Lying on the western foot of the S. do Estreito, radar imagery (it has not been possible to visit the area) clearly shows another sand plain with a much more uniform topography and a less reflectant surface than the dune fields. Shallow depressions, up to 1 km in diameter, have been excavated in this plain and it appears to be a loamy aeolian accumulation, with some deflation basins. Presumably it consists of fine material blown above the sierra by easterly winds. Many other deflation hollows, of greater size and dampness, as a result of their greater depth, can be observed on the western bank of the S. Francisco River from the surroundings of Pilão Arcado, where the dune field is at its greater extent, to as far up stream as Xique Xique.

The extensive dune field of Pilão Arcado seems to have been shaped by aeolian reworking of an extensive sandy flood deposit of the S. Francisco River. Under a semi-arid climate, great quantities of sand were transported by runoff on the scarps of the sandstone plateaux of the upper catchment and accumulated on the pediments. Part of this sand was transported by the river during floods. Some of it has accumulated along the river, but another part, probably during major floods, reached the extensive lowland of Pilão Arcado, where the river waters spread out, infiltrated or evaporated. Between the floods, these deposits were reworked into a dune field. This type of evolution is similar to that which occurred during dry periods of the late Quaternary along the Niger River, in the Mali, in West Africa, around Lake Debo. However, a positive feedback mechanism appeared: the accumulation of the sand led to less channelling of the flood flows and enhanced infiltration, which was further encouraged by dune formation. So, over time, the deposition of sandy flood deposits progressively receded up stream.

Differences in colour of dune sand suggest the existence of distinct periods of aridity. Better evidence of these phases can be gathered from the eastern bank of the S. Francisco River.

On this right bank, the Rios Jacaré and Verde flow with difficulty upon old alluvial deposits laid down upon the pediments at the foot of the inselbergs. It is possible to distinguish two areas:

(a) There is an older accumulation, forming a slightly higher area between the two rivers. This consists of loamy–sandy material and is thus not able to form very typical aeolian landforms. Nevertheless, low ridges and deflation basins can be identified. From these features it may be concluded that a more humid period (outwash deposition of the material) was succeeded by a drier one (aeolian reworking).

(b) At the foot of the inselbergs, mainly along the lower R. Jacaré, numerous deflation basins have been excavated as a result of the eddies generated by these topographic obstacles. They are irregularly shaped, but their fresh forms appear to be more recent than the former landforms. At present near Xique Xique, a good proportion of the streams coming from the hills fail to reach the S. Francisco. Their waters spread out during floods, depositing their sandy loam load. Such hydrological behaviour seems typical of the 'wet' climate periods. This region is, presently, at the climatic limit of exoreism and it is easy to understand that any drier climate would result in endoreic conditions.

Both the upper courses of the Paraguay and S. Francisco rivers, on the drier margins of Amazonia, ceased to run into the ocean at certain periods during the Quaternary. When drought was at its peak, dune fields were formed in the middle S. Francisco valley. This probably occurred twice during the Quaternary. In eastern Bolivia, Jordan (1981) has also described aeolian landforms, including dune fields, some of which are stable and some of which are still active. In spite of a great difference of intensity, these features are somewhat similar to those recognised in West Africa on the southern margin of the Sahara.

Northeastern margin: the Guyana Shield

The Guyana Shield has been identified by biogeographers as one of the refugia where humid tropics biocoenoses were able to survive during the drier periods which affected Amazonia. A rapid geomorphological survey has given us the opportunity to observe the geomorphological evidence for this assumption.

Intense and deep weathering has been observed on the northern slope of the Gran Sabana plateau from Tumeremo to La Escalera. The main road cuttings are not deep enough to reach the solid rock, even on the greywackes, siltites, andesitic and rhyodacitic volcanics of the Caballape Formation. Even in a moderately jointed granite, 57 km south of the road to El Dorado, a cut 10 m deep reveals only rotten granite. In it two completely decomposed corestone boulders 2 m in diameter can still be identified. This intense alteration is of ferrallitic type, with abundant kaolinite. Nowhere along this road has any truncation of the weathering been observed: no exhumed boulders, no stone lines. Only one small outcrop of fresh rock has been noticed; this was found in the River Yuruari bed and consists of a small exhumed pinnacle of a very resistant dacitic rock. Everywhere else, the entire surface is mantled by a thick cap of ferrallitic alteration products. Landforms are those typical of warm and wet climates, with rounded hills (meias laranjas).

Obviously, this northern slope of the Guyana Shield has experienced humid tropical conditions for a long time without any Quaternary interruptions. Geomorphological features support clearly the biogeographical assumption that the area was a refuge for humid tropical biocoenosis.

The lower regions of the north are somewhat different. Around Caicara, El Milagro and Maripa, metasediments, gneisses and granites have suffered intensive alteration phases alternating with drier periods when weathering products were partly stripped and removed. This evolution has resulted in landforms showing differential weathering:

(a) Quartzites, in spite of their close jointing, always form ridges, being highly resistant to weathering with only superficial ferruginisation. The fact that these quartzites act as a resistant rock in spite of their jointing suggests that weathering has been the prevailing process in the past.
(b) Gneisses, on the contrary, have suffered intense ferrallitic alteration. North of Upata, the road to Ciudad Guayana cuts a very narrow syncline consisting of gneiss (20 m broad) between vertical quartzites. These are quite fresh and form ridges, whereas the gneiss is completely soft and excavated. In some places, massive gneisses are able to form slabs on steep slopes, but this is rare and restricted to very small areas, such as west and south-west of El Paso. More frequently, rock pinnacles and outcropping boulders can be observed. These features attest to the succession

of deep weathering periods and phases of denudation. Alteration products have been stripped off during the latter phases.

As a result of these climatic changes, the landforms in the gneiss are varied. Generally they are shaped into rounded hills with a concave, colluvial foot. Less commonly, where joints are scarce, they form residual hills, although never inselbergs.

It can be concluded that this region was characterised during the Quaternary by predominantly humid tropical conditions, interrupted, during one or more short periods, by somewhat drier climates. However, the climate was probably semi-arid at the end of the Pliocene. Many hill-tops seem to be the remnants of a pediplain and are coated with siliceous, mainly quartz, pebbles. One can presume that these gravels are a marginal facies of the Mesa Formation which occupies the eastern Llanos of Venezuela and which, in some places, begins to the south of the Orinoco River. However, in the area of the basement rocks, this dry tropical period has not left important landforms: it just provides a benchmark in the geomorphic timescale.

In the Guyanas, these views are supported by biogeographical evidence presented by Descamps and Alii (1978), who consider that the bare rock hills now form refugia for the xeromorphic species, like *Cyrtopodium andersonii*, which were more widespread during a recent drier period. In French Guyana, at present, the rainforest is gaining ground on heliophic plants and on the niche of the lizard *Tropidurus torquatus hispidus* Spix. The coastal savannas are recent. This can be deduced from the complete absence of endemic species in French Guyana and from their very low percentage (4%) in Surinam. The same authors assume that the humid tropical refuge of the Guyana Shield was discontinuous and consisted of at least three discrete sub-units. An argument in favour of this view is the complete lack of common species amongst the dendrophilous acridians living in the canopy of the forest between Guyana and Amazonian Colombia.

On the north slope of the Guyana Shield, an important humidity gradient can be recognised for the recent Quaternary:

(a) On the margin of the Gran Sabana, no dry period can be recognised.
(b) On the Venezuelan foot of the Gran Sabana, there is an evidence for short drier spells during this period, with a partial stripping of the deep alteration products.
(c) Near the Orinoco delta, Tricart and Alfonsi (1981) have identified deflation features and sand dunes, some of which have been fossilised by the deposits of the Flandrian transgression: a semi-arid climate prevailed there during the last dry period which affected central Amazonia.

This humidity gradient exists at the present day, but it was enhanced during the period from the end of the Würm (Wisconsin) to the very beginning of the Holocene.

From the hillslopes on the south-east margin of the Guyana Shield, Journaux (1975a,b) has described frequent stone lines covered by a yellow earth. These stone lines fade into the alluvium of the valley bottoms which contain rounded quartz pebbles. As in this chapter, Journaux equates these colluvial and alluvial deposits with a dry period, synchronous with the last glacio-eustatic regression. He also agrees that the present period is one of pioneer forest extension along the border of Brazil (west Pará) and Surinam.

Just south of the Guyana Shield, in Venezuela, around Sta Elena de Uairén, important evidence of climatic change has been found. This region, at an altitude of about 900 m, receives at present 1695 mm of rain annually (period 1951–71), but from December to March, inclusive, each month is relatively dry with less than 100 mm (71 mm in January, 65 mm in February), so that shrub savanna grows where edaphic conditions are less favourable to trees. Drier climates during a recent period can be deduced from the following geomorphic features:

(a) About 10 km west of the airfield, volcanic rocks have suffered intense alteration and have been transformed into clay, but the weathering profile, on the slope of a low hill, has been truncated. A pavement of altered volcanic rock fragments, up to 5–10 cm thick, coats the slopes with a thickness of 0.2–0.4 m. Below this pavement and on top of the clay, a lammellar duricrust, 0.01–0.02 m thick, consists of a combination of silica and iron oxide.
(b) Pediments are frequent at the foot of the sandstone cuesta. In their upper sections, fragments of sandstone have been deposited by running water, but these disappear progressively down slope, and on the lower sections only sand and loam can be observed. Along the valley bottoms, the alluvium consists exclusively of silt and clay. It can be deduced that runoff did not have a high competence and was only able to wash away the fine material on the shallow slopes, thus concentrating the stones into a pavement. On steeper slopes, like the cuesta front, runoff was able to transport some stones, but not far from the cliff foot. These facts suggest that the vegetal cover was open but nevertheless relatively thick on the pediments, where water was more abundant.
(c) On the front of the sandstone cuesta, above the small town of Sta Elena, torrential catchment basins have been observed at the heads of the sharp re-entrants, deeply cut into solid bedrock. At the foot of the scarp, typical fans have their head lower than the neighbouring colluvial concavity, when their toe has been built upon it. This torrential episode obviously post-dates the last period of activity on the pediments described above.

The area around Sta Elena de Uarién provides evidence for the existence of erosional landforms subjected to several alternations of wetter and drier climates than at present. Humid environments prevailed for long periods of time, but a drier climate has followed them recently, with two different episodes, one of unconcentrated runoff and one with a torrential regime. Unfortunately, it is not possible to date these episodes precisely.

In Roraima Territory, Brazil, in the same Rio Branco catchment as Sta Elena de Uairén but further south, Radambrasil (1976b) describe extensive accumulations of ferruginous sandstones interbedded with clays, quartz pebbles and lignites. They reach the Casiquiare in the west and have been dated as being early Quaternary in age. These outwash deposits denote a drier climate. They are covered by alluvial sands which cross the Venezuelan border. This more recent flood outwash deposition, erroneously attributed to the Holocene, consists of gravel, sands, silts and clays. It is very similar to the pediment deposits of Sta Elena and probably of the same age; at least, they have been deposited by the same morphogenic system. On the hillslopes, the same association of stone lines and outcropping boulders, typical of the northern foot of the Guyana Shield, has also been observed here. Near Catrimani (1°N), deflation hollows, presently drowned, and low sand dunes can be identified on infrared photographs, radar imagery and slides. The dune orientation is not quite distinct but seems to be NNE to SSW. If this is correct, then these forms represent the southern limit of influence of the Trade winds. A southerly displacement of the tropical high-pressure belt of the Caribbean would obviously explain the evidence for drier climates around the Guyana Shield. Such a change in meteorological conditions at the end of the Würm (Wisconsin) is compatible with the extended ice cap of North America and the lower temperatures that affected the whole world.

Between Amazonia and the Andes: the Llanos del Orinoco

Evidence for Quaternary aridity is not limited to the margins of the Orinoco delta but is also apparent over a great part of the catchment of this river.

Tricart (1974a,b) has described an extensive dune field on the west bank of the Orinoco. It coincides with a tectonic basin at the foot of the Andes, where Tertiary sediments have been covered by Quaternary alluvium brought down by the rivers issuing from the Andes.

The Orinoco River runs along the eastern border of an extensive dune field up stream of the small town of Caicará. The northern limit of the field coincides roughly with the Apure River. In the south, an important flexure, to which the River Meta is adapted, has raised and exposed the Tertiary sediments; these are not favourable to dune formation.

In this extensive area, all the sandy material has been reworked by the wind and shaped into longitudinal dunes, orientated NE–SW, a trend seen more clearly on LANDSAT imagery than in the field. Some finer accumulations, in the east, are characterised by lower dunes. Many dunes form loops open to the south-west and are of parabolic type. This feature demonstrates that the winds responsible for dune construction were the Trade winds and that the transportation of sand was only local. Along the streams at the foot of the Andes dunes are frequently more developed on the south-west bank of the river than on the opposite bank, thus providing further evidence for reconstructing the wind direction. The orientation of the dune axis changes gradually at the foot of the Andes, passing from NE–SW to NNE–SSW westwards, like the direction of the chain itself.

Dunes are shown to have been formed with the alluvial material of the lower terrace of the Andean streams on the piedmont zone of the mountain chain. In the mountains themselves, the same low terrace was built during the last cold period, as evidenced by its relationships with periglacial slope deposits and moraines. This last dry period, as in eastern central Amazonia, coincided with the Würm (Wisconsin) last maximum and the last glacio-eustatic low sea level.

During the Holocene, the climate became warmer and more humid. The Andean streams have been able to reorganise their drainage networks to provide once again an exoreic network towards the Atlantic Ocean. They have succeeded in cutting their way through the dune fields. As a result of the Flandrian transgression and of the higher base level, part of the dune fields has been fossilised by alluvium and is presently flooded during the high-water season.

South of the Meta River, around San Martin and Puerto Lleras, Neogene sediments consist of indurated sands, soft ferruginous sandstones and sandy clays. These are too cohesive for the construction of dunes but, being impervious, are favourable to runoff when the vegetal cover is sufficiently sparse and open. At present, (July 1973), the soil has an approximately 75% cover of shrub savanna and is used for ranching. Runoff is active and cuts small gullies associated with piping. These processes dissect extensive pediments. Hillslopes are covered by a layer of debris, up to 0.3 m thick, consisting of fragments of iron duricrust, sandstone and reworked quartz pebbles. Intense runoff has washed the fine fraction of the sediments from these slopes before eroding the pediments. These hills and pediments result from the dissection of the area which has taken place after the accumulation of the third terrace above the valley bottoms. It can be considered as embracing the time span of the Middle and Upper Quaternary.

10.4 Conclusion

It can now be clearly recognised how much we were mistaken some 20 years ago. South America north of the Tropic of Capricorn suffered important climatic changes during the Quaternary. To some extent, the consequences of these changes can be compared to those which affected Africa. In both continents, between the Tropics, a high proportion of landforms are inherited or relict. The areas where morphogenesis has been polygenetic are largely predominant. Landforms have been shaped successively under quite different climatological regimes and are simultaneously polygenetic and polychronic. Areas where the same bioclimatological environment has persisted during a long period, like the northern slope of the Gran Sabana of Venezuela, are not common and have a restricted spatial extent.

One must be very cautious when trying to define the specific landforms of the present climatic zones. Only accurate studies can enable us to define that part of the landscape which is presently active and that which is inherited from former conditions that were, usually, quite different. Much confusion has resulted from insufficient care being taken over this evaluation. The problem is even more complicated as a consequence of the feedback of landforms upon the processes which model them; through inherited landforms, past climates still influence present processes.

As has already been proposed, climatic influence is mostly an indirect one, strongly dependent on the vegetal cover which is, in part, climate-dependent. This is why, in this chapter, available information on vegetational changes has been applied to, and correlated with, palaeoclimatic geomorphology.

A general summary can be presented. During the last cold period, most of South America north of the Tropic of Capricorn was affected by climates drier than those of the present, when, on the contrary, lower temperatures ought to have made water budgets more favourable. This means that evaporation from the oceans was lower, indeed much lower according to the laws of physics, and that the patterns of atmospheric circulation and meteorological conditions were quite different from those at present. Some landforms, like dunes, can provide clues for their reconstruction.

11
Pleistocene aridity in tropical Africa, Australia and Asia

M. A. J. Williams

11.1 The humid tropics: stable or dynamic?

For long regarded as relatively unmodified Tertiary relicts, the rainforests of the tropical lowlands have only recently begun to reveal their chequered history. The diversity and extent of rainforests led many workers until the 1970s to equate diversity with longevity and present extent with past distribution. If the tropical lowland forests were indeed a stable ecosystem, then they would be one of the few such features in the world at the present time, with important consequences for process studies and geomorphic theory.

The appealing notion that the humid tropics were but little influenced by the events of the Quaternary had been challenged by a number of biologists working in Africa and South America before 1970. Moreau (1963) noted the similarities in the avifauna of the now widely separated montane forests of Ethiopia, Kenya, Uganda and the Cameroon uplands. He concluded that there must have been a recent connection between these now isolated forests and suggested that lower temperatures during the last glacial maximum may have depressed the montane forest belts sufficiently to establish such a link. A difficulty with this particular interpretation is that the last glacial appears to have been significantly drier as well as cooler in the highlands of central and east Africa, with expansion of grassland at the expense of the upland forests (Livingstone 1975, 1980).

In tropical Africa, it was De Heinzelin (1952) and De Ploey (1963, 1965) who first argued for intervals of Pleistocene aridity in the Zaire basin and who thus demonstrated the degree to which even the lowland tropics had been influenced by Quaternary climatic fluctuations. The geomorphic and other evidence of late Pleistocene aridity ran counter to the well entrenched notion that pluvial climates in the tropics corresponded with glacial climates at higher latitudes. One of the reasons for this protracted debate is the inherently ambiguous character of some of the evidence. Was a tropical lake high as a result of more runoff into the lake, or more precipitation onto the lake, or less evaporation from the lake, or an increase in groundwater inflow? Was fluviatile deposition of silts and clays a response to a change in river regime from braiding to meandering as a consequence of more regular rainfall or was it a result of an influx of wind-blown dust? Did the equatorial grasslands in the Nile headwaters expand in the late Pleistocene because it was cooler, or because it was drier?

The problem is to distinguish the geomorphic and ecological effects of temperature change from those of precipitation. Figure 11.1 illustrates the dilemma nicely. If the curve of inferred late Pleistocene rainfall fluctuations at Lynch's Crater in tropical northeastern Queensland (Fig. 2.5) is correct, why does it parallel the palaeotemperature curves from the three deep-sea cores? Is there some causal connection between lower equatorial oceanic temperatures and intertropical aridity? The aim of this chapter is to suggest that there is, and that glacial maxima during the Quaternary were reflected in cooler, drier and windier climates throughout much of the intertropical zone.

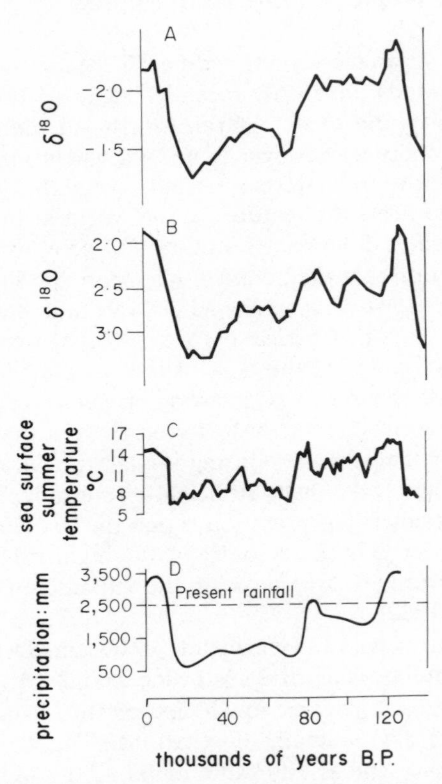

Figure 11.1 The correspondence between low temperature and low rainfall in the tropics during the Upper Quaternary is illustrated in this diagram, adapted from Kershaw (1978). Curve A is the oxygen isotope curve from core V28-238 in the equatorial Pacific (Shackleton & Opdyke 1976). Curve B is from core RC11-120 in the southern Indian Ocean (Hays *et al.* 1976). Curve C is deduced from microfossils in core V23-82 in the North Atlantic (Sancetta *et al.* 1972). Curve D is the precipitation inferred from pollen recovered by Kershaw from Lynch's Crater in tropical north-east Queensland, Australia.

11.2 The duration and amplitude of the Quaternary climatic fluctuations

The late Cenozoic build-up of ice in Antarctica and in the Northern Hemisphere caused changes in the oxygen-isotope composition of the oceans. With the removal of water rich in ^{16}O and its transfer to the growing ice-caps, the oceans became progressively depleted of the lighter oxygen isotope. Hence the increase in ^{18}O in ocean water in the late Tertiary which is a measure of the total volume of ocean water locked up in the form of ice. Specifically the fluctuations in the $^{16}O/^{18}O$ ratio of the calcareous tests of marine foraminifera reflect global changes in ice volume, as well as more local changes in ocean temperature and salinity. By identifying ice-volume maxima and minima in the deep-sea core isotopic record, it is possible to identify the Plio–Pleistocene glacial–interglacial cycles. Using evidence from deep-sea cores and from raised marine terraces in the Caribbean, Broecker and Van Donk (1970) have argued that the primary glacial cycle was 'sawtoothed'; gradual build-ups over periods averaging 90 000 years in length being terminated by deglaciations completed in less than one-tenth of this time. Modulating this primary cycle are secondary oscillations. The glacio-eustatic sea-level curve reconstructed from dated flights of uplifted coral reefs in the Huon peninsula of New Guinea confirms this suggestion, particularly for the past 200 000 years, after which the dates are unreliable (Chappell 1978).

Detailed analyses of well preserved deposits of fossil-bearing stratified loess in central Europe have demonstrated that there were at least 17 glacial–interglacial cycles during the last 1.7 million years since the Olduvai Event (Fink & Kukla 1977). The most recent attempt at a Quaternary timescale places the base of the Quaternary at the top of the Olduvai Event (Berggren *et al.* 1980) and microfossil first-and-last-appearance data in deep-sea cores suggest biostratigraphic subdivisions with a mean duration of about 100 000 years. The marine and continental records thus seem in fair agreement and recall the 100 000 year orbital cycle first recognised by Milankovitch (Broecker *et al.* 1968, Veeh & Chappell 1970, Chappell 1973, Hays *et al.* 1976, Kerr 1981). There are, however, serious discrepancies between the ice-volume changes reflected in the New Guinea and other middle and late Pleistocene sea-level curves, and those apparently depicted in the oxygen-isotope curves from deep-sea cores of similar age. Until the latter have been correctly calibrated, it is best to regard the sea-level curves as a truer measure of global ice-volume change, taking due account of dating uncertainties (J. Chappell pers. comm. 1981, P. Aharon pers. comm. 1982).

To sum up thus far, there were at least 17 glacial–interglacial cycles since the palaeomagnetically defined base of the Quaternary, each of about 100 000 years' duration (Fink & Kukla 1977, Berggren *et al.* 1980). Each interglacial only lasted about 10 000 years (Broecker & Van Donk 1970), and, during the last 400 000 years at least, sea surface temperatures in the tropics were as warm as, or warmer than, today for only one-tenth of that time (Emiliani 1972).

Before exploring the links between low equatorial sea surface temperatures and intertropical aridity ('glacial aridity'), it is worth noting the speed with which certain palaeoclimatic events took place.

During the past 0.7 million years sudden coolings with a probable intensity of 5°C/50 years and a duration of 100 years occurred about once every 10^3 to 10^4 years (Flohn 1979). Kukla (1980) has analysed data from northern high-latitude ice cores, pollen spectra and marine cores, and concluded that the onset of sudden cooling (marking the change from stable non-glacial to immediate pre-glacial climates) can be relatively rapid, perhaps only 2×10^3 years. The cause of these abrupt changes in climate is not yet clear. A possible factor may be changes in sea ice and snow cover, which will strongly influence the global albedo (Kukla 1975, Kukla *et al.* 1977). The problem here is to disentangle the web of cause and effect, a familiar dilemma.

11.3 Intertropical ice-age aridity

Nilsson's pioneering studies of glacial deposits and former lake strandlines in Kenya and Ethiopia aroused interest in Quaternary climatic fluctuations in East Africa (Nilsson 1931, 1940). The four Alpine glaciations proposed by Penck and Brückner (1909) appeared to be matched by the four major glaciations recognised in North America, the last of which was known to correspond to the last major interval of high lake levels in the western United States (Flint 1971 pp. 443–4). If high- and middle-latitude glaciations were synchronous with times of middle-latitude high lake levels, there seemed no good reason why times of high tropical lake levels should not also have been in phase with intervals of tropical and temperate glaciation. This view was tacitly accepted by many prehistorians working in Africa until effectively challenged in the late 1950s by Cooke (1958) and Flint (1959a,b).

Once shells and charcoal from a number of high lake shorelines had been dated, it soon became clear that the 'late Pleistocene' Gamblian Pluvial was of early Holocene age, and that a number of tropical African lakes had risen from very low levels in the late Pleistocene to exceptionally high levels in the early Holocene (Faure 1966, Kendall 1969, Butzer *et al.* 1972). Later studies confirmed the early Holocene rise in lake levels along the southern margin of the Sahara from the Atlantic to the Red Sea (Chamard 1973, Clark *et al.* 1973, Gasse *et al.* 1974, Servant 1974), as well as further afield in Rajasthan and northern Queensland (Kershaw 1970, 1971, Singh *et al.* 1972, 1974).

By 1971 it was clear that the climate had been very dry south of the Sahara roughly 20 000 years ago (Burke *et al.* 1971b) and by 1975 it was reasonable to argue that this late Pleistocene phase of tropical aridity had been synchronous in the two hemispheres. Independent studies from South America, the Caribbean and the Middle East, and cores taken from the tropical Atlantic and Pacific, all pointed to a marked reduction in rainfall throughout the humid tropics during the last glacial maximum (Williams 1975).

Ice-age aridity was not unique to the last of the 17 or more Quaternary glacial–interglacial cycles. Variations in the proportions of Saharan quartz particles blown into the equatorial North Atlantic during the past 600 000 years show that high quartz inputs correlate with cold periods and low quartz inputs with warm periods (Bowles 1975). The same relationship holds good for wind-blown dust of organic origin, such as opal phytoliths and fossil freshwater diatom frustules blown from exposed lake floors (Parmenter & Folger 1974).

The oxygen-isotope composition of planktonic foraminifera collected from the Red Sea and Gulf of Aden also shows that, over the past 250 000 years, glacial maxima coincided with times of extreme aridity, and of high salinity, possibly related to unusually high evaporation from the Red Sea (Deuser *et al.* 1976). Runoff into the Red Sea may also have been lower during glacial maxima, much as in the eastern Mediterranean during times of much diminished late Pleistocene Nile flow (Williams & Adamson 1980, Adamson *et al.* 1980, Rossignol-Strick *et al.* 1982).

In the Indian Ocean the salinity gradient was steeper during the last glacial maximum, as it was over the northern Arabian Ocean. A combination of reduced freshwater inflow from the Ganges, Brahmaputra, Indus and Irrawaddy, and higher evaporation over the Arabian Sea, seems adequate to account for this pattern of glacial sea surface salinity (Duplessy 1982) – particularly as glacial sea surface temperatures over the Indian Ocean apparently were not much lower than those of today (Prell *et al.* 1980).

Sea surface temperatures seem also to have been slightly lower around Australia during the last glacial maximum (McIntyre *et al.* 1976), but not by the amount postulated by Webster and Streten (1972, 1978) to account for a reduction in cyclone incidence over tropical northern Australia during the late Pleistocene.

We are faced here with a paradox. If lower glacial sea surface temperatures produced less rainfall in the intertropical zone, as suggested by Flohn (1953) and later workers (Galloway 1965a, Fairbridge 1970, Flohn & Nicholson 1980), how do we reconcile a reduction in intertropical evaporation with the very compelling evidence now emerging of more vigorous atmospheric circulation during the last glacial? The investigations of Parkin and Shackleton (1973), Parkin (1974), Diester-Haass (1976), Kolla *et al.* (1979) and Sarnthein *et al.* (1981) all indicate stronger trade winds during the last glacial maximum.

In the Southern Hemisphere, Petit *et al.* (1981) have shown that there was a stronger meridional circulation over Antarctica during the late Pleistocene. East Antarctica dome C ice core spans the last 32 000 years. In this core, marine and continental aerosol inputs at the end of the last glacial (*c.* 20 000 BP) were respectively five and 20 times higher than at present (Petit *et al.* 1981). That the sea salt input into central Antarctica was far higher, despite the greater extent of sea ice during glacial time, points to stronger winds. If, as Sarnthein (1978) suggests, the tropical arid zones were five times larger towards 18 000–20 000 BP, then wind speeds 1.3 to 1.6 times higher than the present would suffice to increase the continental dust load over central Antarctica by the required amount (Petit *et al.* 1981). If Sarnthein's map is an overestimate of the glacial arid zone, even stronger winds would be needed to blow 10–20 times more dust into central Antarctica from Australia and elsewhere around 20 000 BP.

Another test of glacial wind velocities is the particle size of aeolian dust in deep-sea cores (Parkin 1974). Sarnthein *et al.* (1981) have used a modified version of Parkin's method in their analysis of glacial and interglacial wind regimes over Saharan Africa and the adjacent Atlantic, and have concluded that the Trade winds at 18 000 BP were much stronger than today – with wind speeds of 20 $m\,s^{-1}$, in contrast to the weakened Harmattan which today brings West African dust to Cayenne in South America (Prospero *et al.* 1981).

The dilemma of the association of stronger glacial trade winds and reduced intertropical rainfall may be resolved by assuming lower rates of evaporation from the tropical seas (though the reverse seems true of the Red Sea and northern Arabian Sea): but a more plausible assumption would be weaker summer monsoonal rain systems over Northern Australia (Rognon & Williams 1977), West Africa (Maley 1981) and India (Duplessy 1982). The evidence is the active linear dunes that developed well beyond their present limits during the last interval of glacially lowered sea level in northwestern Australia (Jennings 1975), Senegal (Michel 1973) and Rajasthan (Goudie *et al.* 1973). The cause may have been persistent summer anticyclones over subtropical continents and a greater influx of tropospheric dry and initially cold air from ice-covered high latitudes that blocked the summer monsoonal rains (Maley 1976, Rognon & Williams 1977, Nicholson & Flohn 1980).

11.4 The evidence of Quaternary climatic change

It is over 30 years since Faegri remarked that 'palaeoclimatology suffers from the disadvantage that those who can judge the evidence, e.g. biologists and geologists, cannot judge the conclusions, and meteorologists who can judge the conclusions cannot judge the evidence' (Faegri 1950 p. 194). With certain distinguished exceptions, this comment seems as true today as it was in 1950.

Table 11.1 lists some of the main lines of evidence used in palaeoclimatic reconstruction. Each, in isolation, has inherent limitations.

Flint (1976) has appraised some of the physical evidence (IA in Table 11.1). To his *caveats* should be added the caution that the horizons of certain duplex or texture-contrast soils may be depositional, not pedogenic. Furthermore, laterite is not peculiar to the tropics; calcrete is not diagnostic of aridity; and podsols can form under a variety of bioclimatic conditions, sometimes very rapidly. Similar palaeoclimatic limitations apply to most soils. In the case of vertisols, the parent material may range from stratified alluvial clays as in semi-arid central Sudan to deeply weathered basalts as in the humid uplands of Ethiopia, and even to wind-blown dust as in the Chad basin. In such circumstances the influence of parent material can far outweigh that of climate.

Table 11.1 Sources of evidence for palaeoclimatic reconstruction.

I. Geology and geomorphology
  A. Continental
    (1) rivers
    (2) lakes
    (3) dunes, sand plains
    (4) loess, desert dust, cyclic salt
    (5) soils
    (6) evaporites, tufas
    (7) glacial and periglacial deposits
  B. Marine
    (1) continental dust
    (2) fluviatile inputs, beaches, shorelines
    (3) biogenic dust: pollen, diatoms, phytoliths

II. Biology and biogeography
  A. Continental
    (1) plant and animal distributions
    (2) fossil pollen and spores
    (3) plant macrofossils
    (4) diatoms
    (5) vertebrate fossils
    (6) invertebrate fossils: mollusca, ostracods
  B. Marine
    (1) diatoms
    (2) foraminifera
    (3) coral reefs

III. Archaeology
    (1) plant remains
    (2) animal remains, including hominids
    (3) artefacts: bone, stone, wood, shell, leather
    (4) rock art
    (5) hearths, dwellings, workshops

IV. Isotope geochemistry
    (1) oxygen isotopes (ice, marine foraminifera, carbonate)
    (2) carbon isotopes (plants, bones, carbonate)
    (3) hydrogen isotopes (ground water, peat)

Inferring climate from alluvium is equally hazardous. River deposition is often strongly time-transgressive and may depend upon local geomorphic thresholds. To complicate interpretation further, charcoal in point bars may be considerably older than the time of deposition of the alluvium. For these and other reasons simple frequency analysis of radiocarbon dates obtained from alluvial deposits cannot be considered an accurate palaeoclimatic tool. Alluvial history inferred from detailed stratigraphic mapping is not the same as that deduced from numbers of dates.

Palaeohydrologic inferences based on lake-level fluctuations may likewise prove chimerical, for reasons noted earlier. Some lakes are insensitive to climatic change, particularly if sustained by hot springs or by ground water.

Only one major Holocene transgression is evident at Lake Besaka in Ethiopia, for instance (Williams *et al.* 1981), but four Holocene transgressions are discernible in other lakes in the same region which depend more upon inputs from runoff (Gasse & Street 1978). Runoff-fed lakes in closed basins in which the catchment is large relative to lake size will tend to amplify climatic oscillations. Groundwater-fed lakes and lakes situated along major rivers may be relatively insensitive even to long-term changes (Street 1980). Depending upon the time-scale used, the Holocene lake-level curves from East Africa may seem broadly synchronous or individually quite variable. Dating uncertainties caused by too few samples, or by sample quality (e.g. partial recrystallisation of shell) impose additional constraints upon interpretation.

The orientation of desert dunes may provide information on former wind directions (Warren 1970, 1972, Jennings 1975, Bowler 1976, Fryberger 1980), but such information is not easy to interpret. We have long known that desert dunes were active far beyond their present limits in Africa, India–Pakistan and Australia. Does such activity imply less rainfall and an equatorward shift of the sand moving winds, as suggested by Grove and Warren (1968) and Diester-Haass (1976) for North Africa? Or does it imply stronger trade winds, with no latitudinal displacement of the Saharan high-pressure belt and its associated wind systems (Talbot 1980, Fryberger 1980, Sarnthein *et al.* 1981)? Is a combination of greater aridity, stronger winds, less vegetation and a pre-existing sand supply all that is needed to activate dunes as near to the equator as latitude 9°30′N in western Sudan and 17°S in northwestern Australia? In the case of the source-bordering dunes of Australia, Sudan and, presumably, of Zaire and Amazonia, the three prerequisites for dune formation seem to be a seasonally abundant supply of river sand, strong unidirectional seasonal winds, and a less dense tree cover than today. Such dunes do not connote extreme aridity.

Aeolian dust mantles are potentially informative, and can be dated by thermoluminescence, as can sand dunes (Singhvi *et al.* 1982). The sterile argument over whether desert dust is loess is not relevant here. Other forms of evidence are perhaps more questionable. For example Livingstone (1980) has shown the limitations of pollen analysis in tropical Africa, and Harlan and de Wet (1973) have critically evaluated the evidence for the origin and dispersal of cultivated plants. Furthermore, Shackleton and Kennett (1975) have commented on some of the precautions needed in using the isotopic composition of fossil marine foraminifera to infer past ocean temperatures.

Criticism can be levelled at any of the sources of evidence listed in Table 11.1. However, each one is useful at a given scale in time and place, and should be used accordingly. When the conclusions drawn from many lines of evidence converge, confidence is increased. Nevertheless, as Harlan and de Wet (1973 p. 54) point out: 'the glaring item of evidence that does not integrate with other sources of evidence may prove to be the most important clue of all'.

Within the limitations we will now assemble some of the evidence already discussed and attempt a reconstruction of glacial and interglacial climates in

India, tropical Africa and Australia, concentrating upon the last glacial maximum (25 000–17 000 BP) and the early Holocene (*c.* 11 000–7000 BP). The amplitude of temperature and rainfall fluctuations between these two intervals is probably comparable to that within any of the 17 or so glacial–interglacial cycles of the past 1.7–1.8 million years of Quaternary time, with the added advantage that the evidence is reasonably well dated.

11.5 The late Quaternary climates

India

The best record of late Quaternary climatic events in India is that provided by the surrounding Indian Ocean (Prell *et al.* 1980, Duplessy 1982). More limited data from on land include the Holocene pollen spectra of some of the Rajasthan lakes (Singh *et al.* 1972, 1974), inadequately dated geomorphic and archaeological reconnaissance studies in Gujarat and Rajasthan (Allchin *et al.* 1978), and a number of river terrace studies, few with adequate time control.

Part of the problem has been the historical connection between archaeological excavations and investigations of alluvial history (Paterson & Drummond 1962, Sankalia 1978). It is not always evident that the Palaeolithic sites are in primary context (Paddaya 1978), and there has been an undue preoccupation with fossil bones and stone artefacts found within river gravels (e.g. Corvinus 1969, Sankalia 1974). All too often alluvial deposits are dated on the basis of associated archaeological occurrences, with attendant dangers of circularity. There are still too few well dated late Quaternary fluviatile sequences to allow more than very general conclusions to be drawn about past changes in river flow and sediment yield, although recent work in the Son and Belan valleys offers hope for the future.

Duplessy (1982) has used differences in the oxygen-isotope composition of planktonic foraminifera from the Indian Ocean to reconstruct the probable climates of the late Pleistocene and Holocene.

Towards 18 000 BP the salinity gradient in the Bay of Bengal was very much steeper than today, reflecting a drastic reduction in freshwater input from the Ganges and Brahmaputra. This inference is consistent with that of Williams and Royce (1982), who argue from sedimentary evidence that the Son was a more seasonal river at this time, with a sparsely vegetated catchment in the now wooded uplands of north central India.

The upwelling that is now a feature of the southern coast of Arabia, and a contributor to aridity inland, had disappeared at this time, indicating that the south west summer monsoon winds were not particularly strong during the last glacial maximum. A weakened south west monsoon and much reduced summer rainfall would account for the aridity evident in north west India during the last glacial maximum.

In contrast to the weak summer monsoon, the north west monsoon seems to have been stronger during the last glacial maximum. Evidence includes the clockwise circulation pattern in the Bay of Bengal, revealed by a tongue of low-salinity water, and widespread loess deposition in China and northern India.

The combined evidence from rivers, lakes, dunes, loess and pollen spectra accords well with Duplessy's reconstruction of a drier, windier climate towards 18 000 BP over much of India, with less rain in summer and a stronger winter monsoon than today.

During the early Holocene the marine isotopic record indicates a reversal of the late Pleistocene pattern of weak summer and strong winter winds (Duplessy 1982). By 15 000 BP the ice had disappeared below 5000 m in the north-west Himalayas (Singh & Agrawal 1976). Early Holocene cold upwelling water accentuated the aridity of the south coast of Arabia in response to the stronger summer winds. The south-west monsoon blew vigorously, bringing summer rain to swell the flood and fill the lakes of India. Runoff from the Indus, Ganges and Brahmaputra increased, the rivers of central India became less seasonal, and loess deposition ceased.

The lakes of Rajasthan began to fill shortly before 10 000 BP. With minor fluctuations they remained full and fresh until about 4000–3000 BP, after which they became saline (Singh 1971, Singh *et al.* 1972, 1974).

The early to middle Holocene climate of India was generally wet and warm, with heavy monsoonal rain in summer and moderate rain in winter. It was at about this time that the early Neolithic populations of the Ganges and Indus Valley began to cultivate rice and wheat, as well as to domesticate cattle, sheep, goats and water buffalo (Sankalia 1974, Sharma *et al.* 1980).

During the past 4000 years a number of the Rajasthan dunes were again active. The late Holocene desiccation of northern India was also evident in tropical Africa, causing a massive exodus of Neolithic herders from the Sahara, and dune reactivation in southern Ghana (Smith 1980a,b, Talbot 1981).

Tropical Africa

Africa has by far the most complete and best dated late Quaternary record of anywhere in the tropics. Every source of evidence listed in Table 11.1 has been used (with varying degrees of success) in recent attempts to reconstruct the late Pleistocene and Holocene climates of tropical Africa. Roughly equal attention has been given to geology, biology and prehistoric archaeology as sources of evidence and stable isotope analysis has also yielded useful insights into climatic change in and around Africa, particularly in relation to the CLIMAP and DSDP projects.

Comprehensive summaries of late Quaternary lake fluctuations throughout Africa are given by Street and Grove (1976, 1979), and much of the recent work on northern Africa, including the equatorial headwaters of the Nile, is summarised in *The Sahara and the Nile* (Williams & Faure 1980).

The information in Tables 11.2 and 11.3 is a selective summary of the late Pleistocene to early Holocene climatic history of intertropical Africa, with

Table 11.2 Climates of tropical Africa between 25 000 and 17 000 BP.

I. Nile basin

A. Headwaters

(1) Firnline 600 to over 1000 m lower on Ruwenzori, Kilimanjaro and Mt Kenya. Treeline roughly 1000 m lower. Temperature probably 4–8°C colder than today in East African uplands (~White Nile headwaters) (Livingstone 1980, Messerli & Winiger 1980)

(2) Firnline 1000 m lower on Bale Mts, Ethiopia. Treeline 800–1200 m lower in Semien and Bale Mts. Temperatures 4–8°C colder than today in Ethiopian highlands (~Blue Nile headwaters) (Williams *et al.* 1978, Messerli & Winiger 1980)

(3) Lake Mobutu (Albert) no outlet between 25 000 and 18 000 BP. Lake Victoria closed from at least 14 500 to 12 500 BP. Flow of White Nile severely curtailed. Lake Manyara (N. Tanzania) high towards 17 000 BP (Kendall 1969, Livingstone 1980)

(4) Widespread montane grassland and heath in uplands of Ethiopia, Kenya and Uganda (Hamilton 1976, Gasse *et al.* 1980, Livingstone 1980)

Climate colder and drier than today

B. Sudan

(1) White and Blue Nile highly seasonal, with high peak flows but reduced annual discharge (Williams & Adamson 1980, Adamson *et al.* 1980)

(2) Dunes active in central Sudan. Source-bordering dunes formed downwind of Blue Nile distributary channels. Aeolian reworking of sandy alluvium in Kordofan and Darfur (Warren 1970, Williams *et al.* 1980, 1982)

(3) Machar Marshes and Sudd swamps of S. Sudan probably dry (Adamson *et al.* 1982)

Climate hot and dry in summer, with much shorter wet season than today. Winters cold and windy. The 250 mm isohyet probably at least 500 km further south than today, and winds stronger

C. Upper Egypt

(1) Main Nile highly seasonal, with high peak flows but reduced annual discharge (Adamson *et al.* 1980)

(2) Runoff from Red Sea Hills into valleys east of Nile. Valleys west of Nile dry (Butzer 1980b, Adamson 1982)

Climate cooler in winter, but very hot, dry and windy for much of the year

II. Chad basin

(1) Lake levels high until about 20 000 BP, very low thereafter, with many smaller lakes dry. Jebel Marra lakes high before 17 000, but low towards 17 000 BP (Servant & Servant-Vildary 1980, Williams *et al.* 1980)

(2) Temperate diatom flora in lakes between 26 000 and 20 000 BP. Tropical associations towards 20 000–18 000 BP (Servant-Vildary 1978)

(3) Progressive increase in Sahelian vegetation elements from 13 000 BP onwards (Maley 1981)

(4) Increase in aeolian dust blown into Atlantic from Sahara and Chad basin between 25 000 and 18 000 BP. Significant accumulation of aeolian clays in humid tropics south of Chad Basin after 15 000 BP (Maley 1981, Sarnthein *et al.* 1981)

Climate hot, dry and windy in summer, with shorter wet season. Winters cold and usually fairly dry, especially after 20 000 BP, when fewer incursions of cold polar air and associated heavy precipitation

Table continued

Table 11.2 *(Continued)*

III. Other localities

(1) Senegal and Niger rivers reduced annual discharge, with aeolian reworking of sandy alluvium in lower Senegal and middle Niger (Michel 1973, Talbot 1980)

(2) Minimal recharge of S. Saharan aquifers between 19 000 and 15 000 BP (no samples of that age out of 77 dated groundwater samples) (Sonntag *et al.* 1980)

(3) Minimal deposition of alluvial and organic sediments in Sierra Leone between 20 000 and 12 500 BP (Thomas & Thorp 1980)

Climate dry and windy, with shorter wet season(s)

Table 11.3 Climates of tropical Africa between 12 000 and 7000 BP.

I. Nile basin

A. Headwaters

(1) Treeline at least as high as today in uplands of Ethiopia, Uganda and Kenya (Hamilton 1976, Livingstone 1980)

(2) Slopes stable and pedogenesis active in Blue and White Nile headwaters (Williams & Adamson 1980)

(3) Lakes high in Ethiopian and Afar Rifts towards 12 000–11 000, 10 000, 9400–8400 and 7000–6500 BP. Diatom flora show that lakes were warm, oligohaline and very productive (high rates of frustule deposition) (Gasse *et al.* 1980)

(4) L. Victoria and L. Mobutu (Albert) overflow after 12 500 BP. Brief closure of L. Victoria towards 10 000 BP (Kendall 1969, Livingstone 1980)

(5) Upward shift of alpine grassland and heath into areas formerly glaciated or prone to intense periglacial freeze–thaw activity (Hamilton 1976, Van Zinderen Bakker 1976, Messerli & Winiger 1980)

Climate warmer, wetter and less seasonal than today.

B. Sudan

(1) Very high Blue and White Nile floods between 12 000 and 11 000 BP. White Nile floods again high towards 8400–8100 BP and 7000 BP. Blue Nile floods high towards 7500 and 6900 BP (Williams & Adamson 1980, Adamson *et al.* 1980)

(2) Dunes of central Sudan partly submerged beneath Blue and White Nile alluvial clays. Dunes of Kordofan and Darfur fixed by vegetation

(3) Interdune lakes west of Khartoum high between >8400 and 6800 BP. Kordofan and Darfur lakes high between 9300 and 6000 BP (Williams *et al.* 1974, Haynes *et al.* 1979, Williams & Adamson, 1980)

Climate warm to hot, somewhat wetter than today, with a longer summer wet season, and some winter rains in far north

C. Upper Egypt

(1) Exceptionally high Nile floods towards 12 000–11 500 BP (Butzer 1980b, Rossignol-Strick *et al.* 1982)

(2) Playa lakes full west of Nile towards 9500–9000 BP and 7000–5800 BP. Neolithic cultivation of barley and wheat west of the Nile and, about 1000 years later, in the Nile valley itself (Wendorf & Schild 1980)

(3) Wadis flowing from Red Sea hills during winter. Seasonal interdigitation of wadi and Nile sediments (Butzer 1980b)

Climate milder and moister than today (i.e. semi-arid and not arid), with rain in winter and possibly in summer also

Table continued

Table 11.3 *(Continued)*

II. Chad basin

(1) Lake levels high towards 11 000 BP and very high towards 9000–8000 BP. Levels low or lakes dry towards 7500 BP (Servant & Servant-Vildary 1980)

(2) Temperate diatom flora in lakes between 12 000 and 7500 BP. Tropical associations after 7000 BP (Servant-Vildary 1978)

(3) Tropical woodland and savanna in present Sahel between 8000 and 6500 BP (Maley 1977, 1981)

Climate mild and wet. Frequent incursions of cold northern air, interrupted by warm dry intervals. Rainfall less seasonal and less intense than today, with tropical depressions bringing fine, gentle rain

III. Other localities

(1) High runoff from the Senegal between about 11 000 and 8000 BP. Very high Niger discharge from 13 000 to 11 800 BP and again from 11 500 to 4500 BP (Michel 1973, Pastouret *et al.* 1978)

(2) Lake Bosumtwi (S. Ghana) high from *c.*12 500 to 11 000 BP, very high between *c.*10 000 and 8300 BP, and again high from 7500 to 5000 BP, with short, intense regressions around 10 500 and 8000 BP (Talbot & Delibrias 1980)

(3) Groundwater recharge of S. Saharan aquifers after 12 000 BP. Recharge of Kufra and Sirte aquifers between *c.*8000 and 5000 BP by tropical rains with lighter isotopic composition than present-day winter rains. Tropical herbivores in south-central Algeria and Libya (Conrad 1969, Edmunds & Wright 1979, Sonntag *et al.* 1980)

(4) Deposition of coarse basal gravels and alluvial sands between 12 300 and *c.*3800 BP in Sierra Leone. Higher stream runoff and higher rainfall (Thomas & Thorp 1980)

Climate wetter along southern margins of Sahara, with longer summer wet season and frequent tropical depressions

special emphasis being given to events in and around the Nile and Chad basins. Events in other parts of tropical Africa are less well dated and are based upon fewer independent lines of evidence.

The evidence tabulated above is consistent with a colder, windier and generally drier climate than today between about 25 000 and 17 000 BP. In contrast to this, the early Holocene climate was wetter than today, temperatures were generally quite high, and rainfall was more evenly distributed throughout the year. Is is probably fair to assume that the inferred 25 000–17 000 BP climate is broadly representative of full glacial conditions in tropical Africa during the Quaternary, and that the 12 000–7000 BP climate is reasonably representative of interglacial times. The truth of this statement needs to be tested by future work.

Tropical Australia

Until the advent of glacial aridity towards 25 000 BP, the southern half of Australia was a land of lakes (Barbetti & Allen 1972, Bowler *et al.* 1972, Killigrew & Gilkes 1974). By about 17 000–15 000 BP, the majority of these inland lakes were dry (Bowler 1975, 1976).

Northeastern Australia was significantly less humid than today well before 25 000 BP. The long pollen record from Lynch's crater may span the last interglacial–glacial cycle (Kershaw 1978). If the dates and interpretation are correct, the pollen spectra from Lynch's Crater indicate a drastically reduced rainfall (from 2500 mm to 500 mm yr^{-1}) in the interval between about 80 000 and 20 000 BP. This long interval of aridity in tropical north-east Queensland probably reflects the impact of generally lower sea levels throughout the 60 000 or so years preceding the glacial maximum (Chappell 1976, 1978). This would have led to increased continentality and the probable closure of Torres Strait to equatorial ocean currents.

Except for the equivocal evidence of fluviatile deposits (Wyrwoll 1979), we have no reliable palaeoclimatic data for northwestern Australia for this time. If, as seems likely from the 50 000–25 000 BP glacial and pollen record of Chile, South Africa, New Zealand, Tasmania and the Southern Ocean, the build-up towards maximum glaciation resulted in large measure from a northward advance of the Antarctic Polar Front (Heusser 1981), then the cold West Australian Current may well have had a greater desiccating influence upon north-west Australia than it does today. The sea would in any event have been at least 30–50 m lower at that time.

The combined evidence from marine cores (Van Andel *et al.* 1967, McIntyre *et al.* 1976, Prell *et al.* 1980), pollen analysis (Kershaw 1981, Walker & Singh 1982, Singh 1982), lakes, dunes and lunettes (Bowler *et al.* 1976, Bowler 1978), vertebrate fossils (Baynes *et al.* 1975) and periglacial deposits (Galloway 1965b) strongly suggests that during the last glacial maximum (25 000–18 000 BP) much of Australia was drier and more windy than today. Winters were colder, but inland summers were probably quite warm.

With the sea up to 150 m lower, Australia was at its most continental. Broad land bridges linked New Guinea to northern Australia, and much of what is now the Arafura Sea and the Gulf of Carpentaria was then dry land. Sea surface temperatures in the oceans west, north and east of Australia were somewhat lower, although data points are few and marine sedimentation rates were often low (McIntyre *et al.* 1976, Shackleton & Opdyke 1976, Prell *et al.* 1980, Webster & Streten 1981). Empirical evidence and theoretical argument both point to a shorter summer wet season between 25 000 and 18 000 BP (Bowler *et al.* 1976, Rognon & Williams 1977, Webster & Streten 1978).

After 18 000 BP the sea rose, rapidly at first, then more slowly, reaching present level towards 6000 BP (Thom & Chappell 1975). The previously exposed Great Barrier Reef became submerged beneath the rising sea, and the land link with New Guinea was severed after 11 000 BP (Jennings 1971).

Whether the early Holocene expansion of rainforest in northern Queensland was largely a result of sudden Barrier Reef flooding bringing more rain locally to the east coast or whether rainfall was generally higher in northern Australia is still open to debate, The wettest parts of Australia today are on or near the coast, and these would tend to migrate inland as the sea rose. Warmer seas and marine flooding of the Sahul shelf would also tend to increase the incidence of tropical cyclones (Webster & Streten 1972, 1978). Whatever the ultimate causes, rainfall was higher and the summer wet season was longer throughout tropical northern Australia between about 12 000 and 7000 BP. The monsoonal summer rains appear, on palynological evidence, to have penetrated as far south as Lake Frome (30–31°S) during the early to middle Holocene, but after about 4500 BP the summer rains became more erratic in southern Australia, and the levels of many lakes fell (Singh 1981). Shortly thereafter conditions were sufficiently dry and windy for sand dunes to form in central New South Wales (Wasson 1976), at about the same time dunes were also active in Rajasthan and southern Ghana.

11.6 Future palaeoclimatic research in the tropics

Although the broad pattern of climatic oscillation from glacial to interglacial seems reasonably clear on the basis of proxy evidence of late Pleistocene and early Holocene age from tropical India, Africa and Australia, there yet remains a sobering number of unresolved issues. Perhaps the thorniest of these is how to distinguish between the effects of temperature and precipitation change, as instanced in Figure 11.1. Some form of reliable palaeothermometer is urgently needed in palaeobotanical studies. At least two lines of enquiry seem worth pursuing in this regard. One, already tested by carefully controlled laboratory experiments, is the demonstrated value of oxygen-isotope analysis in testing the degree to which plants respond to changes in relative humidity and temperature (Ferhi & Letolle 1977a,b). The other potentially useful paleothermometer is based upon the known temperature dependence of C_3 and C_4 grasses, the fossil cuticles of which can be identified to genus level (Livingstone & Clayton 1980).

A second major problem in Quaternary palaeoclimatic reconstruction concerns the impact of prehistoric communities upon the landscape. Did ‘overkill’ and careless use of fire cause the extinction of certain plants and animals? Is a high charcoal count diagnostic of man-made fires, as Singh and Kershaw have claimed, or of natural fires, as Clark has suggested (Singh *et al.* 1981), rather more convincingly? Related to the impact of fire is a third problem, that of hillslope erosion and stream aggradation. How far are the middle to late Holocene changes in sedimentation noted by Maley (1981) in Chad and by Thomas and Thorp (1980) in Sierra Leone really a function of climate? Might they not, like the late Holocene retreat of rainforest on Melville Island in northern Australia (Stocker 1971), be at least in part a consequence of prehistoric burning?

Part D

LANDFORM EVOLUTION

Tectonic style and tropical landforms

Editorial comment

The tropics contain some of the earth's most seismically active and highly volcanic terrains as well as some of the most extensive erosional surfaces. This contrast is reflected in rates of erosion (Table 3.7) and in the relative importance of different processes at the present day (Table D.1). Much of the evolution of the stable areas of the tropics may be related to the denudation of the Gondwanaland continent and the drifting of the fragments following its breakup. While Büdel's (1982) argument that the peritropical zone characterised by widespread planation surfaces is one of the two most effective geomorphic process realms may not be universally accepted, there is no doubting the importance of the processes of double planation surface development or etchplanation described by Thomas and Thorp in Chapter 12. On the stable shield areas, such as those of Africa and Western Australia, alternating phases of stability, or biostasy, and uplift, or rhexistasy, occur. Stability dominates for periods of up to 10^9 years. Biostasy allows for the development of deep, highly leached weathering profiles and duricrust formation. Several ages of duricrust of all types are found in the present-day tropics, but their global distribution is poorly understood. The maps provided by Petit in Chapter 13 pose questions about the relationships between tectonics and duricrust occurrence. Why, for example, do bauxites tend to be found on the highest levels and ferricretes at lower levels? Some possible answers will arise from a more detailed analysis of the tectonic structure of the continents, in particular from the study of intra-plate tectonics.

Africa is shown by Summerfield in Chapter 14 to consist of three major cratons (Fig. 14.1), unaffected by significant regional tectonic or thermal activity for the past 1.1×10^9 years. Between them are more elevated, more seismically active non-cratonic areas overlying relatively thin lithosphere. Africa is far more active tectonically than much earlier geomorphological writing suggests. If periods of rapid erosion and sedimentation have occurred, as seems possible between 12 000 and 10 500 BP (Table B.1 and introduction to Part B), is the explanation that they are entirely related to climatic and ecological changes adequate? Could there have been some tectonic or volcanic cause? The analysis of the dust cloud from the El Chicon volcano (Rampino

Table D.1 Comparison of some pedological and geomorphological features of erosional landscapes in northern Australia, Sri Lanka, peninsular Malaysia, Borneo, New Guinea and Hawaii (after Douglas 1978b)[a].

Feature	Australia	Sri Lanka	Malaysia	Borneo	New Guinea	Hawaii
dominant age of landscape	old	old	moderately old	young	very young	very young
proportion of labile rocks	low	low	moderate	moderate	high	moderate
recent tectonism	practically none	practically none	negligible	some	intense	infrequent
cenozoic volcanism	practically absent	practically absent	practically absent	some	abundant	highly abundant
tropical karst	practically absent	virtually absent	residuals only	abundant	abundant	none
relief	low to moderate	moderate	moderate	moderate (except for Kinabulu)	very high	high
rate of current soil development	slow	slow	slow to moderate	moderate	rapid	rapid
rate of current erosion and deposition	low to moderate	low to moderate	moderate	high	very high	very high
depth of soil on slopes	generally shallow	generally shallow	variable	variable	variable	variable but usually high
lateritic crusts	widespread	widespread	fragments	possible	absent	absent
pallid zone of deep weathering	widespread	widespread	few	unknown	rare	rare

[a]The inspiration for this table comes from Galloway and Löffler (1972, Table 2.1).

& Self 1984) shows how great an impact one event can have on tropical regions. Summerfield shows that there is a golden opportunity for bringing the exogenetic and endogenetic processes together in a truly dynamic approach to tropical geomorphology and environmental change.

One fragment of Gondwanaland which offers particular advantages for geomorphology is Sri Lanka (Figs 15.1 & 3.4), the whole of which, except for the Jaffna Peninsula, north west coast and small coastal strips elsewhere, consists of crystalline Precambrian rocks which have undergone varying degrees of metamorphism, granitisation and deformation (Fig. 16.1). The island has a central highland region around Nuwara Eliya (Fig. 15.1) with a steep slope towards the south and more gentle, but greatly dissected, slopes to the north and east. To the west and south of the highlands is a wet zone, well watered by the south-west monsoon, with mean annual rainfall as high as 4000 mm in such places as Ratnapura. To the north and east of the hills there is a dry zone, dependent on the more erratic and less moist north-east monsoon which brings, in November, December and January, the bulk of a mean annual rainfall of 1500 mm at Trincomalee. This climatic contrast over a relatively short distance and on similar parent material provides excellent opportunities to examine the influence of climate on geomorphology.

Studies of the landforms of Sri Lanka have been dominated by attempts to interpret the land surfaces of the island. Three major surfaces have been recognised (Adams 1929, Wadia 1945): a 'lowest' surface up to about 120 m, a 'middle' surface between about 520 and 580 m, and a 'highland' surface ranging from 1220 to 1830 m. Adams considers the surfaces to be sub-aerial peneplains representing successive stages in the uplift of the island, the lowest being the youngest. Wadia, however, believes they are the result of block uplift of an ancient erosion surface in two stages, the earlier lifting the second surface above the first, while the later stage lifted the third above the second. Later workers still tend to reiterate one or other of these views, Sahni (1982) stating that the last major uplift in Sri Lanka, dated as pre-Miocene, saw the island elevated 460 m above present sea level. Peneplanation followed, resulting in the formation of the present coastal plains which were thereafter eroded. However, Curray *et al.* (1982) map Sri Lanka as a platelet, bounded by faults separating it from the Indian plate as a whole. Whether this is an assumption based on Wadia's interpretation of the origin of the three erosion surfaces is not clear. However, Louchet (1981) has identified traces of Jurassic and Cretaceous uplift affecting the highland blocks of Sri Lanka. He stresses that these parts of the island show a critical interdependence of tectonic and climatic influences on landforms. With such a persistent, long-established view of the geomorphology of Sri Lanka, it is appropriate that the final two contributions to this book by Bremer (Ch. 15) and Späth (Ch. 16) are devoted to detailed studies of Sri Lanka to test not only the applicability of the local evolutionary hypotheses but also the theories of etchplanation and tropical landform evolution developed in other areas and discussed in earlier chapters of this volume.

12

Environmental change and episodic etchplanation in the humid tropics of Sierra Leone: the Koidu etchplain

M. F. Thomas & M. B. Thorp

12.1 Introduction

The formation of 'etched plains' was first described by E. J. Wayland (1934). He recognised that a mantle of saprolite would result from vertical penetration of ground water beneath land surfaces of low relief in a seasonally humid, tropical climate, and that 'this saprolite would be largely removed if and when land elevation supervenes, and the process may be repeated again and again as the country rises . . .' (note 376). By this means an 'etched plain' could be derived from an original peneplain and the land surface maintained at or near base level during spasmodic elevation over long timescales. This concept was applied by Bailey Willis (1936) to an interpretation of the Tanganyika Plateau as an 'etched peneplain', 'widely mantled by residual or transported soils' (p. 135). The stripping of the saprolite mantle was ascribed by both authors principally to tectonic causes, though Willis (1936) recognised the possibility of rainfall variations playing a subsidiary role.

Previously, J. D. Falconer (1911) had adduced similar arguments to explain the relief of northern Nigeria and, as Twidale (1982) has recently recalled, many early authors have shown an appreciation of the role of deep weathering in the modelling of landforms in granitoid rocks.

Similar ideas formulated by German geologists found powerful expression in the concept of 'double surfaces of levelling' advocated by Büdel (1957) and since adopted by many workers. Such general models were soon applied in Australia (Mabbutt 1961, 1965), Nigeria (Thomas 1965), India (Büdel 1965) and later in Guyana for example (Eden 1971). Yet development of the 'etchplanation model' has been slow and not without difficulty. Thornbury (1954) for example considered that although etching 'might contribute to local differential lowering of a peneplain surface . . . it is difficult to visualise this process operating widely enough to produce etchplains of regional extent'

(p. 193). Moreover, writers such as Mabbutt (1961), Bishop (1966) and Ollier (1969) sought to confine the application of the term 'etchplain' to areas of stripped and exposed bedrock. Because such areas are of restricted extent, Bishop (1966) considered such a designation superfluous, while it is not surprising that most instances of etched surfaces *sensu stricto* have been illustrated from semi-arid areas, where the prevailing aridity has promoted inter-stream degradation and the removal of a deep saprolite mantle previously developed during the late Mesozoic or early Cenozoic eras.

Recognition that the situation in the humid tropics of today was more complex led Thomas (1965, 1974) to elaborate an etchplain typology based on experience in Nigeria. Further modifications of this scheme have been applied in southwestern Australia (Finkl & Churchward 1973, Finkl 1979, Fairbridge & Finkl 1980), in India (Demangeot 1975) and most recently in southern Fennoscandia (Lidmar-Bergström 1982). Recent publication of an English translation of Büdel's *Climatic Geomorphology* (Büdel 1982) has now unified the terminology from English and German sources and extended it to embrace relict and active etchplains within Bremer's (1971) concept of 'divergent weathering and erosion'.

In this study the concept of etchplanation will be applied to: (i) landforms that may have been produced by one or more distinct episodes of stripping of a relict saprolite mantle; (ii) landforms continuing to evolve by surface erosion acting on preweathered materials; and (iii) landscapes within which emergent rock forms coexist with more rapidly evolving forms controlled by the shifting balance between weathering penetration and surface lowering within a system involving active or dynamic etchplanation.

All these conditions may coexist within a single landscape and the analysis recognises that in circumstances where etching of the crystalline rocks of the Archaean basement has continued throughout Phanerozoic time, the formation, stripping and renewal of the saprolite mantle will reflect variations in relative and absolute rates of weathering penetration and surface denudation in both temporal and spatial dimensions, to produce a complex of landforms on regolith, on rock and on transported materials.

Specification of an adequate geological framework for the understanding of etchplanation has also proved difficult. The timing and nature of the formative tectonic events have remained uncertain although these have been clarified by application of modern plate tectonic theory, and correlation of phases of saprolite removal with the sedimentary record has not always proved possible. Nevertheless, the work of Millot (1970) and many others has now revealed some of the essential relationships between regolith mantles, tectonics and sedimentation, particularly in intracontinental basins. Such ideas build upon those of Erhart (1956) who advanced the concepts of 'biostasy' or biological equilibrium associated with stability and pedogenesis, and 'rhexistasy' marked by disequilibrium, instability and surface erosion.

Fairbridge and Finkl (1980) have now developed these ideas towards the definition of a 'cratonic regime', based on studies in Western Australia.

According to their analysis, alternating planation and transgression take place without major disturbance over periods of up to 10^9 years. During this time a 'thalassocratic regime' corresponding to biostasy alternates with brief intervals of an 'epeirocratic regime' or rhexistasy. The prolonged biostasy leads to conditions of low suspended and relatively high dissolved loads in streams, removing silica and calcium to the oceans forming limestones and chert and leaving deep ferrallitic soils and weathering profiles on the land mass. This is what Millot described as the 'lessivage' of the continents. Conditions of rhexistasy induced by tectonic uplift are indicated by the stripping of the ferrallitic soils by headward erosion of streams and the flushing of residual quartz during entrenchment. Intervening plateaux, according to this model, will be desiccated by falling water tables leading to duricrust formation while oceanic sedimentation will be dominated by the formation of red beds and quartz sands. Like the etchplain model itself, this simplified and abstracted view of erosion and sedimentation has to be confronted with reality within local frameworks of space and time. But it is readily seen that the two concepts are remarkably congruent.

A further problem concerns the relationships between etchplanation and (Quaternary) environmental change. The effects of the latter on morphogenesis in the tropics, especially the drier tropics, are now well appreciated if not fully understood (Tricart 1972, Fairbridge 1976). Many of the studies on which this appreciation is based focus upon recent stratigraphic history and their findings have yet to be fully integrated into models of etchplanation although their implications are clear (Fölster 1969, De Meis & Monteiro 1979, Bigarella & Andrade 1965, Journeaux 1975b) and are made explicit, for example, by Finkl and Churchward (1973).

Thus there are several aspects of etchplanation theory as applied to tropical landscapes which require further testing and research, These include:

(a) detailed descriptions of the landforms and deposits within specific areas of supposed etchplanation;
(b) information on current geomorphic processes which would allow a more accurate identification of dynamic and relict features in the landscape;
(c) specification of the factors and processes controlling landform development in such areas over different time periods, requiring special attention to be given to process interpretations of datable forms and deposits due to late Quaternary environmental changes, and their use as analogues for earlier events;
(d) investigation in humid tropical areas for which the concept of etchplanation was first advanced, but within which only limited research has taken place.

This enquiry attempts to consider these questions for a humid tropical rainforest area of West Africa, the Koidu etchplain, where recent vegetation clearance and mining activity have revealed clearly the morphology and the surface deposits. These geomorphic details will be placed in a theoretical framework based upon the etchplanation concept.

12.2 Geological and geomorphological background

Sierra Leone, together with the adjacent parts of Guinea, Upper Volta, Liberia and Ivory Coast, occupies an emergent dome (the Leo Uplift) or cymatogen over much of which are exposed Archaean or 'Basement Complex' rocks (*c.* 2700 Ma) of the West African craton, and it is with these areas that this study is concerned. Antecedents of the present landscape may reach back to events which led to the removal of a former cover of Ordovician sediments (Thomas 1980, Hubbard 1983). But the present relief has almost certainly been fashioned from late Mesozoic–Early Cenozoic surfaces of low relief and advanced weathering, that became subject to epeirogenic uplift associated with the opening of the South Atlantic. Dissection of the early Cenozoic surfaces has led to marked differential erosion amongst the basement rocks and to the deposition of ferrallitic sediments and quartz sands in the shallow estuarine and shelf facies now forming the coastal plain sediments (Bullom Group).

Planation surfaces of possible Palaeocene–Eocene age (Gaskin 1975) are preserved at altitudes of 650–800 m by thick duricrusts developed over very deeply weathered (over 100 m) supracrustal relics forming elongate schist belts of amphibolite grade (Rollinson 1978) and collectively called the Sula Group (McFarlane *et al.* 1981). These form bastions enclosing dissected lower ground between 350 and 450 m a.s.l., developed on the basement rocks (see Fig. 12.1). This whole complex area is commonly referred to as the 'Main Plateau' (Dixey 1922b, Hall 1969, McFarlane *et al.* 1981), but this description is misleading. There is a close adjustment of drainage and relief to structure and lithology, and the granitoid basement areas mostly occur as plains and basins of varying extent. These rather planate areas have been described as true planation surfaces (Hall 1969), but may alternatively be viewed as etchplains or etchsurfaces held up at various levels, by local bedrock contrasts, during the regional dissection of a complex early Cenozoic land surface of low relief.

One of the planate areas, called here the Koidu basin, is the object of this study (Fig. 12.2). Lying at 355–415 m, it is cradled below the duricrusted schists of the Nimini Hills to the west and partly rimmed by other local metasedimentary rocks or granite hills. In reality the area is not a basin but a divide of generally low relief for its southern perimeter lies along the upper edge of a major dissection and scarp zone. This marks the watershed between south-flowing strike-aligned river systems and west-flowing strike-discordant rivers of the Main Plateau. The Koidu basin is developed mainly on migmatised granodioritic gneiss and hornblende granite but minor schist belts and certain granites give rise to higher hill groups and inselbergs rising some 150–200 m above the general level, and also to isolated tors, boulder inselbergs and rocky outcrops that seldom rise more than 50 m above the level of

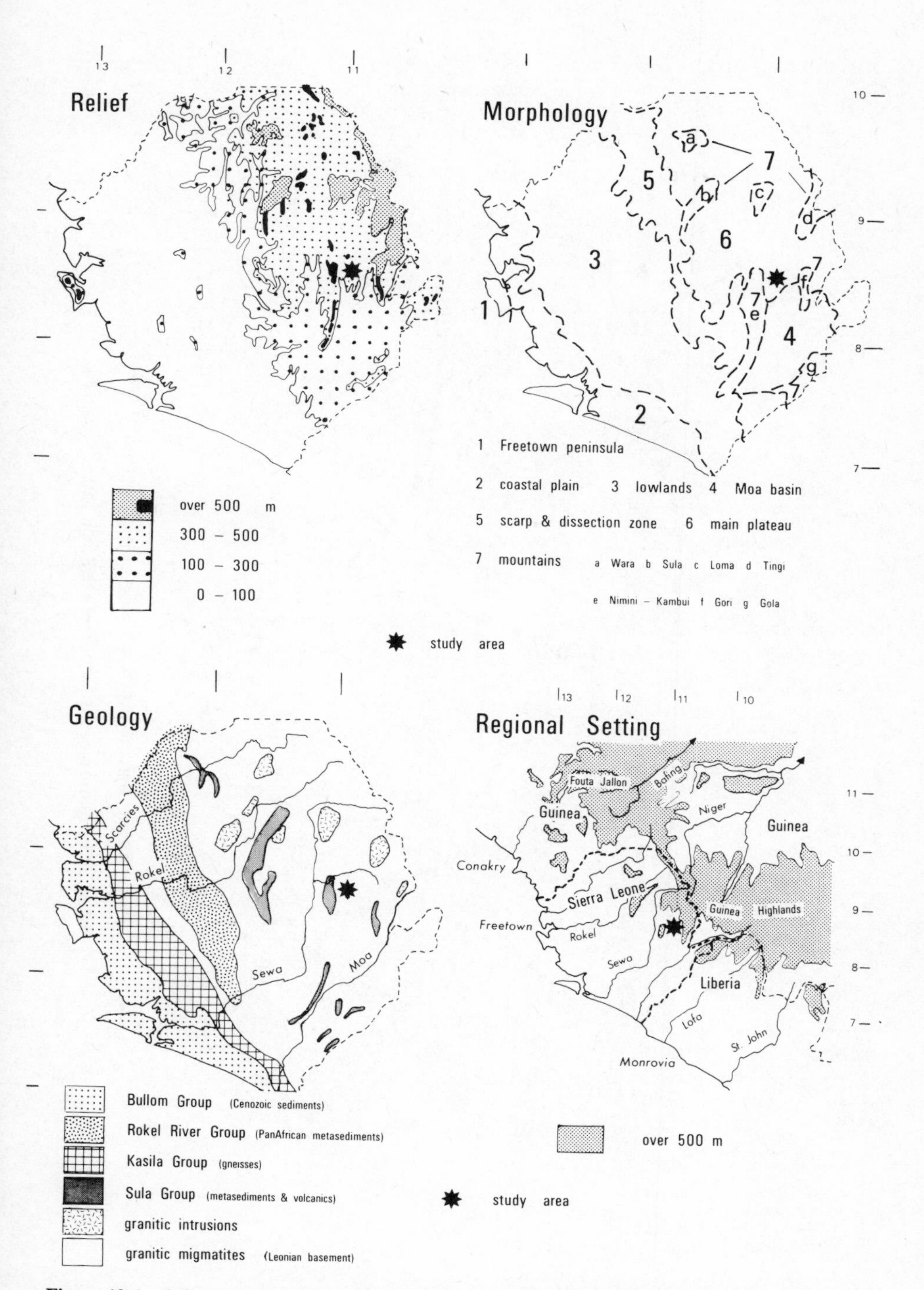

Figure 12.1 Relief, morphology, geology and regional setting of Sierra Leone and the study area.

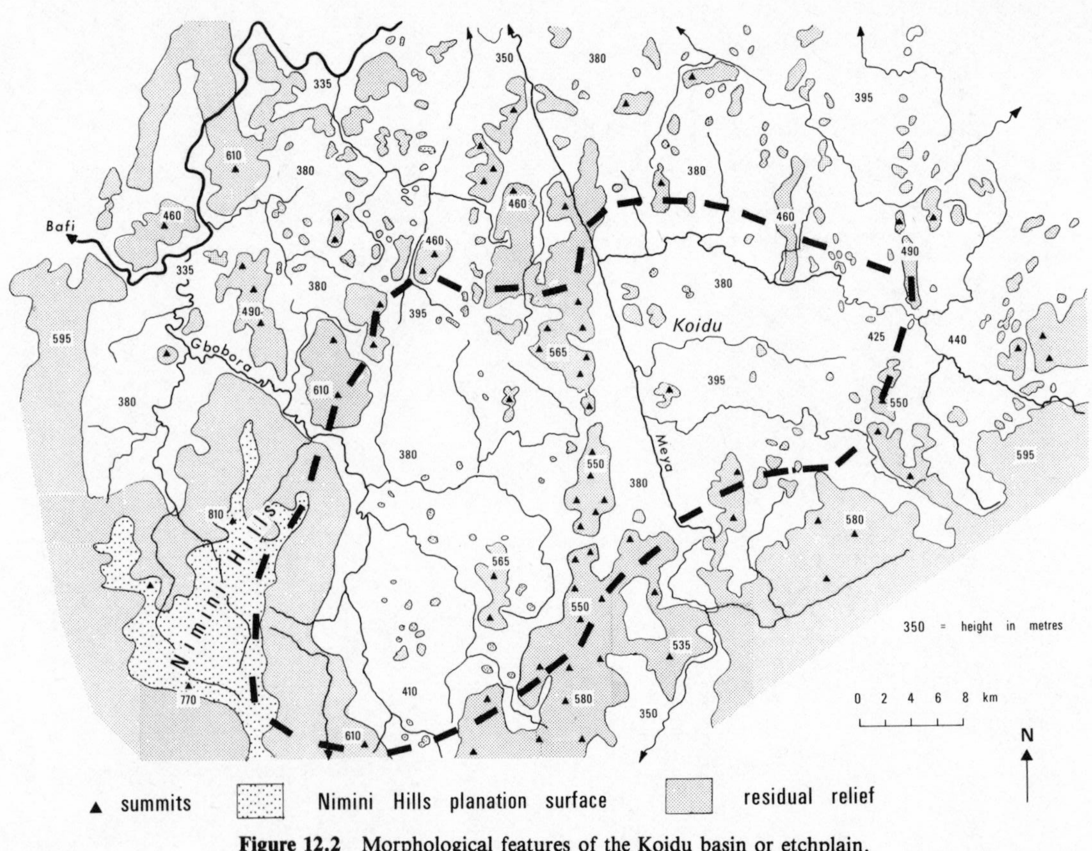

Figure 12.2 Morphological features of the Koidu basin or etchplain.

the interfluves. These higher relief elements are disposed in north-south strike belts which act as local stream and river base levels to the discordant west-trending drainage.

The tall rainforest natural vegetation is now preserved only on the steep slopes, the remaining areas having been cleared for cultivation. The annual rainfall is *c.* 2300 mm with a 'monsoonal' summer season concentration but no month receives less than 25 mm.

12.3 Landform elements and deposits

The landscape comprises the following landform elements (Figs 12.3 & 4): residual hills and steep hillslopes; planar, ramp-like pediment or glacis slopes; planate interfluves; higher terraces in the larger valleys which may merge up slope with the glacis; lower alluvial terraces and the present swampy floodplain flats of the valley floors (Fig. 12.5).

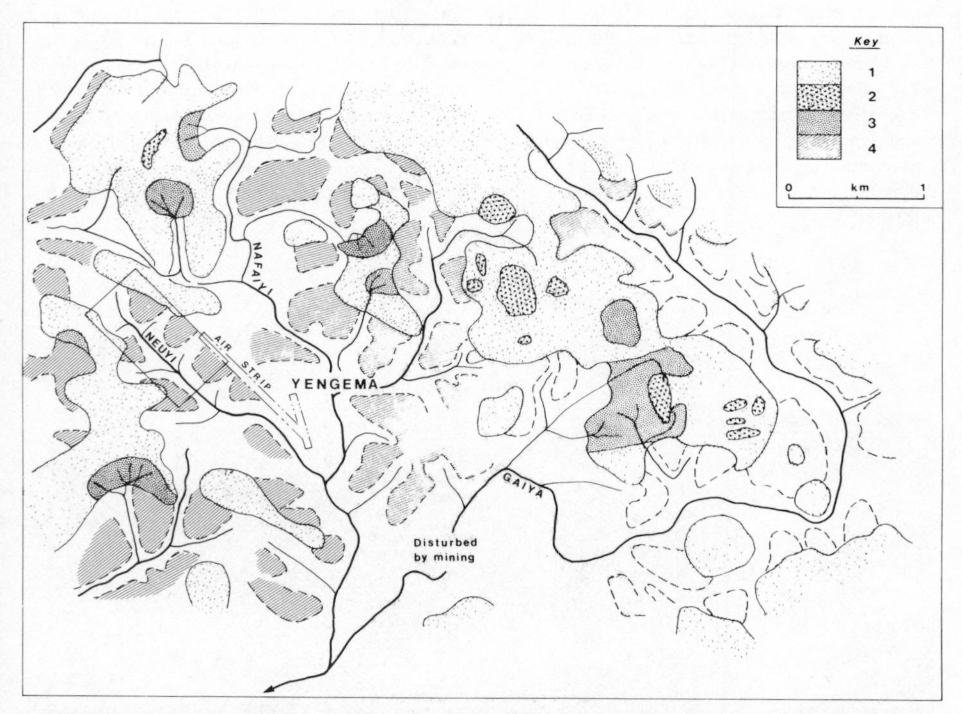

Figure 12.3 Landform elements on part of the Koidu etchplain, near Yengema (rectangle indicates location of Neuyi headwaters shown in Fig. 12.7). Key: 1, residual relief, including 2, rocky areas – exposed tors and rock slabs, and 3, inferred areas of deep weathering; 4, main glacis surface. Low terrace and alluvial flats and floodplains are left uncoloured. (Compiled from 1 : 20 000 panchromatic aerial photographs by Huntings Surveys 1967. Profile relationships are indicated in Fig 12.4.)

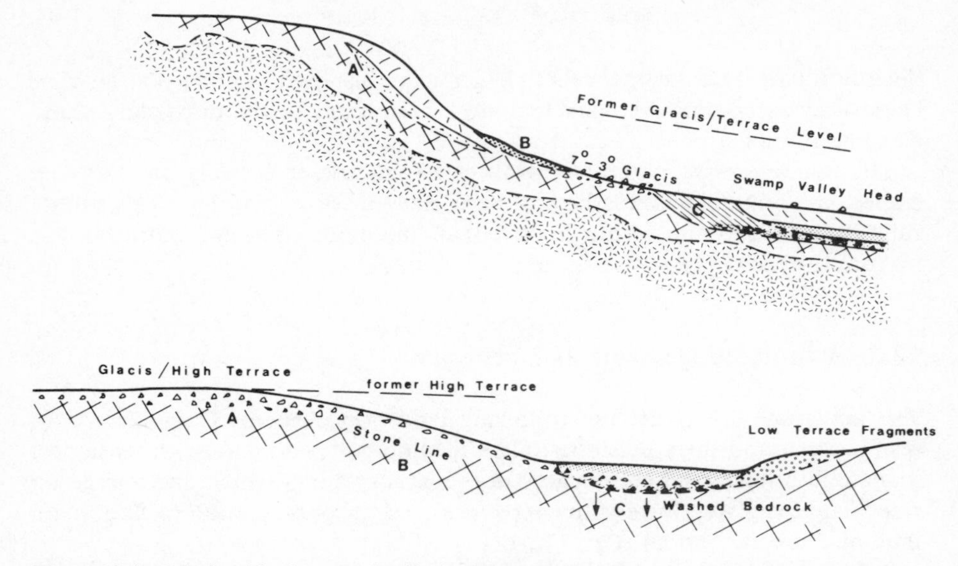

Figure 12.4 Relationships between landform elements of the Koidu etchplain. Upper: Relationships between residual relief (hillslopes), glacis and valley heads, showing some superficial accumulations: (A) gully fill on hillslope; (B) colluvium at hillfoot sites; and (C) valley head colluvium as described in Figure 12.6. Lower: Relationships between planate interfluves and valley morphology and deposits, showing: (A) densely packed residual gravels containing duricrust fragments, stained quartz including pebbles, corrundum and diamond – some appear to be glacis slope deposits, others derived in part from high alluvial terraces; (B) stone line derived from A and continuing below swamp deposits; and (C) valley swamp floors showing sequences of colluvial/alluvial sands, quartz gravel and 'washed bedrock' or saprolite. A low terrace is also indicated.

Figure 12.5 General view of landform elements in the Koidu basin, showing residual hills, the main glacis surface and valley floors in the middle Meya valley north of Koidu.

Residual hills and slopes

These vary in complexity, in dimensions and in morphology, from largely rocky boulder inselbergs and bornhardts to those retaining a weathered mantle. Summits vary in area and in form from small, flat-topped isolated hills to more extensive uplands, but it is not yet clear whether these are of more than purely local significance. Hillslopes, with inclinations of 20–40° are smooth, rectilinear or concave on the schists, but on the granites commonly display an irregular microrelief with core boulders, small tors, benches and slabs. Characteristically, there is a sharp break of slope below the summit, and the hillslopes are dissected by shallow valley heads and dells leading into first-order valleys. The heads of these 'incipient' valley forms are partly infilled with 'colluvium' which may be thick enough to obscure the bedrock form. Where present, the regolith shows an upper, mobile layer 0.1–1.3 m thick, containing a loosely packed gravel-sized material consisting of ferruginised or partly weathered bedrock fragments, quartz pebbles, lateritic nodules and pisoliths, and, occasionally, allogenic corundum and diamond presumably derived from ancient high-level alluvial deposits now largely destroyed. The hillslopes commonly lead into the glacis via a pronounced concavity and here the 'boulder colluvium' shows evidence of mass movement from the slopes above.

Glacis and low interfluves

These constitute an important part of the land surface of the Koidu basin and possess certain features in common though differing in morphology and position. Glacis are gently sloping, more or less planar surfaces developed across unconsolidated materials and have been widely observed in the sub-humid and semi-arid parts of West Africa and of other continents and also in Ivory Coast (Michel 1973, Delvigne & Grandin 1969). In Koidu the glacis slope towards the axial valley floors (from 7° to 1°) but are dissected by the first-order valleys which have gradients that are much flatter than those of the flanking glacis. Where they flank valleys of Strahler order 3 and higher, the glacis frequently merge imperceptibly down slope into truncated alluvial high terraces although the glacis surface deposits pass over the terrace material. The glacis are cut across *in situ* mottled saprolite mantles, partially decomposed and even fresh bedrock in some places (Fig. 12.4).

On low interfluves the summit areas are flat in long and cross section but camber towards the valley sides. They are dimpled by swales and shallow cols at the heads of streams and valleys, and occasional whalebacks, tors and boulders break the surface.

The glacis and interfluves are underlain by deposits which might be termed stone lines but there are important variations in stratigraphy and lithology. These deposits vary from 1 to 2 m in thickness and contain an upper organic topsoil and a lower gravel layer. The gravel layer varies from stringers of single clasts to densely packed gravel beds up to 1 m thick. The gravels contain lateritic pisoliths and nodules; quartz pebbles (sub-angular to well rounded, and variably coated, stained and impregnated by iron oxides); occasional ferruginised bedrock pieces, pebbles and cobbles and heavy mineral assemblages which include worn allochthonous corundum, tourmaline and diamond. The gravel matrix comprises illuviated topsoil and sandy clay. Under forest, the stone line may be deeply buried beneath some 2 m of ochre-coloured sandy clay, but where the soil has been cultivated and eroded a thin dark organic loam rests directly upon or extends into the gravel. The stone lines repose upon decomposed mottled bedrock (Figs 12.6 & 7).

These deposits form a continuous sheet from hillside to valley floor, but exhibit great variability in clast type and proportions, indicating that these lag deposits are polygenetic. Some seem to be residual accumulations arising from the slow removal of fines from weathering profiles by throughflow, eluviation, rainsplash and wash, no doubt assisted by termite translocation as described by Komanda (1978), Roose (1980) and M. A. J. Williams (1978). By such pro-

Figure 12.6 Densely packed gravels associated with planate interfluves (glacis/high terrace), and containing duricrust fragments, ferruginised bedrock pieces, stained angular vein quartz, subrounded quartz pebbles and cobbles, and abundant iron pisoliths.

Figure 12.7 Stone line mainly of quartz but containing iron pisoliths, in lower slope position near swamp margins. These gravels have been traced into the basal gravel below the swamp flat, where they contain no iron concretions or duricrust fragments.

cesses the land surface would be vertically lowered but the mechanism may be self-limiting when a certain degree of clast packing has been achieved. Profile sequences of densely packed over sparsely packed pisoliths as described by McFarlane (1976) in Uganda are common.

The presence in many sample pits of rounded quartz pebbles and diamonds points clearly to an alluvial origin for some of the deposits. These are especially common where the interfluve or glacis is beginning to curve down into the valley-side slope and must be interpreted as a relict alluvial terrace, more or less in place but much thinned and eroded (Fig. 12.4). In the schist areas such deposits are better preserved by indurated oxides. The occurrence of occasional well rounded pebbles together with worn allochthonous heavy minerals suggests a more complex genesis involving the incorporation of older alluvial material. In several respects such gravels are similar to those in the small swamp valleys. Indeed, the presence of diamonds, corundum and rounded quartz pebbles at different topographic levels and in many morphological positions in the landscape is striking (West 1979), and raises fundamental questions as to the precise nature of landform evolution. It suggests that over periods of geologic time ($10^5 - 10^6$ years) and during land surface lowering by 30–100 m a high proportion of the total area may have been fluvially reworked, notwithstanding the present strong structural control over stream-channel position. Crickmay (1974) has discussed this phenomenon in other contexts and, although the evidence here does not support Crickmay's specific model of panplanation, such fluvial working may be a general feature of land surface evolution by etchplanation in deeply weathered terrains.

Subsequent translocations of clasts from these gravels is evidenced by their identification within more recent sedimentary units spanning the last 35 000 years of alluvial deposition in contemporary valleys, while downslope transfer of fragments from truncated quartz veins requires simultaneous lowering of the slope profile. Colluvial accumulations in hillslope gullies and valley-head locations also demonstrate the dynamic nature of the landscape. Collectively, these stone lines form drapes across hillslopes, glacis and valley-side slopes, linking and in part derived from bedrock and ancient alluvial deposits. They therefore integrate a long and complex history of land surface evolution as do the diamonds themselves. These surface gravels are also interlinked by process cascades activated by energy pulses becoming effective at different times and locations in the landscape.

Alluvial terraces

These are best considered within the context of the valley morphology and this will be described in a down-system sequence. Stream heads take several forms but all are very sharply distinguished from the surrounding watershed areas by abrupt convex breaks of slope, and the majority are sharply etched into the low planar interfluves. Some arise on the residual hills and slopes, as steeply sloping chutes formed in weathered material or as rocky gullies following joints in the solid rock, others commence at the junction between the rocky hillslopes and the weathered glacis and may be more deeply incised at the valley head than down stream (Figs 12.8–10).

All first-order valleys rapidly open out to form swampy valley flats defined by abrupt breaks of slope above which rise 5–10 m high, steep convex valley-side slopes. Thus the valleys have the pure 'sohlenkerbtal' morphology. They possess no permanent stream channel, and during the wet season water moves across the whole width of the valley floor. Beneath these flat floors are 1–2 m of sediment commonly comprising some 0.5 m of black sandy clay loam overlying up to 1 m of grey clayey sand, below which is found 0.1 to 0.4 m of bleached angular quartz gravel in a sandy, gritty clay matrix. Alluvial sedimentary structures are apparent only in valleys of Strahler order 3 and higher. The sediments are saturated for most of the year, but water tables below the interfluves, in contrast, are always very low.

The bedrock floor is irregular showing low humps and shallow basins (Fig. 12.11). The rock is commonly decomposed to depths of 15 m and more and is typically a stiff white, gritty clay containing kaolinite, sericite and quartz derived from the granite or gneiss. However, in many of the headwater swamp valleys fresh rock protrudes through the deposits and side slopes as corestones and whalebacks (Figs 12.8 & 9).

Figure 12.8 Headwater swamps etched into low (glacis/high terrace) interfluves.

Figure 12.9 Flat channel-less valley floor punctuated by core boulders, and break of slope at the valley-floor margin.

Figure 12.10 Alluviated valley cut into residual hilly relief. Ubiquitous granite cores indicate widespread and geologically recent stripping. Abrupt break of slope with valley flat is characteristic.

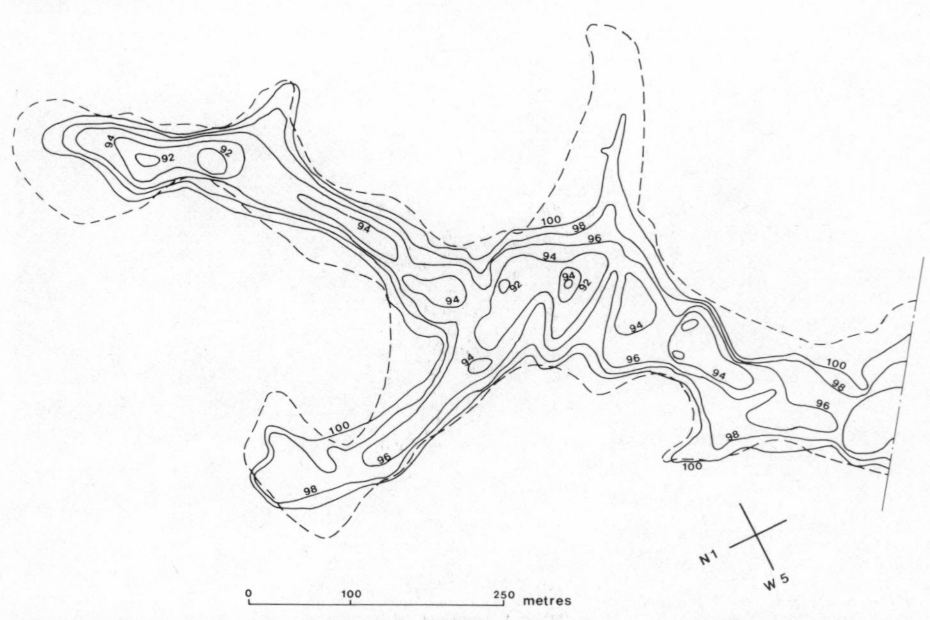

Figure 12.11 Bedrock configuration of the Neuyi valley. All heights are compared to stream level taken as 100. Figures are in feet (taken from work by R. W. Crofts with permission). The broken line indicates the limits of known alluvial deposits.

Frequently the swamp sediments contain organic material and six ^{14}C age determinations on pieces of wood from basal gravels in four swamp valleys gave dates ranging from 8200 to 750 BP (Thomas & Thorp 1980), but the possibility of timbers sinking through the top sediments to rest on the underlying gravels leads to uncertainty concerning the relationship between the age of the wood and the age of the swamp deposits.

Within the valleys there is an older suite of alluvial deposits, partly truncated by the convex valley-side slopes and resting on bedrock floors slightly higher than those of the contemporary swamp floors. The sedimentary character of these older deposits shows a more obvious alluvial origin and the bedrock morphology indicates deposition in valleys broader than the present, while sedimentation also marks previous extensions of present-day valley heads.

Excavation of one valley head (Figs 12.12 & 13) draining a low interfluve revealed a buried gravel stratum apparently once continuous with a stone line linking to the interfluve. The gravel is buried beneath 3–4 m of strongly mottled, structureless sandy clay colluvium wedging out towards the interfluve. The upper surface of this colluvium is strongly indurated and contains larger quartz and duricrust clasts also derived from the interfluve deposits, thus consituting a second stone line. Above this duricrusted and truncated palaeosurface is a second thinner sandy clay deposit coloured yellow-brown. A pisolithic intra-soil gravel within this deposit is also continuous with the gravel on the interfluve surface. The lower gravel, colluvial unit and extended valley head probably mark the extension up valley of the flanking low terrace. This latter feature elsewhere in the area contained organic material giving ^{14}C dates spanning the period 36 000–25 000 BP. The indurated horizon may then correlate with the dry Ogolian period (*c.* 20 000 BP) and possibly formed after a contraction of the drainage network when valley heads becamed infilled with colluvium. The onset of more humid conditions in the late Pleistocene may be recorded by the truncation of the indurated deposit and the later colluvial fill possibly marks subsequent Holocene fluctuations in land surface dynamics.

A further observation from this section may be relevant to an understanding of subsurface processes. In the decomposed granite below the sediments the quartz grains are matrix supported by the kaolinised feldspars but the fabric was porous and contained some large voids. Between this decomposed bedrock and the basal gravels is a layer *c.* 0.3 m thick comprising a more compact quartz grit and fine gravel within a sticky clay matrix lacking visible voids. We refer to this as 'washed bedrock' and it may represent the first stage in the modification of the saprolite towards a swamp gravel by water moving as throughflow beneath the gravels.

Valleys of Strahler order 3 and higher contain permanent open river channels discharging a regular seasonal or even perennial flow. They also contain unambiguous alluvial sediments exhibiting bedding and size sorting. The floodplains are floored by a basal gravel containing sub-angular to well rounded quartz gravel, pebbles and cobbles in coarse sandy matrices and are overlain

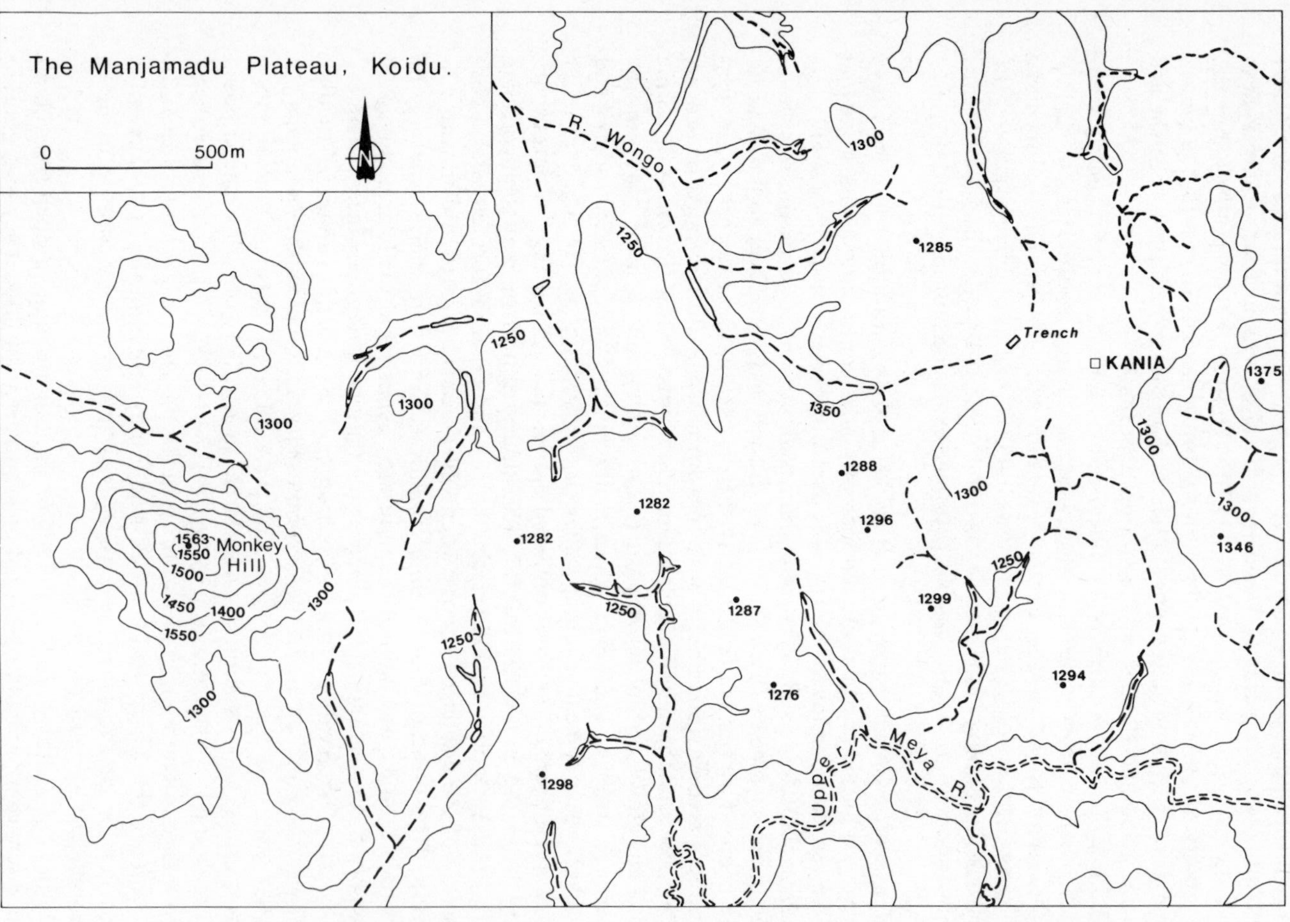

Figure 12.12 Map of the Manjamadu Plateau showing the site of the Kania Trench section (Fig 12.13).

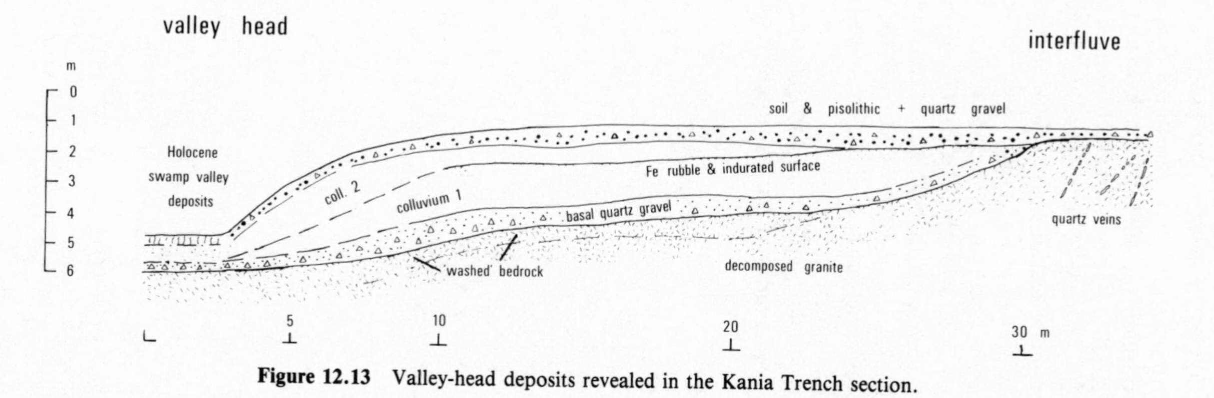

Figure 12.13 Valley-head deposits revealed in the Kania Trench section.

by channel sands and fine-grained overbank floodplain deposits. ^{14}C age determinations of interbedded organic-rich clays and of large pieces of timber in the gravels range from 12 000 BP to Present but group into several distinct age classes (Fig 12.14) which relate to two morphological floodplain units: an inner meander belt floodplain (*c.* 4000 BP to Present) and an older lateral floodplain (*c.* 9600–7800 BP). The present river beds lie above the level of the basal gravel, except in the vicinity of rock bars, and gravels remain saturated throughout the year. Artificial sections often show strong iron precipitation as soft tufas during the dry season and the older suite of sediments are strongly stained and frequently show iron pans in their gravel layers. Large timbers and trash are commonly associated with the late Holocene sediments but common to both suites are clay beds rich in organic matter both beneath the basal gravels and within the channel sand facies. Clearly there have been strong fluctuations in the fluvial systems with distinct episodes of deposition, fluvial stagnation and floodplain erosion.

Early Holocene basal gravels dating from 12 500 to *c.* 10 000 BP are commonly found partially cemented and packed under and around the incurved

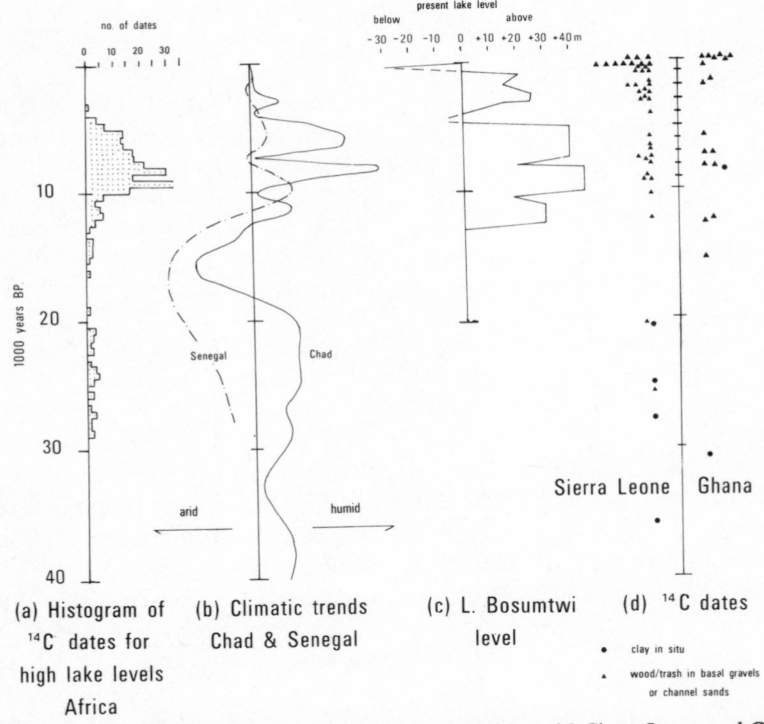

Figure 12.14 Late Quaternary correlation diagram for Africa with Sierra Leone and Ghana ^{14}C dates. Sources: (a) after Street and Grove (1979); (b) after Michel (1980) and Servant and Servant-Vildary (1980); (c) from Talbot and Delibrias (1980); from Thomas and Thorp (1980) and unpublished data.

lower surfaces of large corestones (2–4 m diameter) which lie buried beneath the floodplains in distinct reaches of the large rivers (Figs 12.15 & 16). Earlier deposits, dating from 36 000 to 20 500 BP, constitute a low terrace. Some of these latter deposits are sticky organic-rich clays over coarse basal gravels while others are organic-rich laminated clays resting upon scoured bedrock around exposed corestones and overlain by standard floodplain deposits. Unfortunately, there are too few dates for this period on which to base a stratigraphic history with confidence but, as with the Holocene suites, there is evidence for strongly fluctuating fluvial conditions.

Relationships of stream scour to bedrock weathering appear complex, with incision into the weathering profile during the last 36 000 years to expose the corestone zone and other rock barriers in specific reaches of the larger rivers. But elsewhere in the same valleys the weathering front, as determined by drilling, lies up to 50 m below the alluvial floors. Although partial removal of an irregular, inherited saprolite is indicated, further and continued weathering penetration is proceeding below the alluvium of both small swamps and the larger river valleys.

There are, in addition to the alluvial deposits described above, four or five higher alluvial terraces rising more than 75 m above the present river level in the Gbobora valley where it passes discordantly through a constriction in the Nimini Hills Schist belt (Fig. 12.2). All are heavily lateritised and indurated but some of the exposures show greater thicknesses of gravels and better rounding and sorting as compared with the Holocene gravels. These extend the sequence of fluctuating environmental conditions probably into the early Quaternary or Pliocene, but little can be inferred about their origin, which may involve epeirogenic uplift. Little evidence of these higher terraces remains within the Koidu granitic basin, where the rounded quartz pebbles and allochthonous heavy minerals in the interfluve and glacis deposits already described are all that remain of these earlier topographies (Figs 12.16 & 17).

The surface morphology and deposits thus show unmistakable evidence for episodic pulses of surface processes across the landscape (Fig. 12.18). Those of the swamp valleys also suggest a second group of processes. First, bedrock contours established for several of these headwater valleys show the detailed structural control and the enclosed bedrock hollows that can arise from differential weathering (Fig. 12.11), while reference has already been made to the appearance of corestones on valley floors, sides and interfluves. Secondly, comparison of the granulometry of the decomposed bedrock and of the swamp gravels suggests that the quartz could have accumulated, at least in part, as the bedrock becomes weathered and the clay products are removed from the sediments both mechanically and possibly by chemical degradation or ferrolysis as argued by Brinkman *et al.* (1973). In this view the gravels are residual lag accumulations. Thirdly, artificial sections show that where the valley-side stone lines pass below the dry-season water table in the swamp floors, the lateritic components disappear, being dissolved and leached away by the reducing conditions of the swamp margins, a process described from

Figure 12.15 Core boulders exposed in alluvial diamond workings at different reaches of the Moinde valley. Alluvial deposition appears to have followed scour into partially decomposed bedrock after 12 500 BP. Some timber is found jammed beneath large cores, often associated with lightly cemented gravels.

Figure 12.16 See caption to Figure 12.15.

Figure 12.17 Rounded pebbles from the channel gravels appear to have been repeatedly entrained into the river systems without transport out of the area. They are associated with sub-angular to sub-rounded material of both similar and different calibres.

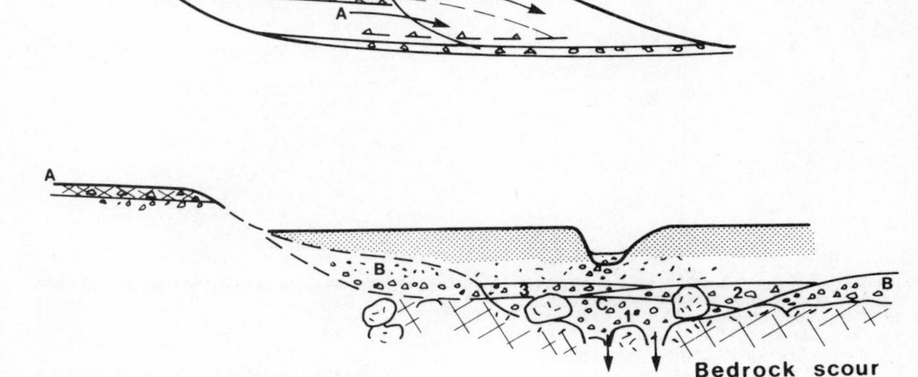

Figure 12.18 Pulsed energy relationships in stream heads and valley floodplains. Upper: Two phases of colluvial fill (A,B) alternating with headward extension of swamp flat. Successive positions of swamp head shown as 1, pre 20 000?; 2,3, present day. Based on Kania Trench section (Fig 12.12) and similar stratigraphies. Lower: Major phases of fluvial activity indicated by high terrace deposits (A), low terrace deposits (B) and channel fill deposits (C), schematically divided into three cut and fill periods 1, 2 and 3. In both cases indications are that slope and fluvial systems have responded to environmental changes.

other areas by Sivarajasingham (1968) and by Mäckel (1979). Moreover, around swamp margins the ground water is often highly charged with dissolved iron (1700 mg l^{-1} in one recorded case; IITA 1975). Clearly, therefore, it is possible to view the swamp valleys as sites of active chemical denudation and perhaps even as the products of such denudation. Hall (1969) considered that these valleys and their deposits evolved by headward extention due to spring and seepage sapping along structural lineaments and joints in the rock. Clays and fines from the sapping areas are removed mechanically and chemically from the valley and a carpet of lag quartz gravel is left behind to become covered by colluvium from the valley-side slopes (Fig. 12.19).

To provide a more complete model for etchplain development it is necessary to integrate this concept of dynamic etching with the proven episodic nature of fluvial activity. It is also important to provide an understanding of spatial linkages from interfluves to valley floors and from headwater swamps into larger river valleys.

The lateral continuity between interfluve gravels, stone lines and valley swamp gravels implies that the quartz in the subswamp gravels has come from upslope sites as well as from the weathering of subjacent rock, but the absence of iron pisoliths in the swamp gravels indicates that the rate of iron leaching exceeds the rate of transport of the pisoliths. However both the densely packed residual gravels of the interfluves and the angular, unworked gravels of the swampy valleys indicate a gradual concentration of lag deposits as solutes and fines are removed in a form of dynamic etchplanation (Fig. 12.19).

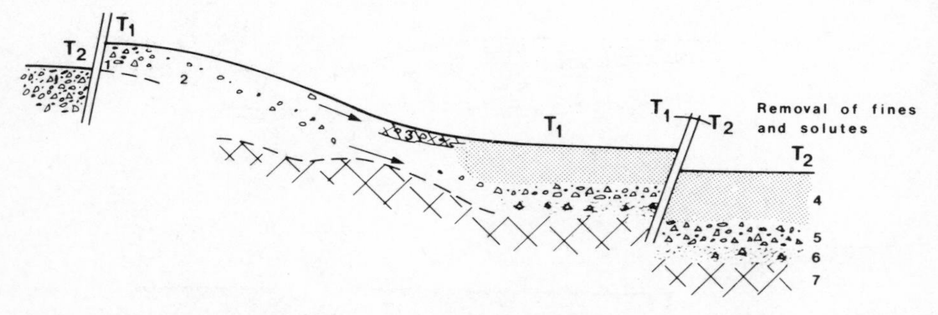

Key:
1 'residual' gravels derived from duricrusts, vein quartz and alluvial terrace remnants, becoming thicker and more densely packed with landscape lowering through loss of fines
2 stone line deposits derived from 1
3 colluvial wedge at foot of some slopes
4 clayey sand of swamp flat (channel-less floodplain)
5 basal 'swamp gravel', mainly residual from lowering of valley floor by chemical weathering and eluviation, but with some elements feeding from valley-side stone line
6 washed bedrock or saprolite, showing effects of lateral seepage through weathered rock
7 saprolite in place

Figure 12.19 Continuous etching model for valleys and interfluves. Generalised stratigraphy based on headwater valleys in the Koidu basin (etchplain) from time T_1 to time T_2. From T_1 to T_2 lowering of both interfluve and valley floor is indicated; similar lowering will affect intermediate slopes. Operation of the model is limited not by accumulation of residual products so much as by intervention of high-energy pulses through the system as shown in Figure 12.9.

On the other hand, sediment stores found as gully fills, valley-head colluvium and discrete sedimentary units in the larger valleys are evidence of episodic pulses of energy operating through the landscape. Moreover the widespread stripping of upland regoliths, emergence of cores within the headwater valleys and their burial by alluvium down valley point to a net loss of material of all calibres from the slopes and swamps and storage in the larger valleys.

There are therefore grounds for considering the entire etchplain as a complex morphogenetic system undergoing more or less continuous 'low-energy' transformations (biostasy) that have been periodically overtaken by 'high-energy' systems of erosion and sedimentation (rhexistasy) (Figs 12.17 & 19), closely related to concepts of ground surface stability/instability or *K*-cycles of Butler (1959) and others. Is it now possible to relate these surface and subsurface processes to known Quaternary environmental changes?

12.4 Episodic etchplanation and environmental change

Although the study of Quaternary environmental change in the lowland humid tropics has suffered from relative neglect, there is abundant evidence for such changes in West Africa, based on interpretations of relict features such as glacis (Delvigne & Grandin 1969), polygenetic weathering profiles and stone lines (Avenard 1973a, Bruckner 1955, Burke & Durotoye 1971, Flageollet 1981, Fölster 1969, Rohdenburg 1970) and river terraces (Vogt 1959). The relative chronologies of these studies have now been largely superseded by chronologies for the late Pleistocene based on radiocarbon dating (Livingstone 1975, Street 1981). However emphasis has centred on the semi-arid and arid zones, and discussions of the chronology and magnitude of environmental changes in the lowland rainforest zone have remained tentative and based on very few chronostratigraphic records.

Many aspects of early views on the desiccation of the rainforest areas (Aubréville 1949, 1962) have been confirmed and dated to the Ogolian period, 20 000–12 500 BP (Fig 12.14). There is also agreement that the period from *c*. 11 000 to 7500 BP was wetter than the present, and that another sub-humid event may have succeeded it around 4500 BP (Talbot 1981, Rossignol-Strick *et al*. 1982). Radiocarbon dates from the study area have also confirmed this late Quaternary chronology (Thomas & Thorp 1980), but assessment of the degree of environmental and ecological change has proved more difficult (Sowunmi 1981) and alternative non-climatic interpretations have been advanced for some landscape features (Jeje 1980).

For the period *c*. 40 000–20 000 BP there are few dates and the geomorphic evidence has not been analysed in detail. In Koidu, the fluviatile and related sediments with dates between 36 000–20 000 BP show three genetic modes: (1) floodplains scoured to bedrock with subsequent deposition of laminated organic clays and floodplain aggradation; (ii) coarse, well sorted and rounded gravels overlain by thick organic clays beds; and (iii) sedimentary strata (basal gravels, sandy clays and organic top clays) identical to those found in the swamp valleys of Holocene age. Fluctuating conditions are indicated but the dates (five) are too few to support a reliable process or environmental history. Elsewhere in West Africa the evidence is similarly enigmatic (Lang 1975, Pastouret *et al.* 1978, Sowunmi 1981) but does not indicate any prolonged or pronounced 'aridity crisis'. There is no reason for thinking that conditions did not favour chemical weathering beneath interfluves, slopes and valley floors although erosive 'events' are indicated by the Sierra Leone data. Sustained morphological stability cannot therefore be said to characterise the only prolonged humid forested period for which we have any data.

The higher alluvial terraces of the study area contain no datable organic remains but possibly resulted from environmental fluctuations between 75 000 and *c.* 40 000 BP. A period or periods of prolonged semi-aridity were probably responsible for the development of the relict glacis which form benches 200–400 m in width bordering the valley flats (Fig. 12.3) and much of the extensive stripping of granite hillslopes to expose core boulders and rock slabs was almost certainly achieved before the Ogolian period (Figs 12.5 & 8).

Fitting a timescale to these features is not possible at present, but Klammer (1981, 1982) has recognised important indications of aridity in south-east Brazil and suggested that they date to the early Quaternary or late Pliocene. The extensive but largely non-indurated glacis are probably not as old as this, but duricrusted deposits preserved within the schist belt and in many other areas of Sierra Leone, including the Freetown Peninsula, possibly are. The Koidu etchplain therefore contained many indications of forms and deposits, resulting from climatic oscillations prior to those of the late Quaternary.

The Ogolian dry phase (20 000–12 500 BP) corresponded to a southward expansion (not a shift) of the zone of maximum Saharan dust fallout by 8° of latitude from 23°N to 15°N at 18 000 BP with significant increases in sand influx occurring as far south as 5°N (Kolla *et al.* 1979) indicating a higher frequency and intensity of the NE Trades. Sahelian conditions could thus have existed in the Koidu area (8°N) for perhaps 4000–7000 years. Rossignol-Strick and Duzer (1979) have suggested that the Tropical Convergence Zone at this time attained only 8°N. Drill cores in the offshore zone of the Niger delta (Pastouret *et al.* 1978) indicate a dramatic reduction in the influx of fresh waters at this time, while palaeobotanical evidence from the Ivory Coast littoral zone (Assemien *et al.* 1970) and from the Niger delta (Sowunmi 1981) record an invasion of savanna tree and grass species into areas previously and subsequently occupied by closed rainforest. Before 12 000 BP water levels in crater lake Bosumtwi, southern Ghana, were abnormally low and surrounding vegetation was savanna grassland and open woodland communities (Talbot & Delibrias 1980, Talbot & Hall 1981).

The total absence of organic remains with dates of 20 500–12 500 BP from the Koidu area (Fig. 12.14) suggests to the authors that riparian woodland there was sparse and rapidly oxidised after death, that runoff in the main streams was reduced to ephemeral or brief seasonal flows and that stream sedimentation was at a minimum. With the onset of Ogolian desiccation, degradation of the forest towards more open communities would have lagged behind changes in precipitation. However, destabilisation of hillslopes would have ensued eventually, leading to enhanced rainsplash erosion, sheetwash and gullying on steeper slopes. Removal of the fine topsoil would leave a coarse lag gravel on the surface which might subsequently become encrusted. Widespread colluviation will have occurred at hillfoot sites, valley heads and along the margins of valley floors. If the precipitation was reduced to 600 mm or less (Lauer & Frankenberg 1980), then this part of Sierra Leone would have experienced a transition through the conditions of maximum sediment yield from slopes predicted by Schumm (1965).

The apparently rapid return of humid conditions after 13 000 BP together with a delayed response in forest recolonisation led into a period lasting some 2000–3000 years and ending *c.* 10 000 BP, during which runoff and sediment yield increased dramatically (Pastouret *et al.* 1978). In the Koidu area, gully-head advance in steep terrain and low-relief areas alike seems likely at this time and high peak discharges in the rivers evacuated pre-existing alluvial deposits and scoured into the decomposed bedrock floors exposing corestone weathering zones.

The early Holocene interglacial 'pluvial' or 'climatic optimum' lasted from *c.* 10 500 BP at least until 7600 BP. In southern Ivory Coast and Ghana closed rainforest appears to have become fully established by 7500 BP. Lake Bosumtwi, like Lake Chad, was periodically well above its present-day level. In the Koidu area, floodplains were built with total alluvial thicknesses 10–20% greater than those of the later Holocene. The available runoff also appeared capable of generating discharges which caused episodic floodplain reworking in contrast to the period between 7800 and 3000 BP.

A sharp fall in Bosumtwi lake levels around 4500 BP but lasting only a few hundred years, and coastal sand sheets and dunes over palaeosoils in southern Ghana (Talbot 1981), suggest another but shorter humidity crisis during the mid-Holocene. A paucity of organic matter between 7800 and 3300 BP in the Koidu area suggests that peak discharges in the rivers were probably reduced in magnitude and the frequency of floodplain erosion thereby diminished (Fig. 12.14). However, it is unlikely that the environmental changes were as profound as those of the Ogolian. After this it becomes increasingly difficult to distinguish the effects of lesser climatic oscillations, shifting cultivation and high magnitude–low frequency geomorphic events in the stratigraphies of the Koidu floodplains and swamp valleys.

During the past 40 000 years, therefore, external environmental controls have generated episodes of sediment entrainment and deposition in the Koidu landscape, the major events having a periodicity of 10^3 years.

It now appears possible to outline a model of etchsurface development that takes account of both continuity and fluctuation in the operation of morphogenesis over the landscape. Fairbridge and Finkl (1980) refer to etchplanation and 'cationic denudation' as the dominant process (more correctly a group of processes) operating during prolonged biostasy with pediplanation and 'mechanical erosion' intervening during brief periods of rhexistasy which would correspond with desiccation during the timescale of the 100 000 year 'glacial cycle'.

In the light of this study, and with respect to events of the late Quaternary, such alternation becomes complex and takes place over shorter timescales. Stability of the ecosystem and of the land surface may have been short-lived if the climatic oscillations occurred with the magnitude and periodicity indicated by available evidence (Fig. 12.14). Moreover, 'activation' of different slopes and drainageways in the landscape will not be synchronous during environmental fluctuation, and the 'complex response' of the drainage system to changed controls (Schumm & Parker 1973, Butzer 1980c) will have ensured that the surviving morphological and sedimentological record will itself be complex.

12.5 Continuous versus episodic etchplanation

Views of landform development that emphasise continuity either refer to long-term development and ignore short-term interruption, as was discussed in a previous study of Sierra Leone (Thomas 1980), or are concerned with very brief periods of time during which conditions remain essentially constant or oscillate about a mean that itself may be changing too slowly to be observed.

The long-term (10^7 years) view of the development of the Koidu etchplain and similar terrain on tropical cratons indicates the importance of geochemically controlled differential denudation from a possible late Mesozoic–early Cenozoic surface of widespread planation and profound decomposition. In contrast to the situation on the Western Australian craton described by Fairbridge and Finkl (1980), the Atlantic slope of the Guinea Highlands, of which Sierra Leone is a part, appears to have undergone repeated epeirogenic uplift. This has led to the accentuation of relief as the more readily weathered areas of granite and gneiss have been lowered by simultaneous reduction of the double surfaces of planation while on the schist belts only the basal weathering surface is lowered, the old land surface being more or less preserved by the resistant duricrust capping. Over the Koidu granites there is little evidence for the preservation of ancient palaeoforms except for the massive inselbergs or bornhardts.

The mean rate of denudation for the Koidu area cannot be less than the amount of lowering of the granite surface below the duricrusted plateau of the Nimini Hills since the Palaeocene (400 m in *c*. 65 Ma or 0.006 mm yr^{-1}). This figure is higher than the rates of denudation cited by Finkl (1982) for cratonic interiors (0.0001 mm yr^{-1}) and cratonic margins (0.001 mm yr^{-1}), based on Western Australia, but falls within the same 'cratonic regime' of denudation. Such generalisation has limited value, when rates must have varied widely during the long period involved. The overall slowness, however, aids understanding of the survival of deeply weathered duricrusted plateaux on the one hand and the selective weathering to considerable depths of otherwise dominantly stripped etchsurfaces on the other.

Thus all parts of the landscape may be accounted for by reference to saprolite formation and removal, and it is the authors' view that this accords with the concept of the 'etched plain' advanced by Wayland (1934) and discussed by Willis (1936). Limitation of this term of land surfaces of advanced stripping as advocated by Mabbutt (1961), Bishop (1966) and Ollier (1969) not only makes it difficult to apply in the humid tropics, where weathering at the basal weathering front is most active, (Gac & Pinta 1973), but it implies that nearly all instances will be relict features in present-day semi-arid or perhaps cold environments. Etching in the humid tropics is by contrast active, as stressed by Büdel (1982, see Figs 40 & 41) and demonstrated in this study. Etchplanation with its reference to double levelling surfaces (Büdel 1957) emerges therefore as an organising principle within which both relict and active features of tropical cratonic areas can be understood.

Arguments continue about a possible 'monogenetic' regime operating within an inner core of the humid tropics (Demangeot 1975, Twidale 1982) throughout the Cenozoic, but there is now wide agreement for a 'polygenetic' development in seasonal tropical areas that range from rainforest to semi-arid grasslands today (Demangeot 1975, Fairbridge & Finkl 1978, Thomas 1974). Alternating biostasis associated with weathering penetration and rhexistasis leading to pediplanation across weathered mantles has characterised such areas though the balance between these regimes has varied greatly, and the suggested periodicity of rhexistatic interruptions every 10^6–10^7 years advanced by Fairbridge and Finkl (1980) may be misleading within the context of the Koidu etchsurface. This is because the Cenozoic era has witnessed the dismantling of earlier planation surfaces with progressive differential erosion and divergent weathering (Bremer 1971, 1979) leading to the present complex of landforms (Thomas 1980) and involving the environmental fluctuations of the Quaternary. The Koidu etchplain is therefore an example of dynamic etchplanation operating at least since the early Cenozoic.

The role of environmental change during this long history has to be viewed on the timescale of the significant environmental fluctuations about which we know. As already shown this leads to a grouping of forms and processes into periods of 10^3–10^4 years. During these, however, etching at the level of the basal weathering front continues but is imperceptible and its morphological effects become apparent only after stripping of the saprolite cover. Surface erosion is probably most active during recovery from semi-arid to humid environmental states. Superficially, the resulting surface forms detract from the etchplain description but the development of slope pediments or glacis across the saprolite layer only serves to emphasise its importance and the distinction between this process and the cutting of extensive rock pediments (Thomas 1965). The latter is not excluded from consideration, but in regions of steep or vertically folded supracrustal and of basement rocks the close control of geology over the morphological skeleton of the landscape indicates the importance of vertical penetration of the weathering systems which was the starting point of Wayland's argument for the development of etchplanation.

12.6 Some questions of geomorphic theory

Firstly, what we see in a given landscape is governed by our ruling theories. This was always very evident in the work of Davis and King, but inevitably it affects our own interpretations. Certain observable facts about the Koidu landscape offer disturbing evidence in partial support of several differing views of landscape evolution. In the first place the presence of flat-topped hills, broadly accordant in altitude, is suggestive of widespread older planation levels. On the other hand, the adjustment of the present relief to structure and lithology echoes aspects of Hack's concept of dynamic equilibrium (Hack 1960, 1975). Furthermore, the detail of this differential denudation indicates the geochemical control over the processes at work and the importance of rock weathering and the development and removal of saprolite, thus supporting some concept of etchsurface development. Equally, the palaeoforms and deposits in the landscape suggest both the development of pediments and the survival of ancient alluvial elements in the landscape.

Secondly, the events described in this paper are contained within a period of the order of 10^5 years, and have led to forms and deposits sculpted and veneered onto a partially stripped landscape resulting from the progressive dismantling over a period of 10^7 years of a land surface of extensive deep weathering formed in the late Mesozoic and early Cenozoic. Such landscapes have been recognised very widely across the shield lands, and they do not appear dependent upon present aridity or tropicality of climate. They occur not only in Africa and India and semi-arid Australia, but also in the eastern highlands of temperate Australia, between the Snowy Mountains and the coast, in Fennoscandia (Gjessing 1967, Lidmar-Bergström 1982) and in Scotland. Indeed the principles adduced in this study have, we believe, a very wide application outside the zones of Neogene orogenic upheaval.

Thirdly, whereas Fairbridge and Finkl (1980) see biostasy with etchplanation and rhexistasy with pediplanation as alternating conditions within the cratonic regime, the evidence from Sierra Leone suggests that these can be concurrent. The Bullom Group sediments which underlie the coastal plain and offshore zone are thought to record some 60 million years of estuarine and shallow marine sedimentation (MacFarlane *et al.* 1981). Detailed stratigraphic information is lacking but feldspathic sands and kaolinitic clays predominate over occasional pebble beds, lignites and intraformational laterites. They repose upon a weathered and duricrusted basement surface and have thicknesses ranging from several hundred to several thousand metres. Derived from the removal of weathering mantles on the uplifted plains of the interior, they suggest that a strict biostasy has not prevailed throughout the Cenozoic; chemical weathering, the erosion of weathered mantles, slope erosion and pedimentation have continued simultaneously inland. Thus, Fairbridge and Finkl's (1980) model for a 'cratonic regime' needs modification for application to the West African craton.

Fourthly, the essence of the Koidu landscape is that few, if any, forms or deposits survive from the early Cenozoic, and possibly only the duricrusts capping the adjacent schists of the Nimini Hills are of this age. The 'age' of the surviving saprolite cover is indeterminate as are the 'ages' of the scattered quartz pebbles, but both preserve traces of earlier morphogenetic phases and pre-date the land surfaces with which they are today associated. These latter, we believe, are dominantly Quaternary in origin and development, and record the effects of major environmental changes that characterised this period together with the 'complex response' (Schumm & Parker 1973) of the denudation system to such external environmental changes.

During the 10 000 years or so that sediments have been stored in present stream channels and valley heads, the nature, if not yet the mechanisms, of etchplain dynamics are apparent. Beneath low-energy swamp systems that occupy as much as 70% of the valley system of Koidu, weathering penetration into bedrock continues to etch into the jointed granites, while solutes, clays and quartz sand produced by weathering are gradually transformed and transferred by subsurface water movement beneath the alluvial sediment. Marginal sapping of the valley sides and heads appears common and residual iron oxides are transferred downslope and consumed by the weathering system of the swamps.

In outline these are the means by which etchplanation continues, and by which stream system is extended in the landscape. They may also indicate the way in which large-scale etchplanation over long timescales can occur. Over such timescales of 10^7 years the rivers can be seen to have worked and reworked most of the land surface, but during any major phase of stream activity such as the present which may have lasted 10^5 years, interstream processes appear important, forming the glacis, colluvial slopes and stone lines across the landscape.

From a possible datum in the early Cenozoic, progressive removal of a deep saprolite mantle has taken place, but selective renewal of the saprolite has also continued, providing the basic mechanism for both dynamic etchplanation and for differential denudation between major lithologies. The formation of pediments and glacis within this system is not inconsistent with this concept, but their effectiveness as instruments of planation has been limited to development of local forms within individual drainage basins. Dynamic etching also provides the mechanism for the development of landscapes that express geochemical equilibria between rocks of different composition. The detailed facetting of the landscape, on the other hand, reflects clearly the role of oscillatory environmental change during the Quaternary era.

13
A provisional world map of duricrust

M. Petit

13.1 Introduction

This cartographic exercise, resulting from a collective effort and patient compilation, albeit of a provisional character, is of both scientific and pedagogic interest. It permits comparison of the areas of duricrust phenomena with major climatic zones, vegetation types, geological structures, great soil groups and relief features. By drawing upon a wide range of sources, a general explanation integrating both climatic change and plate tectonics becomes possible, especially for the oldest indurated crusts which may well have developed before the major latitudinal changes in position of the continents since the Cretaceous. However, before putting such a proposition forward, previous work must be reviewed.

13.2 Critical review of previous work

Basic principles

Today, a geographer wanting to find out about world duricrust phenomena would have only two authoritative sources: the incomplete synthesis by Prescott and Pendleton (1952) concerned only with Africa, India and Australia at an inconvenient scale, 1 : 80 million; and the FAO–UNESCO (1973) *World Soil Map* at 1 : 5 million which distinguishes two facies of duricrust, one continous and one discontinuous. For Africa the sources are improved by J. D'Hoore's (1963) soil map (African Centre for Technical Co-operation CCTA at 1 : 5 million), by sheet 9 of the *International Atlas of West Africa* (1971) (Organisation for African Unity OUA at 1 : 5 million) and by the 1961 East African map at 1 : 4 million from the Overseas Survey. Beyond these general sources, reference must be made to such national surveys as exist and elsewhere to local studies. Since the FAO–UNESCO map alone has world coverage, it is vital that a common terminology is developed. The international agencies recognise two facies common to the tropical world, with Australia having a unique facies. Three phases are therefore mapped.

(a) 'Petric phase': these soils are characterised by a horizon at least 25 cm thick, 1 m below the surface, at least 40% of its volume comprising ferruginous concretions, indurated plinthite, duricrust or other fragments: thus a discontinuous uncemented concretionary layer.
(b) 'Petroferric phase': similar to the above profile but with a continuous layer, cemented by iron with at most a trace of organic matter; the continuity of the indurated horizon may be broken by fractures ten or more metres apart.
(c) 'Duripan phase': the widespread subsurface layer in Australia cemented by either metallic oxides, silica, carbonates or sulphates; in other words, the calcareous, gypsiferous, siliceous or iron-rich duricrusts known as calcrete, gypcrete, silcrete or ferricrete. The duripan phase is too comprehensive to be of practical value.

Overall, there is a general agreement between the FAO–UNESCO classification and that of D'Hoore; the first two phases (petric and petroferric) may be equated with the soil classes Ab (denuded duricrust) and Bc (lithosols on ferruginous crust) acknowledging that areas mapped as Bc include important remnants of broken or partially denuded continuous duricrust (J. D'Hoore, personal communication). The *International Atlas of West Africa* distinguishes lithosols on duricrusts from more or less indurated ferruginous or ferallitic soils – those with a gravel or duricrust fragment horizon.

Since there appears to be a certain consensus between the various sources which distinguish the continuous or discontinuous horizons, whether broken or denuded, it is possible to move to a critical study of the existing general synthesis maps, referring naturally to the FAO–UNESCO world soil map and considering each of the tropical continents in turn.

Critical study of published maps and reports

In view of the limited number of general synthesis maps, frequent reference will have to be made to the FAO–UNESCO map which is regarded as the basic document.

Tropical Africa. The *World Soil Map* reveals a single northern latitudinal band if the isolated spot between Ogooué and Congo astride the Gabon/Congo border is ignored. The indurated sector is broad in the west, occupying some 1200 km between the shores of the Ivory Coast and 17°N, but narrows to the east after extending to the north of the great bend of the Niger but south of Lake Chad, to disappear at 32°E and 6°N. It thus suggests on the one hand that there is no duricrust south of the equator or east of the Rift Valley, and on the other hand, the division of the band into two big pockets, one southwest of the Senegal–Niger line and the other in the east stretching from Lake Chad to the Congo–Nile divide. These authors thus generally restrict duricrust phenomena at the northern fringe of the savanna areas while emphasising a western indurated zone covering 11% of the surface west of the Niger River.

The CCTA map clearly distinguishes denuded duricrusts (Ac) from discontinuous indurations and crusts (Bc) covered by ferruginous or ferrallitic soils, which together form a crescent oriented NW to SE from 20°N to 15°S. On the one hand this suggests a vast extent of duricrust phenomena in East Africa north of a Zambezi–Ruvuma line, but on the other hand the virtual absence of such phenomena in the whole of the Central and West African humid tropical forest zone. Could this particular delimitation represent the cartographic expression of a genetic concept? The fact that the author has indicated the limit of the rainforest might confirm such a view. In this case then, duricrust phenomena would be essentially characteristic of savanna areas, being capable of reaching relatively high latitudes (20°N) north of the great bend of the Niger. In comparison with the FAO map, a reduction by about one-third of the extent of duricrust in West Africa may be noted, but also the extensions into higher latitudes in both hemispheres and two pockets of induration, one each side of the Rift Valley, south of the equator, and the other on the East African high plateaux.

The *International Atlas of West Africa*, which distinguishes lithosols on duricrusts from the indurations of ferruginous and ferrallitic soils, seems to exhibit a conservative view of the extent of duricrust. The whole mapped zone is basically a mantle, with consequential area of buried duricrust, extending from the Sassandra to the Volta. In contrast to the other two maps, there is absolutely no trace of the duricrust in the Mali sector of the Niger bend. For Africa as a whole, the differences between the FAO–UNESCO map and the CCTA map are of greater importance than those of the *International Atlas of West Africa* whose duricrusted area within the Niger–Senegal curve is limited to 6% of the total land area. As for Madagascar, while the FAO map indicates duricrusts on the high plateaux north of Tananarive (the Tampoketas) as well as on the southeastern coastal slopes, that of D'Hoore only mentions a weak induration on the Hautes Terres.

Tropical America. The *World Soil Map*, the only general synthesis available for tropical America, provides very generalised, succinct information on the extent of the petric facies alone. America would not have undergone evolution as far as the duricrust stage, but should have large areas of gravelly, nodular horizons between latitudes 10°N and 15°N. These large spreads of induration are found in several major provinces:

(a) the western Llanos, between the Guaviare and Meta rivers;
(b) the Guianas;
(c) the northern flank of the Brazilian shield to the south of the Amazon; this region can be divided in three sectors, the interfluve between the Madeira and Tapajos, on both sides of the Xingu and the Nordeste area.

Overall, about 6% of this part of America, excluding the Andean zone, should be covered by petric facies indurations.

South Asia and the Far East. Insular and continental, tropical Asia extends over such a range of latitudes and longitudes that it may be considered in terms of major sectors. For India, there are few differences between the FAO–UNESCO and Prescott–Pendleton maps. Duricrust is found around the edges of the Deccan plateau as far north as the Tropic of Cancer in the north-east of Bihar. It is also found in the eastern, dry zone of Sri Lanka.

A second area includes the equatorial islands, the east of Sumatra, south of Borneo, southern Sulawesi and the western peninsula of New Guinea. Here, as in Africa, according to the FAO map, the petric and petroferric facies are both found. Does Indonesia in fact provide a good example of duricrust in a humid tropical area? Even if the limited specialised literature has little to say about such a distribution, it should be recognised that such duricrust is not just a rare, local phenomenon, for, according to the FAO map, about 4% of the Indian peninsula south of the Tropic, 7% of Sumatra and Borneo, 8% of Sulawesi and 14% of Sri Lanka are covered by duricrust. This map also indicates duricrust phenomena in southern Taiwan, Leizhou Bandao, northern Hainan and central Yunnan.

Australia. The Australian continent has the various regional surveys and general syntheses made by Divisions of the CSIRO in addition to the excellent map of Prescott and Pendleton and the FAO–UNESCO map which groups together under the term 'duripan' too many varied indurations to be practically useful. In the *Handbook of Australian Soils*, (Stace *et al.* 1968) is a map distinguishing two soil categories – podsolic lateritic soils (ironstone gravels) and those with the same gravels but with a sandy, red earth or yellow matrix. Even though well endowed with excellent soil studies, Australians have the problem of varied interpretations in the literature. Whatever the correct interpretation may be, according to Prescott and Pendleton, 13 or 14% of the surface of the continent is duricrusted over a great crescent deployed around the north of the central desert, with the western tip reaching Pointe d'Entrecasteaux in the extreme south-west. There are also duricrust remnants at 41°S in Tasmania.

As for Africa and India, the Prescott and Pendleton map remains of greatest value, requiring only local corrections.

13.3 Mapping of the duricrusted areas at the 1:20 000 000 scale

The different duricrust maps reflect well the diversity of both definitions and interpretations of such phenomena. In presenting the new maps (Figs 13.1 & 13.4) it is appropriate to commence by explaining the objectives and methodology adopted.

The approach adopted

Most of the data used were derived from existing publications, especially the FAO–UNESCO soil map, which remains the only world synthesis, and from consultations with a variety of organisations (ORSTOM and BRGM essentially) and individual experts on duricrusts. While the multidisciplinary organisations set up during the colonial period have created a rich set of scientific work, nothing comparable is available for Latin America south of the equator. India, thanks to the map by Delacourt and Troy (1971), and Australia, through the work of the CSIRO, have knowledge beyond the basic survey stage, while in Indonesia little official information is yet available.

By adopting a simple proposition, a consensus of the views of all advisers may be obtained: the concept of duricrust formation [cuirassement in the original–Editors] lies fundamentally in a continuous and permanent induration with metallic oxyhydroxides, excluding any soil profile layer which becomes compacted seasonally by desiccation, such as plinthite or Buchanan laterite, certain gravelly or nodular horizons or colluvial layers with a clay matrix. The notion of duricrust can be restrictive or extensive according to whether a genetic sense or simply a descriptive sense is implied. Should the term be reserved strictly for a mobilisation or immobilisation of oxyhydroxides of iron or aluminium during pedogenesis or be used for any such concentration whatever the origin? Indeed, is it necessary to reserve the term 'duricrust' [cuirasse] exclusively for continuous and permanent indurated layers of pedological origin and not for the geologic strata often called 'cuirasse de nappe' in the African terminology, or for iron cappings, or certain lake deposits or the indurations due to human activity such as forest clearance? A broad position is adopted here, the intention being to delimit the tracts which have experienced or now experience tendencies to concentrate metallic oxyhydroxides; in other words to demarcate several great tropical ecosystems which have left clear traces in the present-day landscape. As the viewpoint is geomorphological rather than pedological, only denuded or covered, continuous, massive or broken duricrust which has left its traces on the slopes or divides is considered. The scale chosen does not permit distinction of nature, composition, morphological expression or age of the duricrusts. However, comparison of this map with other documents should give rise to ideas for working explanatory hypotheses.

Thus a broad outline of the ecological conditions favouring indurations expressed by pseudo- or true concretions, by debris or blocks of nodular duricrust has been retained. Within these broad limits, the areas of homogeneous duricrust are indicated by shading, showing whether it is continuous, stripped, denuded or buried. Asterisks and full circles are used to indicate occurrences which cannot be mapped accurately but which are either noteworthy or highly localised. Although indurations are not directly excluded, some significant aspects of them have been emphasised. Since the aims of this mapping are basically geomorphological, it is thought that a simple mapping scheme will not only reveal both the latitudinal extent and general distribution but also suggest the diversity of genesis and evolution.

The lessons

The implications of the maps should be considered at two scales, that of the continents where inadequacies of previous maps can be corrected, and that of the world view where an overall picture can be obtained.

While leaving the reader to make comparisons with the *World Soil Map* at 1 : 5 million, some fundamental features may be emphasised:

(a) Duricrust phenomena in tropical America (Fig. 13.1), although restricted, are of considerable importance on the Guyana shield and its detrital margins (Llanos).

(b) In Africa (Fig. 13.2), duricrusts are not merely found north of the equator, but occur on various erosion surfaces from Angola to Mozam-

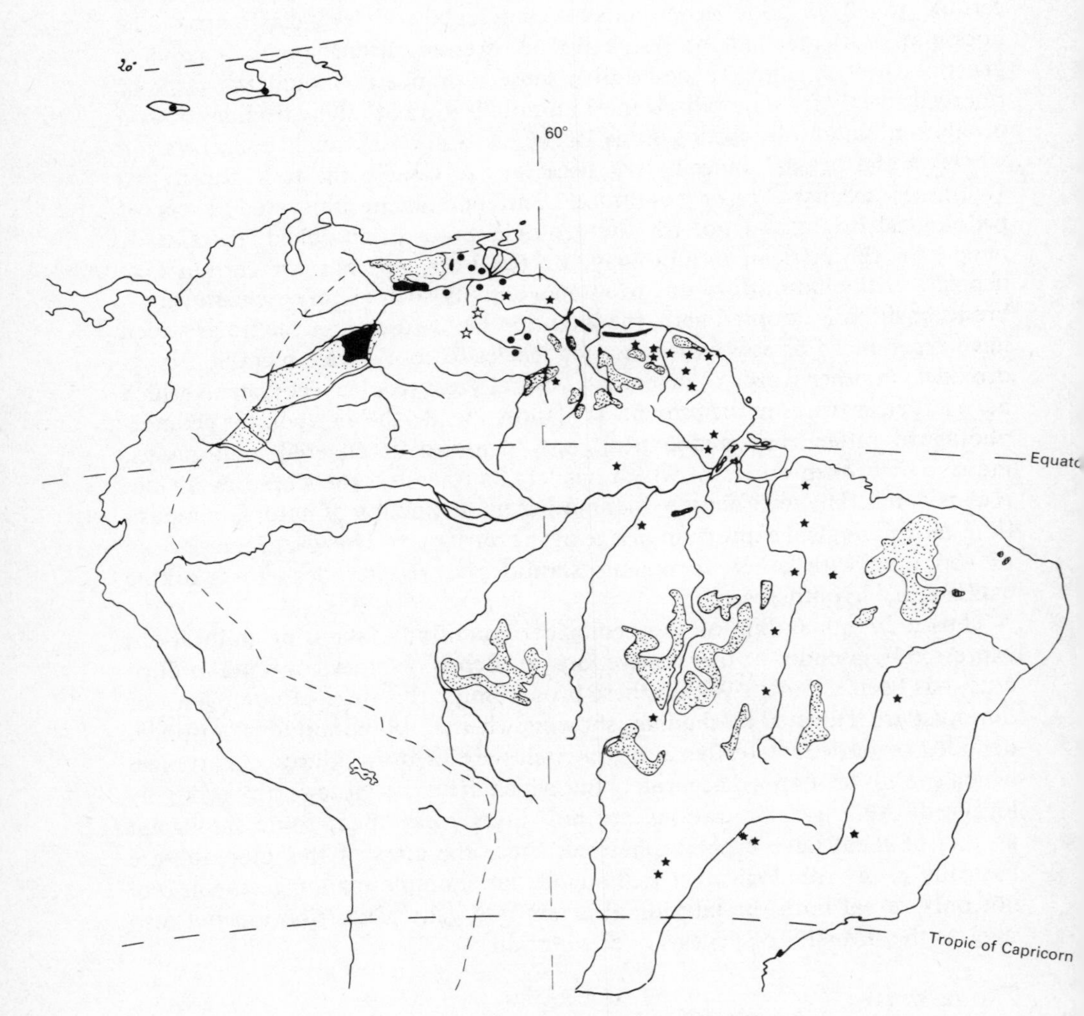

Figure 13.1 Duricrust in tropical America.

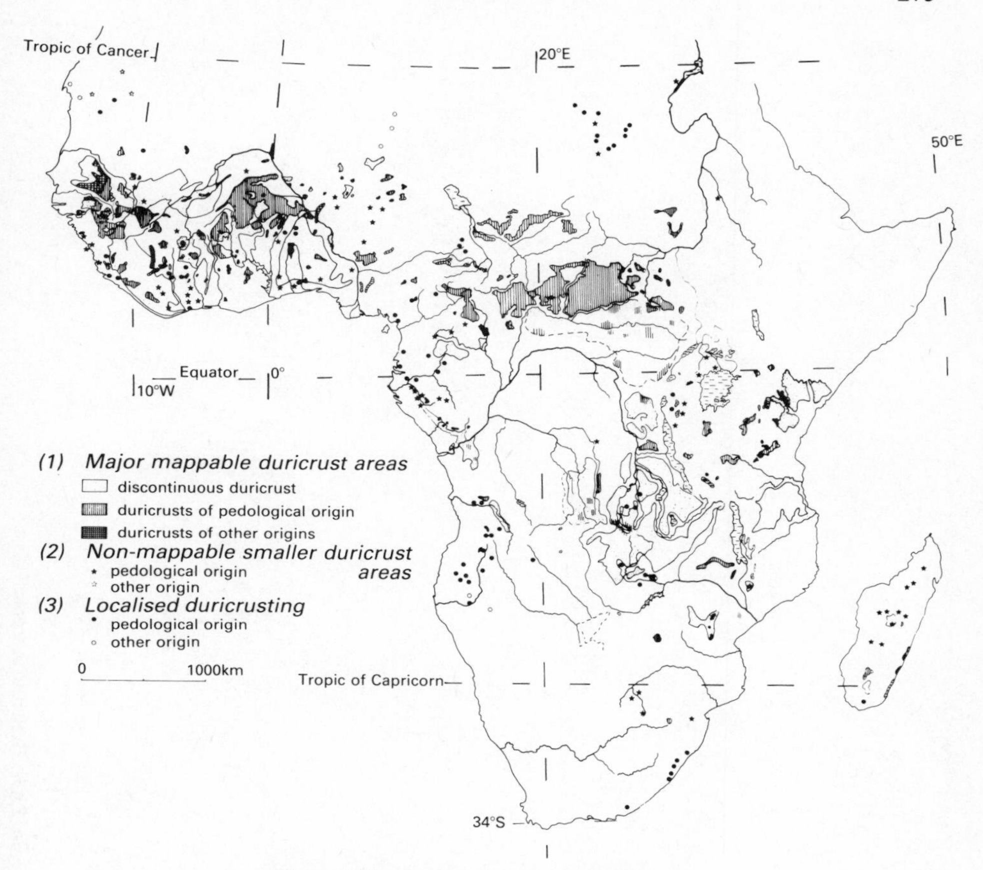

Figure 13.2 Duricrust in tropical Africa.

bique, especially on the Zaire–Zambezi interfluve, and also in the eastern part of the tropical area.

(c) In equatorial Asia (Fig. 13.3), duricrusts seem to be closely linked to the prolonged and significant human impact on the landscape.

(d) In peninsular India (Fig. 13.3) and in Australia (Fig. 13.4), these phenomena covering 4 to 6% of the area appear to be peripheral: edges of the Deccan plateau, margins of the Australian deserts.

However, beyond these anecdotal spatial impressions is a much more rewarding global view.

The first lesson to be learnt from the four maps relates two fundamental facts to the vertical and lateral movements of the Earth's crust. The southern limit of duricrust is at a much higher latitude than the northern limit which barely reaches the Tropic of Cancer. The greatest latitudinal extent of duricrust is over 65° in the Far East and Australia, from 25°N to 41°S. This

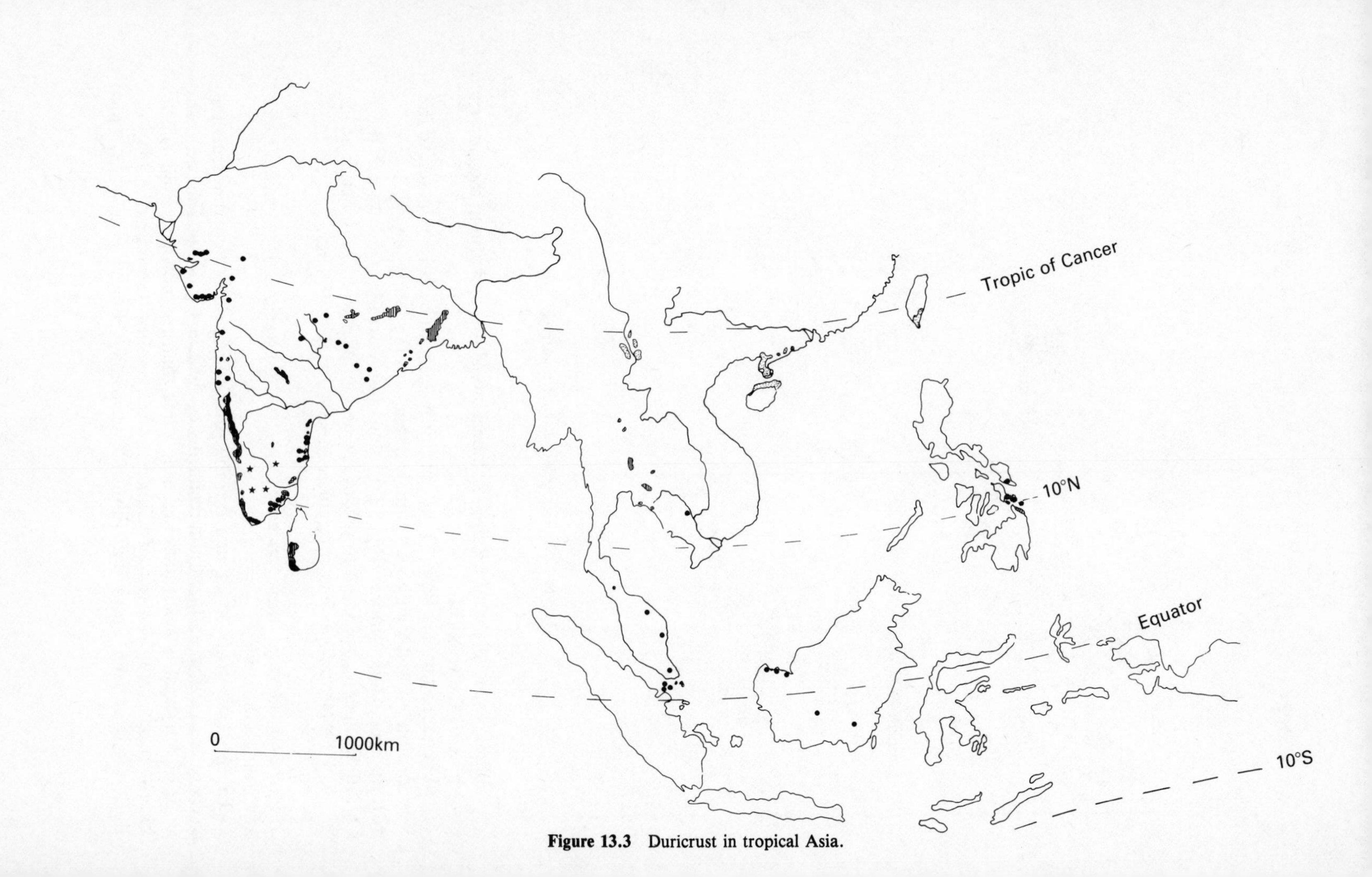

Figure 13.3 Duricrust in tropical Asia.

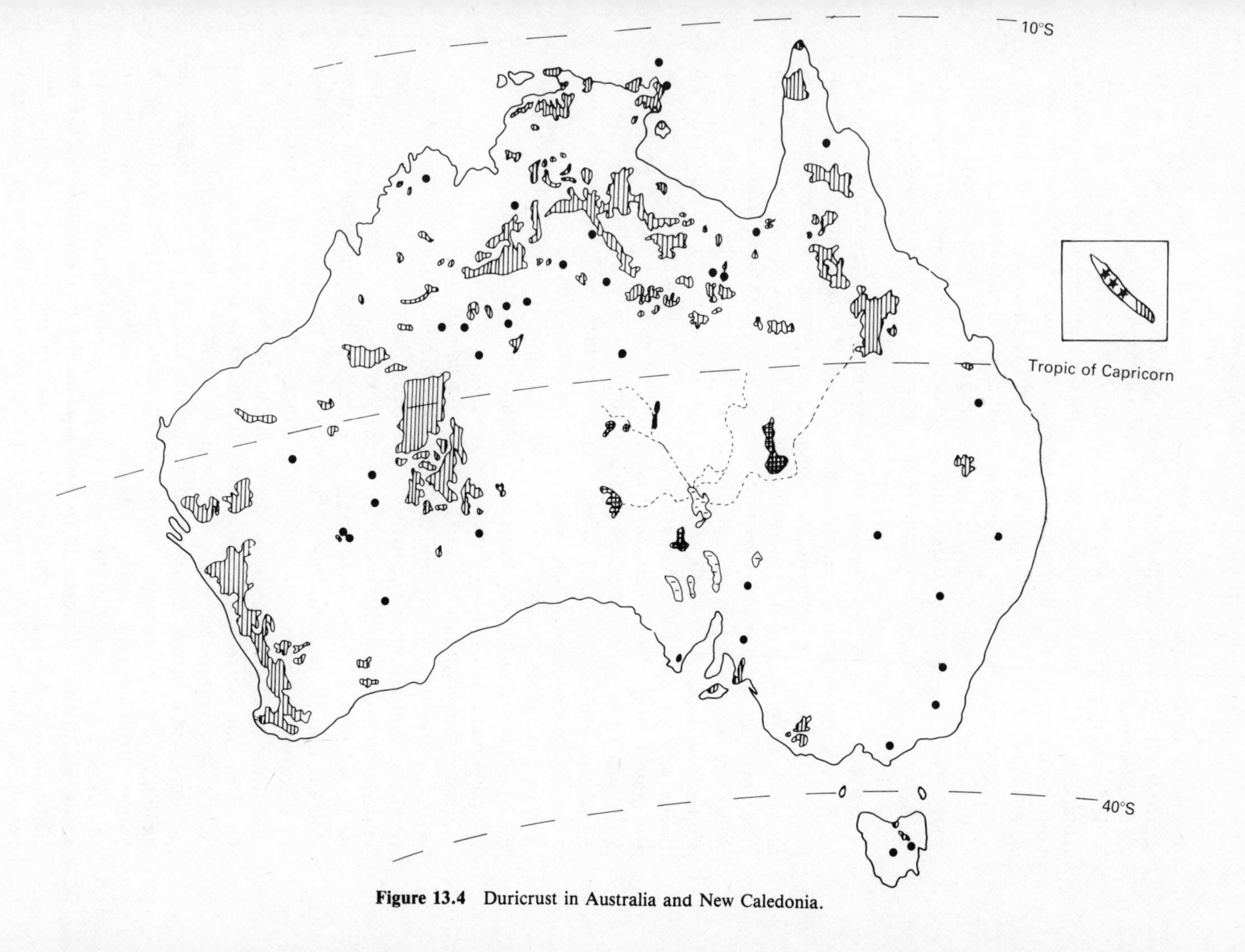

Figure 13.4 Duricrust in Australia and New Caledonia.

distribution may be related to the breakup of Gondwanaland; the most mobile plates would have moved to high latitudes away from the tropical area of duricrust development. From Africa, through Madagascar to Australia, a wide distribution is found, while in Latin America duricrusts are strictly intertropical. This spatial standpoint related to plate tectonics leads to two further considerations: the first is temporal, suggesting a phase of duricrust development before the breakup of Gondwanaland in the Cretaceous and early Tertiary; the second is climatic and invokes the delicate problem of the relative positions of each plate to the thermal equator at any given time (Bardossy 1973). These observations suggest that the simple explanation of a sequence of global climatic changes is inadequate. The dynamics of the Earth's crust must also be considered.

Not only lateral but also vertical displacement must be considered when the map is compared with the distribution of relief. These relationships are not fortuitous but have a recurrent geographical pattern. Certain structural domains seem to favour duricrusts: the marginal uplands of the continents (the Guyanas, Madagascar, western and eastern peninsular India, Western Australia, and the extreme west of Africa); the borders of the rift valleys similarly, although with slight differences (East Africa around the Congo–Nile divide and western Kivu, eastern Brazil on either side of São Francisco), the main continental divides (of West and Central Africa and in Shaba and Zambia). These favoured zones also have a common altitudinal zonation of duricrusts, bauxites on the highest levels, essentially ferricretes at the lower levels. Such a widespread distribution throughout the tropics poses the question of the age of the indurations. Is it a question of a dislocated very old duricrust, raised to higher altitudes by subsequent tectonic movements, or did the movements occur before duricrust development?

Even if the majority of research workers would support the first of the suggestions, it should be recognised that our knowledge of the climatic controls of the mobilisation of oxyhydroxides of metals as a function of parent materials is still limited. Workers such as Troy for the Nilgiri Hills or Gense on La Réunion have provided precise information about allitisation as a function of altitude on well drained basic rocks. If the general synthesis map reveals some general conclusions, it also suggests that some areas have abnormal features. On the one hand there is the absence of duricrust between the Orinoco and the foot of the Roraima massif, even though local conditions would appear favourable: the iron available in the itabirites, extensive level surfaces and a seasonally wet tropical climate with 1000 to 1400 mm annual rainfall. On the other hand, there is much duricrust development on the sandstones of the Roraima itself and on the Plio–Pleistocene detrital deposits of the Uanos. Africa south of 20°S and Madagascar have little duricrust development, a situation which could be explained in the first case by smectites of neoformation developed on basic rocks, and in the second case by active tectonics; while in West Africa, as in the Guyanas, the most developed duricrusts are situated on amphibolites and gabbros.

Finally, there is a considerable duricrust development on Plio–Pleistocene depositional surfaces such as the continental terminal of Africa, the Barreira Formation in Brazil, the red sandstones of Cuddalone (Mio–Pliocene) between the Cauvery and Godavari in India and the sediments around Lake Eyre in Australia. Essentially these detrital deposits have duricrusted horizons with hardpan layers which locally extend onto the adjoining shield rock outcrops. This type of distribution suggests mobility of iron and aluminium of allochthonous origin, while the duricrusts at higher altitudes are strictly dependent on parent rocks.

13.4 Conclusion

If the narrowly restricted approach involved in these maps does not permit a detailed evaluation of the different forms of duricrust, the absence of such a distinction may be excused because there is still a need to relate the nature and genesis of duricrusts to morphostructural conditions. Such work would be far better undertaken at a different scale. However, comparison with relief, climate, vegetation and other world maps has suggested several relationships or apparent contradictions which suggest tentative hypotheses that would not emerge from the detailed analysis of a few exposures.

As this map of world duricrust phenomena still has many imperfections, the author would be most grateful for any suggestions on how it may be improved.

List of authorities consulted

Latin America

Colombia	J. P. Tihay, Université de Pau
Venezuela	D. & M. Pouyllau, CEGET Bordeaux–Talence
Brazil	J. P. de Queiroz Neto, Université de Sao Paulo

Africa

General information	Messieurs Maignien and Chatelin
Cameroons	S. Morin, Université de Yaounde
Niger	J. P. Karche, Université de Besancon
Upper Volta	B. Kaloga, ORSTOM
RCA	Y. Boulvert, ORSTOM
Zaire	J. Alexandre, Université de Liège
	J. D'Hoore, Université de Louvain
Sudan	E. Motti, BRGM, Service de télédetection, Orlèans

Asia

India	J. P. Troy, IGREF, Paris
Sri Lanka	J. Louchet, Université Paris IV
Indonesia	P. Brabant, Mission ORSTOM, Bogor
	F. Blasco, Université de Toulouse

Australia

C. R. Twidale, University of Adelaide

14

Tectonic background to long-term landform development in tropical Africa

M. A. Summerfield

14.1 Introduction

Tropical geomorphology can be considered in at least two distinct ways. First, it can be regarded as the study of landforms in relation to a particular assemblage of climatically determined sub-aerial processes. Ideally it is possible to relate these to characteristic forms, in spite of the increasingly appreciated significance of Quaternary climatic change in the tropics (Street 1981). Secondly, tropical geomorphology can be viewed simply as the analysis of landscapes occurring in the intertropical zone, examining both exogenic and endogenic processes, especially when considering large-scale features (Ollier 1981, Scheidegger 1979, Summerfield 1981).

Over much of the tropics the combination of ancient and relatively stable continental platforms with a predominance of particular lithologies has contributed to a recurrent pattern of landform development. Elsewhere, especially around the margins of the Pacific, a much more active tectonic environment associated with plate convergence has produced a quite different type of landscape. Clearly the role of tectonics in the creation of landforms in the tropics must be evaluated if we are to avoid misleading notions of 'typical' tropical landscapes. Indeed it has been suggested that large-scale tectonic factors probably constitute the major control over the geomorphological diversity of the humid tropics (Douglas 1978b).

The landscapes of Africa have been important in formulating our ideas about tropical landform evolution but in recent years have seldom been considered by geomorphologists in terms of tectonics. There is consequently a need to examine some of the advances that have been made over the past decade or so in our understanding of the macroscale tectonics of tropical Africa as a basis for interpreting its geomorphological development.

Morphological and structural outline

The distinctive morphological and structural features of the African continent include a high mean elevation (nearly 50% being above 500m; Bond 1978), basin-and-swell topography, depositional and erosional surfaces typically of low local relief, and extensive rift valley systems and triple junction structures (Fig. 14.1). Although unaffected by post-Mesozoic orogeny, except in the extreme north-west, Africa has experienced a long history of epeirogenic uplift extending back to the Palaeozoic and producing a gross surface morphology of extensive basins separated by broad swells (Holmes 1965). These are locally surmounted by smaller domal uplifts which attain elevations of around 1000 m in the north and west, but up to 2000 m in the south and east. Superimposed on these structures, and intimately related to them, are sequences of depositional and erosional surfaces, covered extensively by duricrusts and other weathering deposits, and separated by spectacular escarpments which constitute one of the few features producing high local relief on the continent.

The line of swells and domes extending from the Afar region southwards to the Zambesi is bisected by a series of rifts radiating from triple junctions on dome crests and collectively forms the East African Rift System (McConnell 1972). Other major rifts and associated triple junctions include the Benue Rift of Nigeria and the complex structure along the southeastern coastal margin. A number of these rifts are currently active structures, but others are inactive, or reactivated, features dating back to the Mesozoic and beyond.

The three major cratons of Africa (Fig. 14.1) comprise areas which have been unaffected by significant regional tectonic or thermal activity for the past 1100 Ma. They are of low mean elevation and are underlain by relatively thick lithosphere virtually devoid of contemporary seismicity (Gass *et al.* 1978, Sykes 1978). The intervening non-cratonic areas are more elevated, more seismically active, and overlie relatively thin lithosphere.

Chronology of uplift and denudation

The tectonic and erosional history of the African landscape has received much attention yet remains uncertain (Bishop & Trendall 1967, Dixey 1938, King 1962, 1976, Veatch 1935). Figure 14.2 provides a chronological outline of the breakup of Gondwanaland and subsequent major rifting events along the African continental margin together with the timetable of uplift and erosion envisaged by King (1962, 1976). The late Jurassic–Cretaceous tectonism related by King to the fragmentation of Gondwanaland is confirmed by more recent stratigraphic, palaeontological and palaeomagnetic evidence, while the cessation of the prolonged 'African' planation approximately 25 Ma ago correlates closely with the continent-wide increase in volcanism and uplift documented by stratigraphic evidence and K/Ar dating (Burke & Wilson 1972, Briden & Gass 1974). The late Pliocene–Quaternary uplift which was concentrated along the coastal margins of Africa is also supported by the reduction during this period in the proportion of land area below 200 m indicated by hypsometric studies (Bond 1978). Comparison of palaeohypsometric curves also indicates that Africa has been rising relative to the other continents since

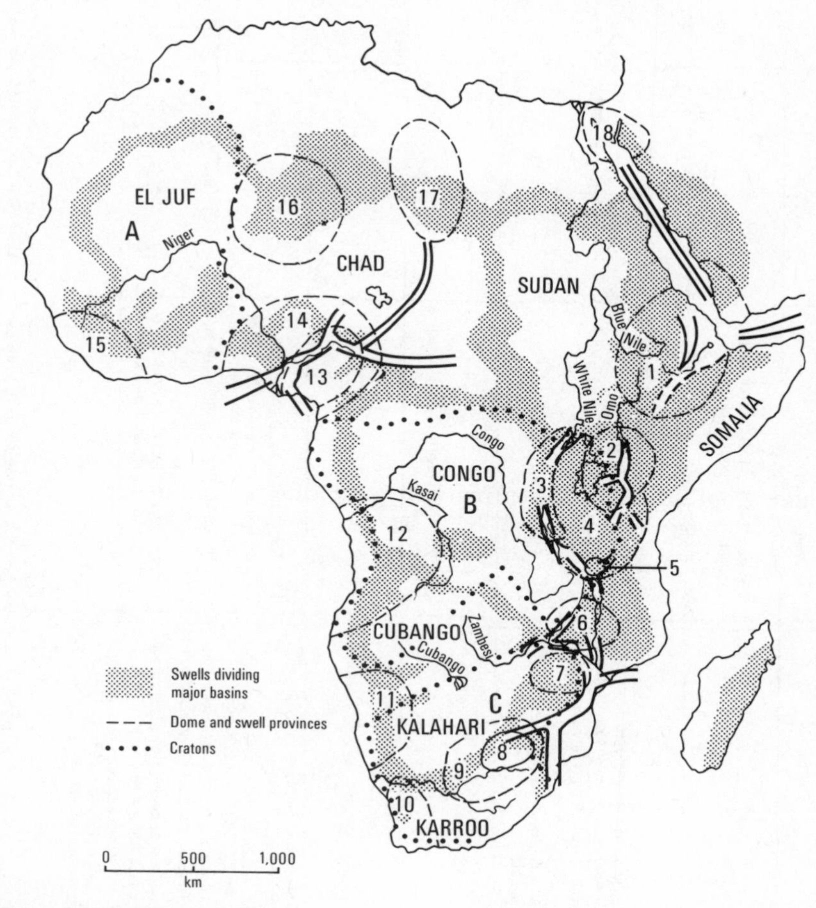

Names and approximate age

1 Ethiopia-Aden (Cenozoic)
2 Kenya (Cenozoic–Holocene)
3 Western Rift (Uganda–Zaire) (Late Cenozoic–Holocene)
4 East African (Late Cretaceous–Cenozoic)
5 Rungwe (Late Cenozoic–Holocene)
6 Chilwa–Zambesi (Cretaceous)
7 Zimbabwe (Early Mesozoic)
8 Transvaal (Precambrian)
9 Kimberley–Pretoria (Mesozoic)
10 Bushmanland–Sutherland (Late Cretaceous–Early Cenozoic)
11 Namibia–Uruguay (Mesozoic)
12 Angola–Southern Brazil (Cretaceous)
13 Cameroon (Cenozoic–Holocene)
14 Nigeria–north-east Brazil (Cretaceous)
15 Guinea–Guiana (Cretaceous)
16 Hoggar (Cenozoic–Holocene)
17 Tibesti (Cenozoic–Holocene)
18 Suez–Sinai (Cenozoic)

Abbreviations

L.C. Lake Chad
L.R. Lake Rudolf
L.A. Lake Albert
L.V. Lake Victoria
L.T. Lake Tanganiyika
L.M. Lake Malawi
O.D. Okavango Delta
S The Sudd
B Benue River
A Awash River
S Shire River
L Luangwa River

Figure 14.1 The morphological and structural elements of Africa. Basin nomenclature and delimitation from Holmes (1965, Fig. 763, p. 1054). Domes (swell provinces) after Le Bas (1971, Fig. 1). Rift systems and triple junctions primarily from Burke and Whiteman (1973, Fig. 4).

Epoch	Era	DEVELOPMENT OF AFRICAN CONTINENTAL MARGIN[1]	DENUDATION/TECTONIC CHRONOLOGY FOR SOUTHERN AND CENTRAL AFRICA[9]	DISCONTINUITIES IN SEA-FLOOR SPREADING[10]: CENTRAL ATLANTIC	SOUTH ATLANTIC	INDIAN OCEAN[11]
QUATERNARY			'Congo' cycle of deep gorge-cutting in eastern and western coastal margins ('Youngest' Cycle)			
			— Strong Cymatogeny —			
PLIOCENE	CENOZOIC		Second phase of Late Cenozoic valley planation ('Widespread' Landscape)		▶10 Ma	
			— Slight Uplift —			
MIOCENE	CENOZOIC	▶20 Ma Rifting of Red Sea accomplished[2] ▶25 Ma Beginning of major phase of post-Gondwana intra-continental rifting–continues to present[3]	Broad valley-floor pediplains developed in 'African' surface ('Rolling' Landsurface)	▶25–21 Ma		▶20–18 Ma
			— Widespread Epeirogenic Uplift —			
OLIGOCENE	CENOZOIC	▶35 Ma Initiation of igneous activity associated with Red Sea rift				
EOCENE	CENOZOIC		'African' landscape of extreme planation ('Moorland Planation')	▶53 Ma	▶53–50 Ma	
PALAEOCENE	CENOZOIC			▶63 Ma		▶55–53 Ma
CRETACEOUS	MESOZOIC	▶95 Ma Final break between Africa and S. America indicated by faunal evidence[4] ▶112 Ma South Atlantic not yet opened appreciably according to palaeo-magnetic evidence[5] ▶127 Ma South Atlantic spreading ridge axis formed[6]	Mid-Cretaceous Disturbances 'post-Gondwana' erosion concentrated in upwarps ('Kretacic Planation') — Fragmentation of Gondwanaland —		▶80 Ma ▶112–110 Ma ▶140–130 Ma	▶80–75 Ma ▶140–125 Ma
JURASSIC	MESOZOIC	▶180–130 Ma Major phase of intra-continental rifting ▶180 Ma Separation of Antarctica from eastern margin of southern Africa – India rifts[7]	'Gondwana' landscape of extreme planation ('Gondwana Planation')			
TRIASSIC	MESOZOIC	▶200–180 Ma N.W. Africa begins to separate from S. E. margin of N. America[8] Pre 200 Ma Gondwanaland intact				

Ma: 0, 20, 40, 60, 80, 100, 120, 140, 160, 180, 200

Key

1 Largely derived from Windley (1977)
2 Kinsman (1975b)
3 Burke and Whiteman (1973)
4 Reyment and Tait (1972)
5 Gidskehaug *et al.* (1975)
6 Ladd *et al.* (1973), Larson and Ladd (1973)
7 Dingle (1973); Dingle and Scrutton (1974), Scrutton and Dingle, (1976)
8 Noltimier (1974)
9 data from Schwan (1980) (Fig. 5), based on various sources
10 excludes data for eastern Indian Ocean
11 According to King (1962), Table VII p. 242 and Table VIII – later modified terminology (King 1976) in parentheses

Figure 14.2 Outline chronology for the formation of the continental margins of Africa, denudation and tectonic in southern and central Africa, and discontinuities in sea-floor spreading around the African Plate.

the late Cretaceous, although currently available data are consistent with continuous uplift since the Eocene as well as more episodic epeirogeny (Bond 1978).

14.2 Domal uplifts and rift systems

The often close spatial and temporal association of uplifts, rifts and volcanism in Africa implies some genetic relationship between them. Two scales of epeirogenic uplift can be identified. Swells comprise broad areas of uplift 1000 km or more across with crestal elevations of 1–3 km. Such features have been termed 'lithospheric domes' by Le Bas (1980) since they may be a result of phase changes deep in the upper mantle. Domes are less extensive features typically less than 500 km across but with elevations similar to swells. They appear to be related to anomalously low-velocity upper mantle at a depth of around 30–50 km and are termed 'crustal domes' by Le Bas (1980). As exemplified in East Africa, domes (such as the Kenya, Rungwe and Western Rift domes) may be superimposed on swells (the East African plateau). Dome crests are commonly broken by rifts and are also characteristically associated with alkali basaltic volcanism.

The location of the rift systems of Africa appears to be related to some extent to ancient structural trends. The East African Rift System, for instance, parallels the trend of Precambrian mobile belts and tends to follow the margins of older cratons (McConnell 1972). Ancient sutures and other deep lineaments are apparently more readily reactivated than adjacent crust and consequently provide more favourable sites for subsequent faulting and rift development. Nevertheless, in the case of the East African Rift System, Precambrian structures have been cut across in a number of places, suggesting some control by the stress pattern initiating rifting.

Rather than comprising a single phase of uplift and rifting, the dome and rift structures of Africa are at various stages of development. Two major periods of formation during the past 200 Ma have been identified; a Jurassic–Cretaceous rifting phase initiated between 180 and 130 Ma ago and chronologically related to the breakup of Gondwanaland, and a more recent period of rifting and volcanism starting 35–25 Ma ago and continuing to the present (Burke & Dewey 1973, Burke & Whiteman 1973).

The history of 16 uplifts (of the size of domes) and 29 rifts of Mesozoic–Cenozoic age in Africa has been documented by Burke and Whiteman (1973) and provides the basis for their evolutionary sequence of uplift and rift development. Three stages are recognised: (1) uplift; (2) uplift with alkaline volcanicity; and (3) rifting and triple junction formation. Within the African continent, development has in many cases failed to progress through all three stages and none has yet developed into the final phase of continental splitting.

The lack of associated rifting and volcanism may make the first stage difficult to identify. Burke and Whiteman (1973) cite parts of the Cameroon uplift in Adamawa as a possible example where elevations in excess of 2 km are apparently not associated with volcanism. The second stage is well illustrated in North Africa where Neogene volcanism has occurred on domal uplifts such as the Tibesti and Ahaggar plateaux. Although there is no direct evidence of these uplift stages in earlier Cretaceous features, structural evidence from the Atlantic continental margin of Africa suggests that uplift preceded the opening of the South Atlantic (Burke *et al.* 1971a).

The third stage is characteristically initiated by the development of three rifts radiating outwards from dome crests forming triple junctions. These are common along the East African Rift System (Fig. 14.1). They are of Jurassic–Cretaceous as well as Neogene age, and in the south Cretaceous structures have been reactivated by Neogene rifting. Burke and Whiteman (1973) recognise 11 triple junctions in Africa and a key problem relates to their possible subsequent development into spreading centres. Spreading can apparently only occur when continental separation can be accommodated into the motions of the global plate system. Where this is not possible rift evolution is halted, the rifts become filled with sediment and the uplift subsides. Although the triple junctions identified in Africa have variously developed by spreading along one, two or all three rift arms, the most common sequence has been for one arm to remain inactive, while spreading occurs along the other two (usually those most nearly orientated north–south). Of the 29 rifts considered by Burke and Whiteman (1973), five have led to continental separation and a further three have caused the rupture of Madagascar away from mainland Africa. The two rifts in the Benue Trough went through a phase of opening followed by later closing. Of the remainder, ten became inactive before the spreading stage and nine are currently active and may spread in the future.

The evolutionary sequence of Burke and Whiteman (1973) represents a useful first approximation but requires some qualification. For instance, much of the Western Rift and some of the Eastern Rift lack volcanics, while the volcanic centres represented by Mount Kenya and Kilimanjaro are situated some 100 km to the east of the Eastern Rift. In spite of numerous studies of individual rift development, such as those of the Neogene rifts of the Nakuru area which, on the basis of elevated erosion surfaces on rift shoulders, show that domal uplift preceded rift propagation (Baker & Wohlenberg 1971), agreement is lacking as to the timing of uplift and rifting. On the basis of a worldwide survey of continental margins, Kent (1980) has argued that upwarping does not invariably pre-date rifting and continental rupture. Indeed, recent work has questioned the widely held view that the formation of the Red Sea followed rather than preceded uplift of the newly rifted marginal crystalline basement (Gass 1977).

14.3 Mechanisms of uplift and rifting

In spite of the abundance of seismic, gravity and heat-flow data that has become available in the past decade there is still major disagreement as to the processes that have generated the warped and rifted surface characteristic of Africa. Prior to the recently presented tectonic models, King (1962) had placed great emphasis on the great crustal arches which he saw as typifying the African landscape. He referred to them as cymatogens and attributed their formation to the process of cymatogeny which he defined as 'a mode of major vertical (radial) deformation of the earth's crust' (King 1962 p. 638) related to the occurrence of tectonic gneisses at intermediate crustal depth. He provided rather generalised notions as to the ultimate cause of uplift, referring to the possible role of 'anomalous levitated subcrust' or 'activated upper mantle'.

Widespread acceptance of sea-floor spreading has led to the suggestion that this has produced overall compressive stress throughout the continent. McConnell (1977) has argued that lateral compression generated during Mesozoic sea-floor spreading has inhibited spreading along the East African Rift System, notwithstanding the presence of upwelling underlying mantle. Similar ideas have been proposed by Bailey (1964, 1974) who has suggested that the basin-and-swell topography and the major rift systems of Africa can be explained by compressive stress exerted from the mid-oceanic ridges of the Indian and Atlantic Oceans. According to this view doming is regarded as a mechanical response to lateral compression which, once initiated, can lead to subcrustal melting through localised decompression, and consequently localised regions of tensional stress with associated alkaline magmatism and rifting.

The marked increase in intra-plate volcanism in Africa 25–35 Ma ago has been linked to the coming to rest of the continent with respect to the underlying mantle (Burke & Whiteman 1973, Burke & Wilson 1972). Thermal anomalies (hot spots) in the upper mantle have consequently become effective in generating uplift and volcanism in the overlying stationary continent. Taking this idea further Thiessen *et al.* (1979) see Africa as possibly providing a 'window' to underlying mantle convection patterns. They have used the distribution of hot spots (areas of Neogene and Quaternary volcanic rocks) and high spots (uplifted areas lacking volcanism presumably attributable to the same process) to construct a pattern of polygonal cells. The size distribution of these cells is seen to match fairly closely the dimensions of mantle convection patterns predicted by experimental models (McKenzie & Richter 1976, Richter & Parsons 1975) scaled to a hypothesised depth of convection of 500 km. A major limitation here is the designation of hot spots and high spots, which Thiessen *et al.* (1979) acknowledge is conjectural.

Gass *et al.* (1978) have attempted to define more specifically the conditions under which uplift, volcanism and rifting in Africa are likely to occur. They suggest that one crucial factor is the velocity of the lithosphere over underlying thermal anomalies. A second factor is the time required to engender thermal effects. This is related to how heat transfer occurs, whether by conduction or by the upward movement of masses of magma (penetrative magmatism), and to the thickness of lithosphere through which the heat flow must pass. Heat transfer by conduction is too inefficient to be significant. Their general conclusion is that the vulnerability to penetrative magmatism and consequently the occurrence and intensity of volcanism and uplift decreases with increasing plate velocity and lithospheric thickness. Vogt (1981) has evaluated this claim both with respect to Africa and on a global basis and is not convinced that it provides an adequate explanation of the distribution of hot spots. Gass and his coworkers have responded with a more explicit modelling of the vulnerability of the lithosphere to penetrative magmatism (Pollack *et al.* 1981). They define a dimensionless vulnerability parameter $(\pi u/kd)^{1/2}\,l$, where k, the thermal diffusivity of the lithosphere (32 $km^2\ Ma^{-1}$), and d, the mean spacing of sublithospheric thermal perturbations according to the McKenzie *et al.* (1974) model (1500 km), are assumed constant; l is lithospheric thickness (km), and u is plate velocity (km Ma^{-1}). Variations in lithospheric thickness and plate velocity for Africa are shown in Figure 14.3A and the resulting vulnerability parameter contours for Africa together with hot-spot locations are illustrated in Figure 14.3B. Lower vulnerability parameter values indicate more vulnerable crust.

On a global scale the correlation between hot-spot distribution and lithospheric vulnerability is fairly convincing and additional support has come from other recent independent work (Sahagian 1980). However, when the African plate is considered separately the relationship actually observed is the opposite to that predicted, with higher hot-spot frequencies associated with less vulnerable lithosphere (high parameter values) than would be expected if hot spots were randomly distributed across the African plate (Fig. 14.3B). Although Gass *et al.* (1978) point out that Cenozoic volcanism is almost completely confined to the non-cratonic regions of Africa, thus suggesting some lithospheric control, there is by no means a perfect correlation between cratonic and non-cratonic areas and lithospheric thickness (compare Fig. 14.3A and 14.3B). As Vogt (1981) has pointed out, the resistance of cratons to penetrative magmatism may have more to do with the mechanical strength of cratonic lithosphere, and its consequent lower vulnerability to rupture and magma injection. This would also accord with the relatively low heat flow and low mean elevation of the African cratons.

In areas of cratonic crust with lithospheric thicknesses of the order of 200 km, even large thermal perturbations are unlikely to produce significant heat-flow anomalies and associated volcanism. In such situations the main surface manifestation will be uplift. The Gass *et al.* (1978) model gives an uplift of 500 m for a temperature increase of 300 K at the base of the lithosphere if the lithosphere is stationary. At a plate velocity of 5 cm yr^{-1} this would be reduced by a factor of about 2. The uplift effect of thermal expansion of the lithosphere alone will in fact be greater in absolute terms for thicker lithosphere but even where it is moving slowing or is motionless this expansion cannot fully account for surface relief of the order of the basin-and-swell topography of Africa. Clearly other factors are operating. These may include expansion in the relatively hot sublithospheric mantle and phase changes, which together with lithospheric thermal expansion can apparently account for the amount of surface uplift observed (Gass *et al.* 1978).

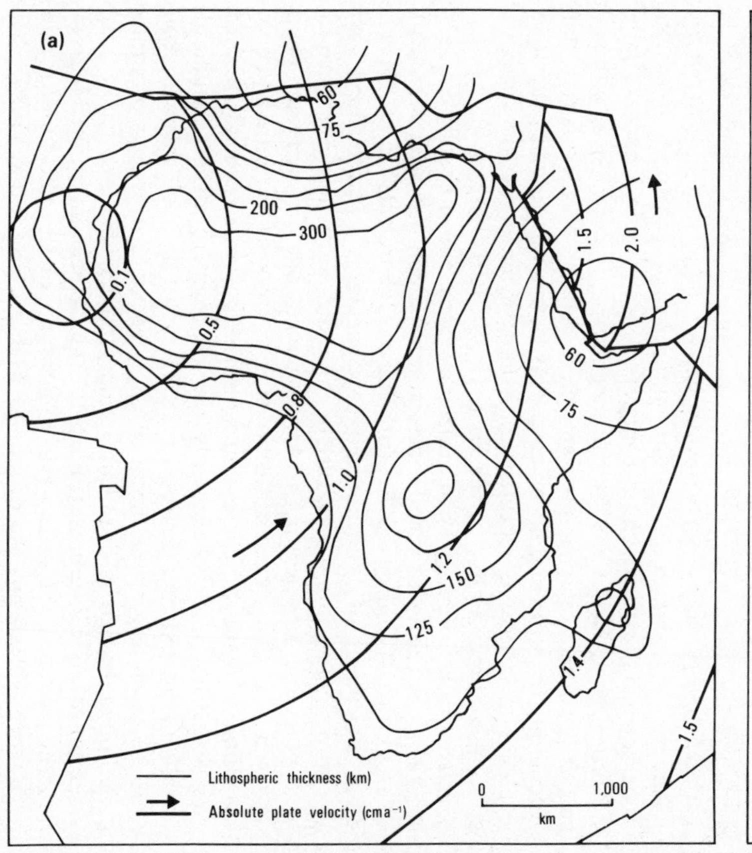

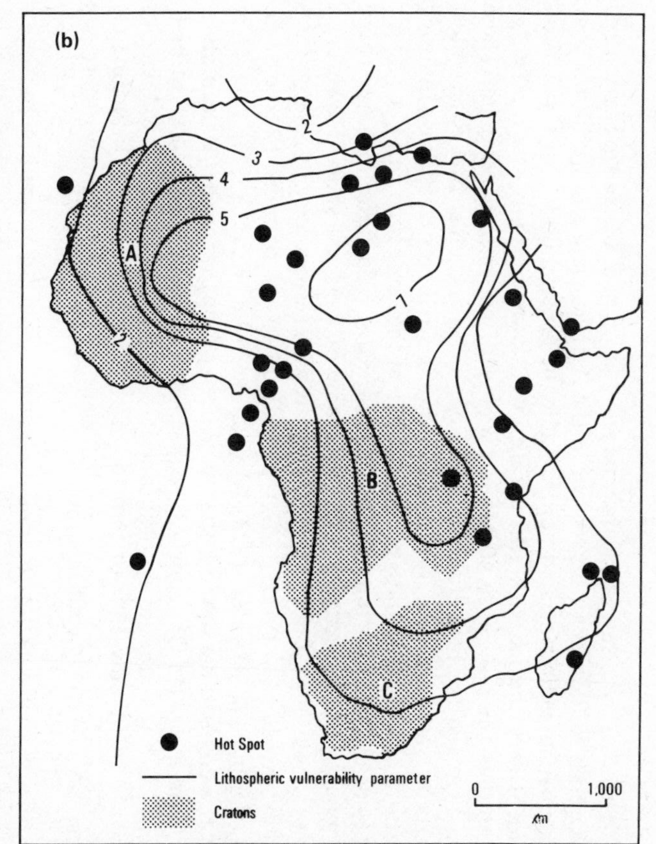

Figure 14.3 Lithospheric thickness, plate velocity, lithospheric variability and hot spots in Africa.

Two general criticisms of the thermal anomaly models of the kinds discussed above relate to the rate of movement of the African plate and the chronology of magmatism and associated uplift. While the general motion of the African continent over the past 200 Ma is fairly well known (Smith *et al.* 1981), its motion over the past 30 to 40 Ma is a matter of considerable dispute. There is some circularity in the Burke and Wilson (1972) argument that Africa must have been stationary for the past 25 Ma or so in order to have allowed volcanism and uplift to have occurred while at the same time citing magmatic activity as evidence for a motionless Africa. Most palaeopositions have been determined with respect to a reference frame of fixed hot spots. However, it is now recognised that hot spots may migrate with respect to each other at rates possibly exceeding 2 $cm\,yr^{-1}$ over periods of 40 Ma or more (Burke *et al.* 1973, Molnar & Atwater 1973, Molnar & Francheteau 1975), that is at velocities comparable to those that have been proposed for the African plate. The Gass *et al.* (1978) model is very sensitive to small differences in plate velocity with the amount of lithospheric thinning predicted at a velocity of 1 $cm\,yr^{-1}$ being only 50% of that for motionless lithosphere. Unfortunately palaeomagnetic methods are not yet sufficiently accurate to identify differences in plate velocities of this order.

Further evidence against the mantle plume origin for volcanism and uplift lies in the age trends of some intra-plate volcanic chains – which show no systematic change in the age of volcanic activity along the chain (Oxburgh 1978, Turcotte & Oxburgh 1978) – and in repeated volcanism over long periods of time along rifts which have migrated considerable distances (Bailey 1977). According to Bailey (1977) rift magmatism is permissive; although magma can potentially be generated anywhere below the lithosphere it only penetrates along fractures and lineaments. This view is supported by the abundant evidence in Africa of repeated uplift and rifting along old structural trends, notably along the line of the East African Rift System.

The active, rather than passive, role played by the lithosphere is also emphasised in the theory of membrane tectonics which has been specifically applied to Africa (Oxburgh 1978, Turcotte & Oxburgh 1973, 1978). The latitudinal movement of lithospheric plates generates stress through their adjustment to changes in the curvature of the geoid, particularly between the 'flattened' poles and the equator. As Africa moved equatorwards from about 160 Ma ago it would have been subject to tension towards its centre and compression along its margins. There would also have been a tendency towards east–west extension and consequently the propagation of predominantly meridionally orientated rifts (Fairbridge 1979). Uplift and volcanism are considered to be secondary effects initiated once tensional stress has promoted rifting. Although the East African Rift System (Oxburgh & Turcotte 1974) and the Cameroon volcanic lineament (Freeth 1978) have been interpreted in terms of membrane tectonics, the theory fails to account for all uplift and for non-linearly aligned volcanism. According to Maguire and Khan (1980) there is insufficient evidence to choose between the hot-spot and membrane tectonics models at present. Indeed Turcotte and Oxburgh (1978) themselves acknowledge that both mechanisms might operate.

14.4 Morphotectonics of continental margin development

Apart from the Mediterranean periphery, Africa is entirely surrounded by trailing-edge continental margins facing spreading ridges which, with the exception of the Red Sea coast, are now thousands of kilometres from the continent. Phases of crustal uplift and oceanward flexure have been widely documented along these continental margins both from the morphological evidence of warped erosion surfaces and, perhaps more convincingly, from the stratigraphy of coastal marine sediments (King 1962, Scrutton & Dingle 1976). The history of the continental margins of Africa extends back to the breakup of Gondwanaland and can be considered in the light of general models of divergent (trailing-edge) continental margin evolution.

Kinsman (1975a) has identified three types of divergent continental margin: (1) plume (dome) margins; (2) interplume (interdome) margins; and (3) transcurrent margins. Dome margins experience the maximum uplift, interdome margins moderate uplift occurring later than at dome sites (although Veevers (1981) suggests uplift here may be minor or absent), and transcurrent margins are associated with little or no uplift. Examination of the coastal margins of Africa shows that they conform in broad terms to the Kinsman (1975a) model (Fig. 14.1). Domes alternate with interdome areas, the former being the major sites of rift development.

Thermal subsidence of continental margins after lithospheric separation probably extends some 500 km inland from the boundary between continental and oceanic crust (Veevers 1981). Because of its lower initial elevation the continental rim in interdome areas would subside below sea level soon after rifting. Additional subsidence would occur as a result of sediment loading and this would be highly dependent upon the location of major drainage basin outlets. Considerable stresses have been developed in this manner in the Niger Delta (Walcott 1970, 1972). Downward flexure of the crust on the oceanward side of the area of sediment accumulation would also engender uplift along a hinge line inland, which would move oceanward with the progradation of the continental slope.

A further implication of this model is for a matching history of erosion surface uplift, denudation, sedimentation and subsidence on either side of a spreading rift. Such a history has already been documented in outline for the Atlantic margins of Africa and South America (Neill 1973, Torquato & Cordani 1981), and is further illustrated along the Red Sea margins of Arabia and north-east Africa where oceanic crust now divides the Ethiopia–Aden Dome in two (Veevers 1981).

A key factor in the rate and amount of continental margin subsidence after lithospheric separation is the amount of crustal thinning accomplished by subaerial erosion during uplift and the initial subsidence phase. In the model developed by Kinsman (1975a) an erosion rate of 0.05 mm yr^{-1} thins 30 km thick continental crust to 27.5 km before it reaches sea level in 50 Ma. With a higher erosion rate of 0.17 mm yr^{-1} crustal thickness is reduced to 25 km and sea level is attained in only 30 Ma. Kinsman (1975a) assumes spatially and temporally uniform erosion rates based on 'typical' global values, though he acknowledges that they would in fact be variable at least spatially. It is a widely held assumption in continental rifting models that areas of domal uplift, particularly the most elevated parts, experience enhanced rates of erosion. Such an assumption cannot be applied without qualification to Africa as extensive areas of duricrust and weathering deposits on Mesozoic and Cenozoic domes testify to minimal rates of erosion over large areas. Elevation of an originally low relief surface does not guarantee increased erosion rates over the entire uplifted area in proportion to the increase in relief. At least in the case of much of Africa such uplift generates highly localised increases in erosion rates in the vicinity of rejuvenated river courses and along newly formed retreating escarpments. Recently Veevers (1981) has argued along similar lines pointing out that erosion has been spatially limited in the case of the Ethiopia–Aden Dome. Even 20 Ma after the formation of the Red Sea the present-day continental rims are morphologically similar to those of the East African Rift System, with erosion largely confined to deeply incised canyons (Veevers 1981).

14.5 Some more general implications

Consideration of some of the tectonic controls over long-term landform development in Africa raises a number of more general issues concerning the interrelationship between exogenic and endogenic processes. First, the tectonic models considered provide a genetic framework for the interpretation of the planation surfaces so extensively developed across the African continent, whether these owe their immediate origin to pediplanation, etchplanation or some other mode of denudation.

Secondly, and crucial to such an approach, is the more precise dating of erosion surfaces and their associated deposits. For instance, the development of domal uplifts in Africa would be expected to generate a complex series of modified base levels to which regional landscapes would be related, rather than the continent-wide base level control to be expected from widespread synchronous epeirogenic uplift or eustatic changes. Evaluation of such effects will only be possible with reliable geochronometry. A major difficulty is, of course, the complex diachronous nature of erosion surfaces. Although possibilities for dating surface deposits employing, for instance, palaeomagnetic and fission track techniques, have yet to be fully explored (Netterberg 1978), recent developments in seismic stratigraphy (Vail *et al.* 1977) hold out the prospect of utilising more fully the complementary records of continental erosion and marine deposition to accurately pinpoint periods of accelerated denudation related to uplift.

Thirdly, advances on this front would enable a fuller examination of the question of the cyclicity and global contemporaneity of uplift and denudation over geological time periods. In their evaluation of the denudation chronology of King (1962, 1976) and the geodynamic model of Sloss and Speed (1974) which relates sea-floor spreading rates to cratonic evolution, Melhorn and Edgar (1975) have argued in support of recurring phases of worldwide denudation. Further evidence of episodicity in endogenic activity has come from the dating of global magmatic episodes (Vogt 1972, 1979). Although there is now convincing evidence of a genetic association between orogenesis and reorganisations in the rate and direction of plate motion (Schwan 1980), there is only an uncertain temporal correlation between such sea-floor spreading discontinuities and intra-plate uplift and rifting in Africa (Fig. 14.2).

Finally, the large-scale morphology of Africa provides a basis for the verification of continental intra-plate tectonic models. Perhaps because of the minimal interest of many geomorphologists in such questions, those currently developing and utilising such models accept somewhat uncritically the denudation chronology and landscape evolution models of Dixey, King and others (e.g. Le Bas 1980, Maguire & Khan 1980, Scrutton & Dingle 1976). Similarly, the assumptions concerning erosion rates and their spatial and temporal variability employed in continental rifting models such as those of Dickinson (1974) and Kinsman (1975a) are oversimplified. The comparison by Veevers (1981) of the morphology of the East Africa–Arabia region and the southern margin of Australia with elements of the Kinsman (1975a) model of continental rifting provides an example of the kind of approach that is required.

The idea that stationary lithosphere, as proposed for the African plate, might provide a means of investigating mantle processes (Thiessen *et al.* 1979) places great emphasis on macroscale surface morphology. Continental uplift and erosion may even provide a means of tracking hot spots in relative motion with respect to the lithosphere (Crough 1979, 1981). The spatial scale of surface relief may reflect distinctive lithospheric processes (Le Bas 1980) and there is a need to identify the underlying 'wavelengths' of topography if endogenic processes are to be linked to surface forms.

14.6 Conclusions

The essential message presented here is that tectonic processes are a vital element in the understanding of the landscapes of the tropics except over small areas and short timescales. This qualification perhaps explains the neglect of the tectonic dimension of morphogenesis by many geomorphologists over the past decade or so as attention has been focused on unravelling the tempo and mode of small-scale sub-aerial processes. It is ironic that the pre-eminence of this research orientation has coincided exactly with the most significant advances this century in our understanding of global tectonics. Nevertheless the integration of small-scale surface process studies into models of long-term landform development necessitates the incorporation of endogenic processes.

Although far from fully substantiated the various tectonic models discussed here provide an initial genetic framework within which the long-term development of the African landscape can be placed. They provide a basis for predicting the general nature of landscape evolution but geomorphological evidence is crucial if they are to be empirically substantiated. Estimation of the spatial and temporal variability of erosion and deposition across developing basin-and-swell topography, and dating of the resulting erosion surface warping are just two examples of the kind of geomorphological contribution required. In spite of its obvious morphological variety, associated primarily with lithological and climatic differences, Africa exhibits a marked tectonic unity. Its landscape bears the impress of this uniformity and at the same time provides a means of elucidating it.

15

Soil and slope development in the wet zone of Sri Lanka

Hanna Bremer

15.1 Introduction

The geomorphology of slopes, particularly the study of slope evolution, has been somewhat neglected by German geomorphologists, some of whom linked slope studies with the hypotheses of Walther Penck which were almost universally rejected by most German geomorphologists. Instead, slopes have been considered to be part of a general development of relief, part of the process of landscape evolution termed 'relief generation' in Germany.

Developing this relief generation concept, Büdel (1970) distinguished between active slopes, those evolving by processes operating on the slope itself or influenced from upslope, and passive slopes, which are mainly controlled by processes at the base of the slope, such as undercutting by a river. Wirthmann (1977) spoke of 'erosive slope development' in the tropics. In German, the term 'erosion' is used to indicate fluvial incision. Wirthmann's concept envisages a dense network of V-shaped valleys dissecting a landscape where the slopes are either valley sides or steep interfluves. Indeed, German studies of slopes in the tropics may be said to be at the stage of collection of new data and the development of explanatory hypotheses. No general concept of slope evolution, let alone a generally accepted hypothesis, has so far emerged.

Rather than present a theoretical discussion or present detailed slope measurements, which would be difficult to justify for the diversity of landforms found in Sri Lanka, this chapter presents tentative conclusions based on field observations.

15.2 Steep slopes and caves

Sri Lanka has, like many other upland areas of the tropics, many smooth, strongly convex, bare rock slopes. Caves are often found at the foot of these whaleback, inselberg forms. As they occur in relatively insoluble crystalline rocks, their presence may be taken as proof of a stable slope above the cave and to a certain extent also of stability of the slope below the cave. No slope retreat or decline would have occurred at least during the existence of the cave, although the lower slope might have become steeper.

At Sigiriya Rock (Fig. 15.1), where overhung recesses at least 2000 years old house the famous paintings dating from AD 485 and since untouched by further weathering, the back wall of the recess or niche has been untouched for a long time. The same probably applies to other such hollows which are most likely to have been formed when the surrounding plain was more or less at the level of the base of the recess. If this were so, the amount of subsequent lowering of the plain would suggest an age of several million years for the Sigiriya niches.

Sigiriya Rock is not a special case, for there are many other rock walls carrying sculptures over a 1000 years old. Another indicator of slope stability is provided by solution features of large rounded karren on bare crystalline rock surfaces on valley sides and inselbergs. Although it is usually difficult to date

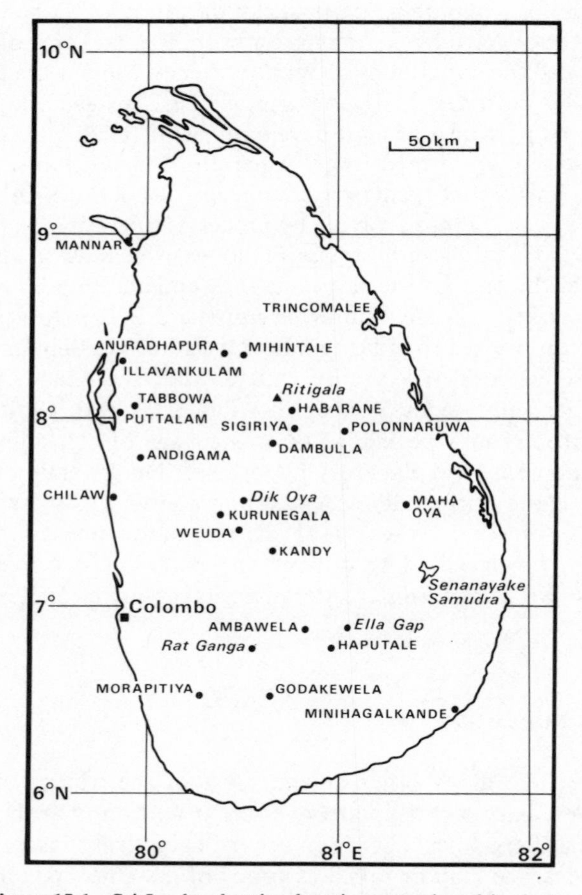

Figure 15.1 Sri Lanka showing locations mentioned in the text.

any of these solution features, their ability to persist from the early Tertiary has been demonstrated at Ayers Rock in Australia (Bremer 1965).

The longevity of rock slopes in the tropics seems to contradict their steepness. The provisional explanation relies on the virtual absence of physical weathering in the tropics. As rain water runs off these smooth rock surfaces very rapidly, there is a lack of moisture for chemical weathering. Although particularly applicable to the smooth, convex, whaleback slope elements, it is also valid for other rock surfaces. Even when such bare rock areas are irregular, few rock fragments are to be found at the base of the slope. Such large blocks as are found form a thin dispersed cover. The well rounded boulders among them probably originated in the regolith and have been left behind after finer material has been removed. However, while the rock surfaces appear to be resistant to erosion, seepage at the base of rock outcrops provides moisture for weathering and denudation and leads to planational lowering. These contrasts have been called 'divergent weathering and degradation' (Bremer 1971).

On the whole, the steep rock slope seems to represent a final stage of slope evolution in the tropics. To the author's knowledge, few such slopes have been accurately dated. In eastern Australia, Eocene basalt flows fill valley bottoms such as those of the upper Shoalhaven (Young 1977), indicating little slope retreat since then. At Ayers Rock, different weathering forms occur in a way that enables older and younger forms to be recognised. By analogy with forms in the present-day humid tropics, the older forms have been ascribed to the early Tertiary (Bremer 1965), correlating with the early Tertiary sediments in the shallow valleys of the plain surrounding Ayers Rock. Evidently, at that time the rock already stood above the plain (Twidale 1978). The slopes have not retreated since the domed form evolved. Although Ayers Rock has been affected by several climatic changes, it has generally experienced wetter conditions than the present semi-arid climate, thus enabling us to consider the rock as an example of slope evolution under a wet tropical climate.

15.3 Moderate slopes and regolith

Further examples of lack of slope retreat may be deduced from morphogenesis. Older phases of landscape evolution or 'relief generation' may be taken to have a great age, even if absolute dating is not possible. Slopes greater than 45° are usually bare rock surfaces, even though in Sri Lanka some slopes of over 60° support a regolith. More commonly, slopes with regolith have gradients around 30°. The slope profile is usually straight with occasional flats. The microtopography is quite irregular. Variations in the regolith reflect differences in geomorphic processes.

Near a small tor on a 40° slope close to Morapitiya in the south-west of Sri Lanka (Fig. 15.1), where the rainfall averages about 4300 mm yr^{-1}, an autochthonous soil profile has been found on a khondalite (quartzite, schist, crystalline limestone) bedrock. The profile contained well rounded corestones in a fine matrix. Surprisingly, laboratory analyses revealed that the matrix is completely weathered, only haematite, kaolinite and a few quartz minerals being found in the thin sections. This mature weathering indicates that the slope has been stable for a long time. The explanation is similar to that for the stability of the rock walls: rapid runoff, with storage of moisture and seepage only at the valley floor or footslope where weathering and erosion occur. This mature weathering on a steep slope may be termed a divergence profile.

The second type of regolith comprises an irregular, bouldery mass, likely to have been transported by slipping of shallow slope surface material, not more than 5 m deep. Sometimes old scars can be found within which new slips have developed. Such slips seem to be spatially and temporally irregular events. The boulders and soil particles cohere, leaving large voids through which water moves rapidly.

A third regolith type is the boulder-strewn slope, with boulders about 1 to 3 m apart protruding from the soil. This regolith appears to evolve from either of the first two types by either surface or subsurface removal of the fine material.

A fourth regolith type consists of a fine-graded soil with few or no stones. Sometimes a stratification or layering is apparent. Some small stones may be arranged in a line. This can be explained by subsurface removal of material as is shown by a fragmented quartz dyke near Godekawela (Späth 1981). Other stone lines suggest slope transport by either wash or creep.

In conclusion, the regolith slopes have a variety of material and gradients ranging from 10° to 45° or more, but averaging around 30°. Slope processes range from localised shallow mass movements and surface work to subsurface lateral eluviation under a stable cover. Probably the steepest slopes are the most stable.

15.4 Shallow slopes

The third type of slopes are the long 2° to 10° footslopes with gradients greater than those of the plains but less than those of the scarps. Usually they are longer than other slopes in the area, commonly several hundred metres, but up to 2 km in places. In general they are straight, but they may have benches or steps. The latter are sometimes broad enough to be taken to be a remnant of a planation surface. In fact, these long footslopes are features of planational lowering, which proceeds more rapidly at the generally wetter lower end of the slope. As there is already a distinction between planation processes on divides and in the lowest parts of the terrain, where wash plains may exist, these long footslopes, or 'stretch' or 'extended' slopes, have been called a form of restricted planation (Bremer 1981a).

These 'stretch' slopes are common in semi-humid areas, where they form the slopes of the flat wash floors or troughs, and the wash pediments beneath escarpments. However, they also occur in the humid part of Sri Lanka and in other tropical areas like southern Nigeria and Amazonia. They are found around escarpments, such as at the base of spurs. The steepest escarpments, however, often head embayments in the mountain front.

From their spatial distribution, one may conclude that both types of slope are the result of planation lowering. The steeper slopes are due to the extension of the younger plain, while the long footslopes occur in positions where there is less seepage and planational processes are thereby reduced. Sometimes they may be a Quaternary infilling of the valley floor connected with these 'stretch' slopes (Mäckel 1974, Thomas 1974).

15.5 Slope development

The three slope types are associated vertically as well as horizontally, although there is no obvious pattern of combinations along a valley side. Quite often the base of the valley side is a long footslope, but the uppermost part may also have the same gentle angle, giving a convex appearance to the transition to the higher surface. Sometimes benches or gently inclined surfaces appear in any section of the slope, possibly indicating how planational processes have left the overall slope as a passive landform,

The bare rock surfaces are most common in the upper half of the slope, although bare outcrops may also occur near the foot. As the bare rock types are considered to be the last of the three slope types to develop, they ought to be the oldest of, or at least as old as, the other slope types. In the planational lowering process concept, the higher relief features are the oldest. Thus the greater frequency of bare rock surfaces on the upper slopes might be explained by age.

Regolith slopes do not occupy any special position in the landscape, except that they are often found below whaleback forms, especially around domed inselbergs where they mantle the base. In broad valleys, regolith slopes form the lower parts of the valley sides, giving the characteristic trough, or almost U-like, shape.

While some slope types appear more frequently in one position than another, their irregular arrangement must be stressed even more strongly. Different moisture conditions can help explain some of this, but there is also an element of chance. Once a slope or a slope element has been isolated from its surroundings, in that it has both better and more efficient water-shedding characteristics than its neighbours, it is stable.

The general concept of slope evolution put forward here is that of planational lowering. Two main types occur:

(a) the extension of an existing plain into an area of higher relief, by steepening regolith or bare rock slopes at their base; and
(b) planation developing more rapidly in the lower and therefore damper

parts of the landscape and so transforming a pre-existing plain into a long, footslope or stretch-slope surface.

In the first type, slopes become shorter, in the second they are extended. In all cases the upper end of the slope remains more or less constant. It is fixed at the moment of the initial lowering of plains, intramontane plains, plain passes, embayments of plains, broad valleys or even narrow valleys.

Completely independently, a fixed position for the upper limit of slopes has been demonstrated for both valley sides and escarpments (Bremer 1971, 1981a). In both Nigeria and Sri Lanka, inselbergs occur more frequently in the rims of escarpments than any other position. Features similar to the inselbergs on valley shoulders that Jessen (1936) observed have developed in Sri Lanka from rock outcrops which once protruded from the slightly steepened parts of the plain. This protrusion has led to faster runoff, thus decreasing the rate of weathering. The position of the eventual escarpment, therefore, was determined to within 4 km early in the planational lowering process. Clearly, there has been no significant scarp retreat, otherwise the inselbergs at the top of the escarpment would have been destroyed. A succession of inselbergs in front of an escarpment does not necessarily indicate scarp retreat (as implied by King 1962) but can be explained equally well as having evolved as long footslopes or 'stretch' slopes continued to be lowered by etchplanation from which the inselbergs were relatively immune as stable water-shedding areas.

Other indicators of past morphogenesis are found on these slopes. The small benches or steps have often been mentioned. If these benches can be linked with planation remnants in the same area, they can be taken to be parts of a former surface. Planation passes, from one valley to another, may be cut off by further lowering, as in the case of Haputale in the great Southern Wall of Sri Lanka (Fig. 15.1). Small rivers running oblique to the main slope, for example near Ambawala or east of the Rat Ganga, are an unusual feature which may be explained by inheritance from an upper surface. This then confirms that the slope has not retreated, for otherwise benches and inherited features would have been removed by erosion.

15.6 Conclusion

The concepts developed here are hard to combine with other ideas. In terms of the parallel retreat of slopes, the foot or base of the slope may be said to retreat, but not the summit. General lowering of divides is extremely slow near the summit of the slope. There may be lowering away from the escarpment edge, further back on the upper surface, giving rise to wall-like features such as the great Southern Wall of Sri Lanka.

It is particularly difficult to comprehend the notion that steep slopes are stable. Gradient is usually used as another expression for gravity in terms of sediment transport. However, in the tropics chemical, rather than physical, processes are dominant. It must be assumed that ultimately every piece of land standing above sea level will be worn down. However, the concept advanced here does help to explain how there can be prominent, steep-sided landforms in an area of rapid lowering and recent uplift. The notion of steady land surface lowering implies continuity. While spatial differences in erosion and sediment supply have been illustrated with reference to varied fluvial conditions in the Amazon earlier in this book, the evolution of these tropical landforms in Sri Lanka demonstrates the existence and importance of temporal discontinuities.

16
Relief generation and soil in the dry zone of Sri Lanka

Heinz Späth

16.1 Climatic characteristics

Although dominated by the monsoons, the climates of Sri Lanka are greatly affected by the central highlands. In the south-west of the island, heavy rains fall from the south-west monsoon from May to September, producing as much as 5000 mm towards the central highlands. The north and north-west of the country are a dry zone, protected by the highlands from the south-west monsoon, but deriving rainfall from the north-east monsoon between December and March.

The climatic boundary between the wet and the dry zones was defined in terms of 'effective dry period': at least three successive months with less than 100 mm rainfall (Dömros 1971, 1976), a boundary coinciding roughly with the 200 mm isohyet.

From the geomorphological and pedological viewpoint the dry zone of Sri Lanka may be defined as having less than about 1750 mm annual rainfall and at least three arid months. This corresponds quite well with the climatic boundary of recent latosol formation in Sri Lanka (Bremer 1981a, Späth 1981; compare also De Alwis & Panabokke 1973). In the dry zone, mean annual temperatures are between 27 and 28°C.

16.2 Geological overview

More than 90% of Sri Lanka consists of Cambrian and Precambrian metamorphic rocks (Fig. 16.1), which are divided into two series (Cooray 1967):

(a) The Highland Series is composed of khondalites (quarzites, schists, crystalline limestone) and greyish-green to greyish-blue charnockites (mostly fine-grained, granitic to gabbroitic rocks). These rocks, mainly in the south-west, are folded, forming narrow, more or less parallel anticlines and synclines which greately influence the landforms of the

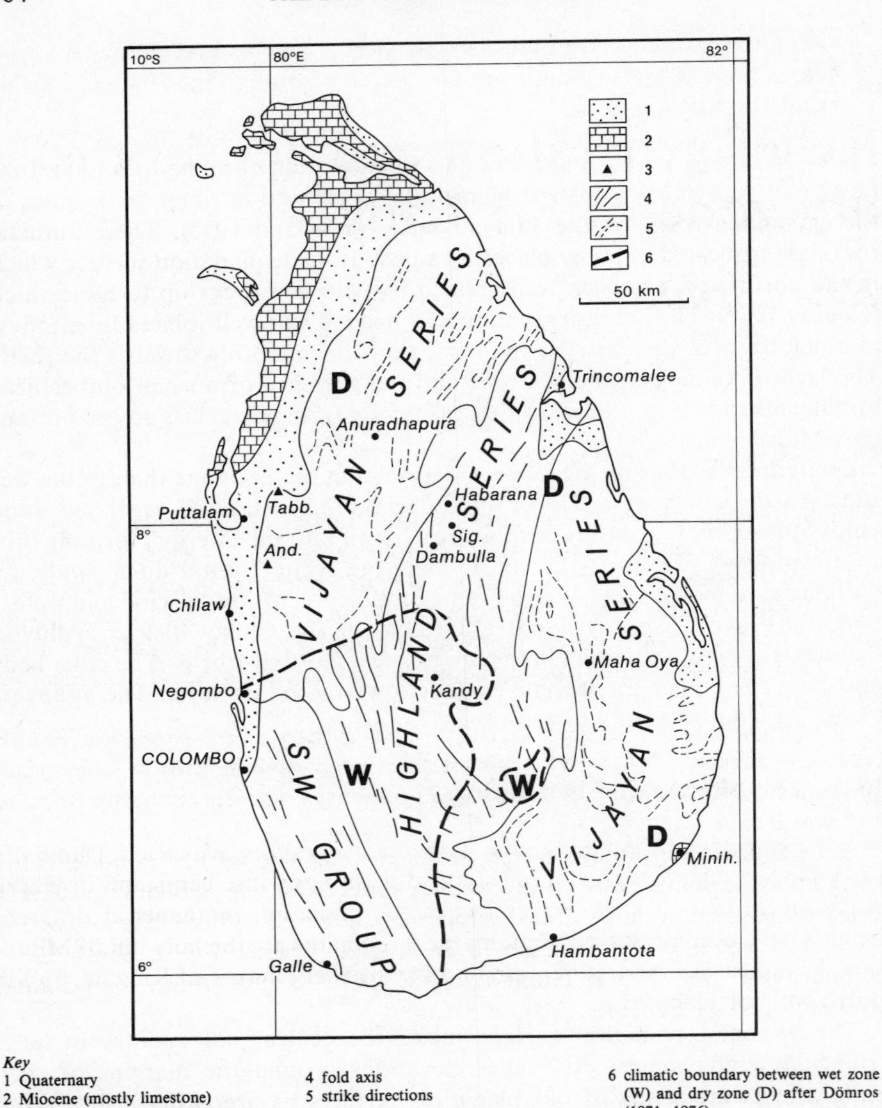

Key
1 Quaternary
2 Miocene (mostly limestone)
3 Jurassic arkose, sandstone and siltstone
4 fold axis
5 strike directions
6 climatic boundary between wet zone (W) and dry zone (D) after Dömros (1971, 1976)

Figure 16.1 Geological structure of Sri Lanka (after Cooray 1967 and Katz 1971).

area. As the bedrock of the dry zone, the Highland Series forms a 50 km broad strip which runs from the central highland to the northeastern coast near Trincomalee.

(b) The Vijayan Series is the most common besement rock of the dry zone. It consists mainly of gneiss with some granites, pyroxenites and amphibolites. The anticlines and synclines of the Vijayan Series are wider than those of the Highland Series and also possess greater strike variations (Fig. 16.1) The Southwest Group is a mixture of the Highland Series and the Vijayan Series.

Upper Jurassic arkose, sandstones and siltstones outcropping locally at Tabbowa and Andigama (near Puttalam) were developed in small depressions in the crystalline rocks of the Vijayan Series (Wayland 1925). These Jurassic rocks are truncated in many places by a post-Jurassic planation surface which in the north-west has been transgressed by Miocene rocks up to 80 m thick (Cooray 1967). The latter are mainly well bedded and well jointed limestones, thinning towards the east. To the west, they dip smoothly towards the shelf. The lack of folding in the Miocene and the presence of a small outlier near Minihagalkande in the far south-east (Fig. 16.1) argue against any important post-Miocene dislocation.

Quaternary sediments are more widespread in the dry zone than in the wet zone. In the north-west, Upper Pleistocene reddish brown, vegetated dune sands are common. To the east, they reach well into the interior, forming thin covers of sand. Towards the coast, there are white coastal dune sands, of postglacial or modern age, which are susceptible to erosion by wind and water. Lateritic gravels (Cooray 1967) are found in the Chilaw district. Alluvial sediments are quite common in the lowest, nearshore riverbeds. The older beds post-date the Upper Pleistocene eustatic lowering of sea level. The youngest loams are flood deposits of recent age.

16.3 The oldest relief generations

In the area between Anuradhapura, Trincomalee, Polonnaruwa and Dambulla the youngest planation surface is studded with many inselbergs and inselberg massifs. On many of the bigger inselbergs, planation remnants at different heights have been preserved. Among these remnants are the holy hill of Mihintale, 10 km east of Anuradhapura, and the inselberg massif of Ritigala, 16 km north-west of Habarana.

The youngest planation surface meets the central hill country in large triangular embayments. At higher elevations around the margins of such embayments are strips of old planation surfaces as are easily visible from Sigiriya Rock and the temple at Dambulla Rock. Inselbergs with such planation remnants are not restricted to the slightly more resistant Highland Series, but are also found in the rocks of the Vijayan Series (e.g. south of Maha Oya and in the vicinity of the Senanayaka Samudra reservoir).

From the Ella Gap resthouse several levels of planation remnants can be seen to the east and south-east. Here, above the youngest surface at about 200 m a.s.l., Bremer (1981a, Fig. 25) has mapped four older levels at 400, 550, 800 and 1100–1400 m. A similar geomorphic arrangement can be seen in the transition zone between the central highlands and the southern plain. In front of the southern wall there are several such planation surface remnants.

In the dry zone, many old planation remnants are preserved. They correspond to the etchplain stairways of the hill country. There Adams (1929) found three such levels, while Timmermann (1935) and Cooray (1967) found four. Recently Bremer (1981a) has suggested more than seven planation surfaces.

In the area of Anuradhapura, Sigiriya and Dambulla are large hollows, or caves, cut in the sides of bare rock inselbergs. The most famous ones contain a temple at the domed Dambulla Rock and spectacular paintings at Sigiriya and sacred sites on the holy inselberg of Mihintale. Preliminary investigations suggest that these caves occur at particular levels which can be correlated with planation remnants. The caves thus document phases of stability during planation lowering. The caves are the products of intensive humid tropical weathering and can be used as evidence of phases during planation lowering as were similar features on Australian inselbergs (Bremer 1965, 1980, 1981b, Twidale 1978, Twidale & Bourne 1975). The lowering of the planation surfaces and exposure of the inselbergs were episodic. Although further detailed investigations are needed there is no doubt that the bigger caves can be connected with planation remnants. The preservation of the caves also means that the inselbergs and their bare rock surfaces are old, stable forms (Bremer 1965, 1980, 1981b, Späth 1977, Wilhelmy 1977). Faced with a lack of correlated sediments or evidence such as datable volcanic rocks, no exact dates can be given for the age of the old planation surface remnants. However, they must date from before the Jurassic as the Jurassic sediments are situated at a much lower altitude and are truncated by the younger planation surface.

16.4 Landform development in the Upper Tertiary

Little is known of landform evolution betwen the Jurassic to Upper Tertiary. The youngest planation surface, being covered by Miocene sediments, must have been in existence by the Upper Tertiary. Following the deposition of the Miocene sediments intensive weathering with terra rossa formation in karstic pipes (Fig. 16.2) and silicification elsewhere took place in a climate wetter than the present. This weathering, which must be older than the Upper Pleistocene sediments, probably occurred during late Tertiary or early Quaternary. After that, the Miocene was partly lowered to the level of the youngest planation surface (for example south-east of Mannar). In the vicinity of Illavankulam (north of Puttalam) this erosion has formed a 20 m high eastward-facing cuesta, giving a mesa-like appearance to the Miocene outcrops. At the same time further erosion affected the crystalline bedrock of the youngest planation surface which in pre-Miocene time was almost at the level it now occupies.

16.5 The youngest planation surface

The youngest planation surface cutting across the Jurassic sandstones already existed in pre-Miocene times and also truncates the Highland Series and the

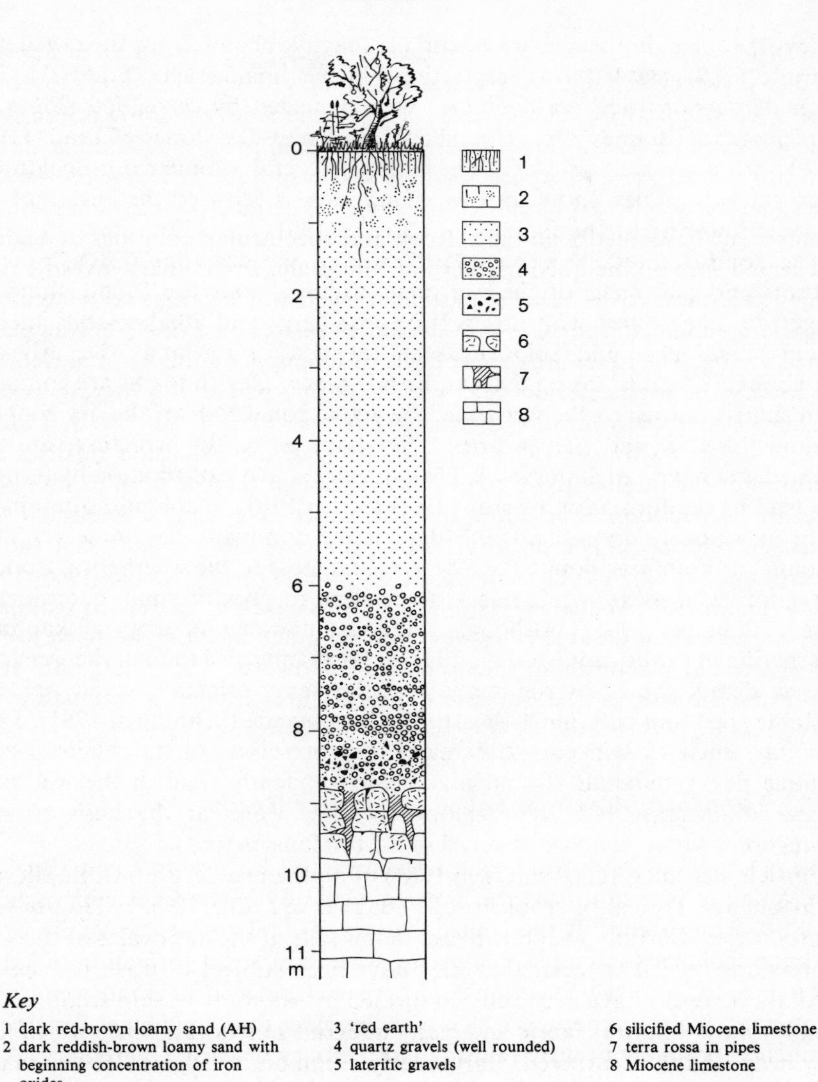

Figure 16.2 'Red earth' over Miocene limestone north of Puttalam (after Späth 1981)

Vijayan Series. It rises very smoothly with a gradient of 0.2% from the coast to 200 m a.s.l. near the centre of the dry zone on the northern, eastern and southern sides of the island. The lowermost planation surface is a typical etch-plain, studded with inselbergs, rock domes and residual core boulders, and divided into wash depressions and wash divides [like the Tamilnadu plain of south India described by Büdel (1982) (Editors)]. The lowest parts of the wash depressions are marked by flat swales, sometimes 2000 m wide, that have a valley-like form, but which are essentially inactive channels for the evacuation of runoff [Bremer (1981a) terms these kleine Spülflächen (Editors)]. The wash depressions and wash divides are connected by 'extended slopes' or 'Streckhänge' (Bremer 1981a), similar to the ramp-like slopes of Louis (1964, 1979) with a gradient of 1–7°. They can be several kilometres in length and their smooth slopes show narrow planations which are the result of the episodic lowering of the flat swales (Bremer 1981a, p. 132).

The dominant reddish-brown earth soils of the dry zone differ from the latosols and plastosols of the wet zone. Alluvial soils are found along the bigger dry-zone rivers with lithosols on inselbergs and alkaline soils such as solonised solonetz and solonchaks in the coastal lowlands (De Alwis & Panabokke 1973). In the padi areas, brown-black, gleyish loams are common.

In sharp contrast to the wet zone, the active zonal soils of the dry zone are shallow, skeletal and rich in grus. They have an earthy structure, are well drained and have a high porosity. The typically active bioturbation by termites can lead to the formation of stone lines (Späth 1981). Kaolinite, often in the form of a poorly crystallised fine clay, is the dominant clay mineral. Small amounts of montmorillonite have been found close to the weathering front on the more basic rocks, while micas decompose to provide small quantities of illite (Tillmanns 1981). Although there is undoubtedly tropical kaolinitic weathering in the dry zone today, it is much less intensive than in the wet zone. This is clearly shown by the analysis of the heavy minerals which yields an unstable spectrum with much garnet and hornblende (Schnütgen 1981). Light minerals, such as feldspars and micas, are preserved in the whole profile. Opaque heavy minerals, found much less frequently than in the wet zone, appear to increase in number along the flat swales at the base of wash depressions where conditions stay humid for long periods.

Particle size analyses (Schnütgen 1981) of all the profiles show little silt; this is, in general, typical of tropical soils (Bakker & Müller 1957). The soils are poorly sorted. Sorting is slightly better in the soils of the flat swales of the wash depressions, which suggests that they have been subject to wash transport.

All these results have been fully confirmed by the study of thin sections. The samples show a porous fabric and many minerals of low resistance which have only been slightly weathered. Particularly common is a brown clayey matrix with many small angular quartz grains and splinters of other minerals (Schnütgen 1981, Späth 1981). The porous brown matrix consists of a fine-grained, thin skin covering the mineral grains and grain-complexes. The thin sections also demonstrate the downward transport of clay. This is part of the subsurface removal of material pointed out by Bremer (1981a). The amount of sesquioxides is lower than in the wet zone. On the wash divides and on the extended slopes haematite is dominant, whereas in the wetter wash depressions there is more goethite.

Figure 16.3 shows a typical cross section and catena between a wash divide and a wash depression. The wash divides and plains in the watershed areas are covered with inselbergs, tors and exposed blocks. In these geomorphological

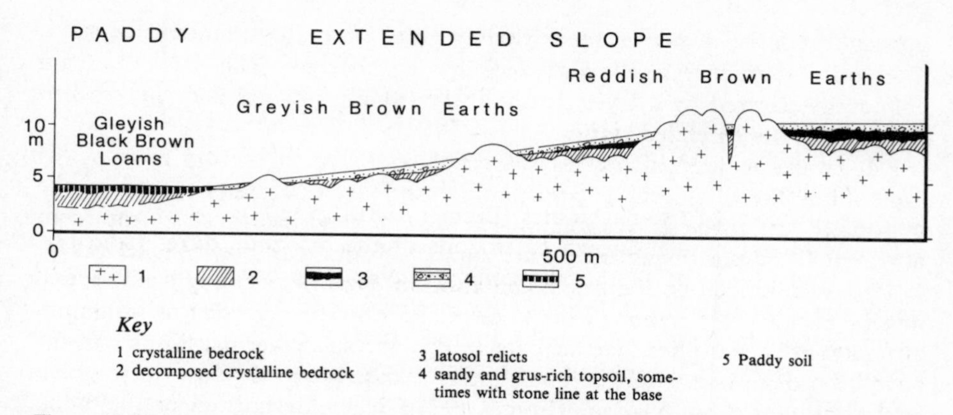

Figure 16.3 Cross section between a wash divide and a wash depression in the dry zone of Sri Lanka.

positions the soil profiles are complex (Fig. 16.3, right part). Above the crystalline bedrock the decomposed zone can be several metres thick. There then follows a mottled zone with a dense red-brown (5YR-3/4) loam with pisolites and quartz grus. This horizon shows all the features of an altered latosol of the wet zone, an interpretation also suggested by the stable spectrum of heavy minerals and by the thin sections. Obvously, this is relict soil from former wetter conditions, as described for many tropical savanna areas (Thomas 1974 p. 15). In our profile, on the top of the relic latosol there is an enrichment of quartz grus, sometimes forming a stone line. The upper layers of the soil consist of a loamy–sandy reddish-brown earth. Pisoliths and pisolith fragments suggest a reworked latosol. On the divides, the stone-line formation cannot be regarded as a former erosion surface with the subsequent accumulation of finer material. Here, the stone lines must be explained by bioturbation as has been reported by Williams (1968) from the Australian Northern Territory and by De Dapper (1978) from the Manika Plateau, Zaire.

From the wash divides, extended slopes fall at angles of 2–3° to the wash depressions. In the upper parts of the slopes there are many outcrops and shallow reddish-brown earths which grade down slope into greyish- and yellowish-brown earths. In the middle and lower parts of the extended slopes, the soils are grus-rich. The spectrum of the heavy minerals is unstable, with many garnets and hornblendes. Micas and feldspars are easily seen in the field. Locally the soils are covered by thin hillwash deposits.

Where the upper parts of the extended slopes give way to wash divides, relics of latosols occur at the surface. As these relics thin out down slope, it would seem likely that the latosol was once continuous to the wash depressions and has since been truncated by the lowering of the wash depression. Such a process produces the extended slopes (Bremer 1979, 1981a). The flat swales at the base of the wash depressions are almost totally occupied by rice paddy cultivation which has greatly altered the soils. Natural soil profiles in these areas appear to consist of brown-black, loamy gleys with a high amount of sand. The soils cover brown-yellowish decomposed bedrock. The flat swales are commonly covered by a layer, less than 1 m thick, of sand and silt, washed down from the extended slopes.

Flat swales in wash depressions are not restricted to the dry zone of Sri Lanka but are a typical geomorphological feature of large parts of the seasonally wet tropics. Ackermann (1936) first described such forms from northern Rhodesia, using the term 'dambo'. From Zambia, Mäckel (1974, 1975) has described episodic sedimentation in dambos, yielding a fill several metres thick. These sediment thicknesses can only be regarded as minimum estimates as they do not take account of any surface lowering. Similar forms have been described from Malawi (Young 1969), where they are surrounded by 2–8° slopes, from Nigeria (Thomas 1974), West Africa (Ledger 1969) and Sierra Leone (Thomas & Thorp 1980). Furthermore, Thomas (1974) has suggested a deeper weathering beneath these flat swales.

The formation of planation surfaces in the dry zone of Sri Lanka today is restricted to the flat swales in the wash depressions (Bremer 1981a, Späth 1981, 1982). However, this does not mean that there is no erosion elsewhere. On the extended slopes surface wash effectively erodes material. Panabokke (1959) found surface sheet erosion even under dense jungle south of Anuradhapura (see also Burns 1947, Joshua 1977). After extreme monsoonal rains Deraniyagala (1958 p. 10) observed alluvial sand and silt in tanks of the dry zone.

The relief of the dry zone is evolving towards a 'tropical ridge relief' ('Tropisches Rückenrelief') as first reported by Büdel (1965, 1982) for southern India. The idea of a planation surface currently forming by the concept of 'double surfaces of levelling' (Büdel 1957: 'Doppelte Einebnungsflächen') is not applicable to the dry zone of Sri Lanka because weathering is not intensive enough to compensate for variations in the resistance of bedrock. The widespread Upper Pleistocene sediments also suggest that there has been no recent lowering of the yougest planation surface of the dry zone. Here, we have planation formation along traditional lines (Büdel 1977 p. 186: 'Traditionale Weiterbildung'). Therefore the lowest planation surface of Sri Lanka today does not belong to the zone of 'seasonal tropics with extreme planation' (Büdel 1980 p. 4).

The catena between wash divides and wash depressions is only partially the product of the present climate. On the wash divides the soil profiles are usually complex and polygenetic. There is no doubt that the latosol relics derive from a wetter climate. In the absence of evidence for tectonic activity during the Upper Pleistocene one has to rely on climatological explanations for the change in soils and relief. This can only mean a phase of a drier climate affecting erosion, mainly on the slopes. Closer dating of this dry phase requires the study of fossil sediments and palaeosols on the lowest planation surface.

16.6 Fossil sediments and palaeosols and their implications for landform evolution

Allegedly Pleistocene fossil red sands and gravels were described by Wayland (1919) while palaeontological studies led Deraniyagala (1958) to suggest several Pleistocene climatic changes which are discussed in more recent research by Cooray (1968), Swan (1979), De Alwis (1980) and Späth (1981). Four examples may be demonstrated here.

'Red earth' in the north-west of Sri Lanka

The north-west and the northern coastal area is covered by red-brown (2.5 YR-3/4 to 3/6; locally 10R-3/4) sands. They extend over a zone more than 30 km wide and form ridges up to 20 m high. A close vegetation cover means that the sands are currently immobile. There are no signs of surface drainage of fluvial erosion on the sands, which show no layering and also no horizon development; they are uniform. Particle size analysis of material from the quarry of Illavankulam, north of Puttalam (Fig. 16.2), reveals a composition of 75% sand, 6% silt and 19% clay. Most of the quartz grains are rounded and slightly polished. The heavy mineral composition consists of 39% sillimanite, 50% garnet, 3% rutile with 6% of minerals undifferentiated. There is also some ilmenite and magnetite. The proportion of opaque minerals is high at 80%, mainly in the form of grains of ilmenite covered by a haematitic skin. The presentation of plagioclase and muscovite throughout the profile (Schnütgen 1981) rules out any intensive weathering. The thin and fine-grained skin of haematite covering and cementing grains, seen in thin sections, proves that the grains have not been transported since being weathered. The dominance of kaolinite in the sands shown by detailed investigations by De Alwis (1980), together with an absence of gibbsite, led to an exclusion of the possibility of a wetter climate during weathering.

At Illavankulam (Fig. 16.2) the sands are underlain by a 2 m thick layer of very well rounded quartz gravels. In the basal part of the profiles there are some lateritic gravels (Cooray 1967, p. 150: 'basal ferruginous gravel'). Both Wayland (1919) and Joachim *et al.* 1938) regarded the sands as aeolian deposits and the gravels as fluvial sediments. More recently, Cooray (1967) suggested these were a coastal accumulation; Swan (1977) considered them to be a coastal feature with dune sands; while De Alwis (1980) wrote of a 'regressive beach/near-shore deposit', partly reworked by wind. As such well rounded fluviatile gravels do not occur elsewhere on the island, the gravel must at least have been reworked and rounded by coastal processes.

The sand material derives mostly from the Vijayan Series although additional old weathered material must also have been included. A great part of the sand has been derived from the beach which must have been greatly widened during eustatic sea-level lowering in the Upper Pleistocene (Späth 1981, p. 216).

Although Deraniyagala (1973) found artefacts dating from about 15 000 years ago and all authors agree that these sands are of Pleistocene age, distinction should be made between their age and the age of their weathering. In south-east India Gardner (1981) described similar sand dunes, which appear to be 30 000 years old. She regards the weathering of these sands as of recent age. The iron comes from iron-rich minerals such as horneblende. Similarly, De Alwis (1980) favours a more recent origin for dune reddening. However, evidence from Upper Pleistocene red dunes in Australia and elsewhere (Gardner & Pye 1981) sugests that the possibility that the main reddening took place in former times should be considered. While there is no doubt that the weathering of the sands is going on today, it may now be less intense than previously.

Lateritic crusts

Laterites east of Negombo, typical of those found in several parts of the dry zone, were studied. On the lowest planation surface there are mesa-like plates and inselbergs up to 25 m high. They consist of totally decomposed crystalline rocks topped by a dark, violet lateritic crust up to 3 m thick. The iron oxides are mostly nodular, have a high porosity and include quartz fragments. Being well jointed, these crusts break up when exposed in road cuttings and small quarries. Thin sections suggest that the crusts are similar to the mottled zones found in the present wet zone. Erosion of the upper parts of complete latosol profiles has led to the exposure, and subsequent induration, of the mottled zones. As this erosion and hardening seems to be impossible under rainforest, it probably occurred under sparse vegetation in drier conditions.

Some slight dissection of the laterites precludes a recent origin. It is more likely that the laterites date from an Upper Pleistocene dry period.

'Ironstone gravels' north-east of Chilaw

Occurring mainly in the vicinity of Erunwala ('Erunwala gravel', Cooray 1963) and Attangane, north-east of Chilaw, these extremely poorly sorted gravels (up to 6 m thick) consist of a sub-angular badly rounded quartz in a sandy grus matrix (Schnütgen & Späth 1978, Späth 1981) and cover decomposed crystalline rocks of the Vijayan Series. The sediments are dissected, forming terrace-like strips along the small rivers (Fig. 16.4). On the top, the gravels are cemented by a resistant limonitic laterite crust (Cooray: 'nodular ironstone').

Formation of those gravels, either in the wet or in the dry zone, is not observed at present. The gravels must, therefore, be fossil because they have been dissected in post-sedimentary time. This could have been related to Upper Pleistocene eustatic sea-level lowering. The sedimentation, perhaps by torrential floods (Cooray 1963), could have taken place during an Upper Pleistocene dry period, probably during the 'Palugaha turia phase' of Deraniyagala (1958). The sediments appear to have filled broad, flat wash depressions where a 'low level laterite' (Hamilton 1964) formed and was subsequently dissected

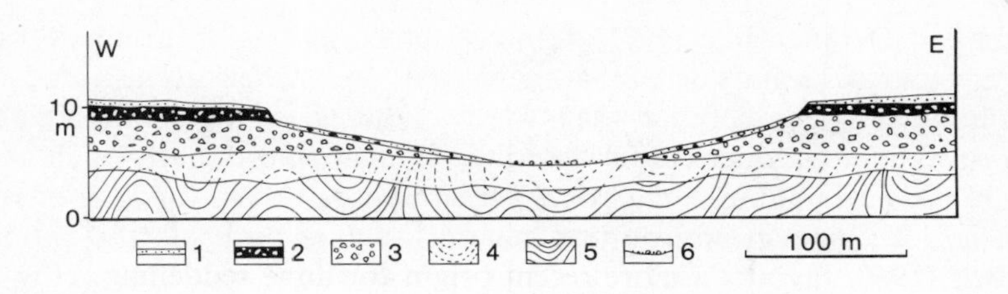

Key

1 sandy brown earth
2 ironstone gravels (nodular ironstone)
3 quartz gravels in sandy matrix
4 decomposed crystalline bedrock
5 crystalline bedrock (gneiss)
6 sand and grus in recent creek bed

Figure 16.4 The 'ironstone gravels' near Attangane, north-east of Chilaw (after Späth 1981).

and indurated. In some places this had led to relief inversion, similar to that decribed by Seuffert (1973) from southern India.

Each of the above three examples of sediment and palaeosol development argues against a recent origin. Conversely, however, much evidence suggests a drier climate accompanied by eustatic sea-level lowering. Many features indicate an Upper Pleistocene age. To obtain more detailed information on this arid phase, the sediments of Weuda (13.9 miles along the national road A10 between Kurunegala and Kandy) have been analysed.

The sediments of Weuda

Along the Dik Oya there is a triangular embayment of the youngest planation surface into the central hill country. The embayment is studded with some inselbergs and is surrounded by old planation remnants at 700 m a.s.l. (Bremer 1981a). Sediments, in part more than 6 m thick, can be traced from the rear of the embayment for *c.* 7 km towards the north-west. The sediments, now dissected and forming terraces, consist of unrounded, partially layered quartz gravels and grus in a sandy matrix (Fig. 16.5). Between the gravels are gleyed, black-brown, organic clayish horizons which become hard and cracked on desiccation. The heavy minerals show an unstable spectrum with many hornblendes, feldspars and micas in the whole profile (Schnütgen 1981). From old place names Deraniyagala (1958, p. 24) and Cooray (1967, p. 175) argued for the existence of a lake at about AD 300.

No clear unconformities or phases of soil formation occur in the profiles, suggesting quite rapid sedimentation. At present, such sedimentation is not found under the dense vegetation and, therefore, sedimentation must have occurred under a drier climate. The dissection of the sediments on terraces up to 10 m above the river level mitigates against a Holocene age. Radiocarbon dates of the first and the fourth fossil humus horizons have shown an age of 22 100–24 300 BP for the sediments, much older than the lake postulated by Deraniyagala and Cooray. This period probably corresponds to a dry climatic phase of the Upper Pleistocene in Sri Lanka (Späth 1981, p. 221).

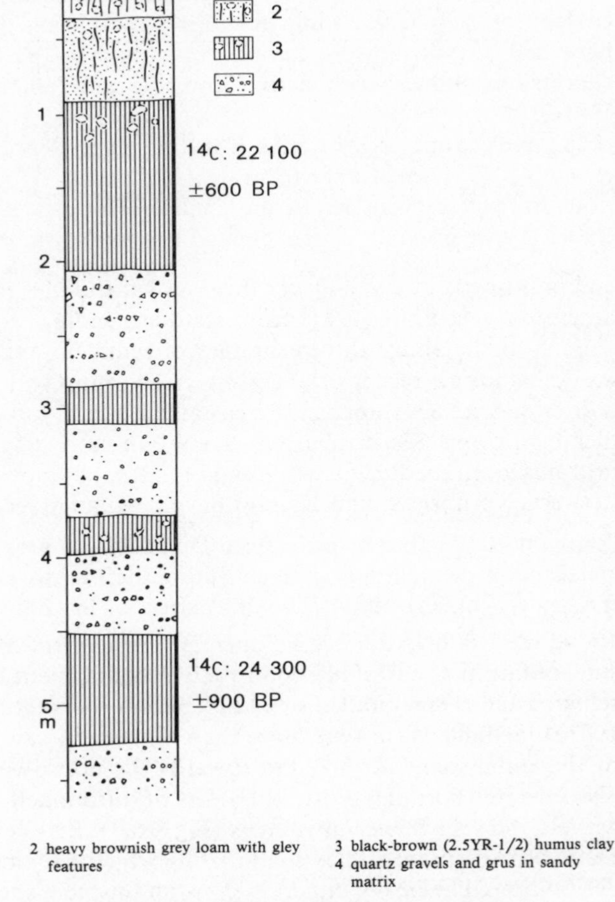

Key
1 dark brown humus loam
2 heavy brownish grey loam with gley features
3 black-brown (2.5YR-1/2) humus clay
4 quartz gravels and grus in sandy matrix

Figure 16.5 Sediments near Wenda, north-west of Kandy (after Späth 1981).

16.7 Discussion

After describing the old planation remnants, this chapter has concentrated on the development of younger landforms in the dry zone of Sri Lanka. There can be no doubt that the lowest planation surface already existed in pre-Miocene time, because of the subsequent transgression. In places, the Miocene has been eroded to the level of the lowest planation surface and thus, clearly, must have undergone a polygenetic development. Swan (1977) has also referred to 'several epicycles'. The intricate evolution of this surface is very well proved by the presence of (a) heterogeneous soils with relics of a wetter climate and (b) diverse sediments and palaeosols from a drier period of the Upper Pleistocene. This complex youngest planation surface is in many places covered by sediments, while elsewhere, mainly on the extended slopes, it is bare and stripped. More humid conditions in the flat swales of the wash depressions induce such active lowering and planation as occur (Bremer 1981a).

In the dry zone of Sri Lanka it is difficult to define the 'normal landscape' (Douglas 1977) because the forms which have developed under previous conditions influence recent landscape change (Douglas 1980b). For this problem Büdel (1983, pp. 180–3) has created the term 'traditional development'.

It is instructive to compare results from Sri Lanka with studies in adjacent areas of south-east India. Büdel (1965) and Demangeot (1975) have closely studied the Tamilnadu surface for which the latter author suggests a polygenetic development, evidence for several changes in climate being shown by incision features and pre-Neolithic 'laterites alluvionnaires'. Such laterites may well correspond to the ironstone gravels of the Chilaw district. The fossil red dunes of coastal Tamilnadu (Gardner 1981) seem to be similar to the 'red earth' dunes of north-west Sri Lanka.

The influence of the Upper Pleistocene dry period on the wet zone of Sri Lanka needs further investigation. Doubtless an area of rainforest must have persisted in the south-west, even during the Pleistocene dry periods (compare Tricart & Cailleux 1965). The dry period of Sri Lanka fits very well into the pattern of Upper Pleistocene climatic change from other tropical areas. Climatic fluctuations have been described from Nigeria (Moss 1965, Fölster 1969, Rohdenburg 1969), Chad (Servant & Servant-Vildary 1972) and Sierra Leone (Thomas 1974, Thomas & Thorp 1980).

Rohdenburg has pointed out that planation surfaces are not actively forming between the Gulf of Guinea and the Sahara at present and Mensching (1970) has observed in Tanzania that with increasing aridity the relief intensity also increases. Thus the area of active planation surface formation may have been overestimated (Louis 1964).

Amongst the evidence of dry phases in the Upper Pleistocene in Latin America is that for Brazil (Birot 1958, Bigarella 1975, Semmel & Rohdenburg 1979), for Central America and adjacent areas (Zonnefeld 1968) and that reviewed by Tricart in this book and by Damuth and Fairbridge (1970) and Klammer (1981). More recently, Bibus (1982) has described Upper Pleistocene superficial deposits and buried accumulations in central Brazil which would not have been associated with etchplain development.

As more information on the complexity of climatic changes and their impacts on landform evolution in the seasonally wet tropics during the Quaternary becomes available, the likelihood of similar changes in the humid tropics increases. However, further investigations are needed to get a clear picture of what happened in present-day rainforest areas.

Part E

CONCLUSION

17 Findings, answers to questions and implications for the future

I. Douglas & T. Spencer

In the first chapter of this book the following questions about tropical landforms were posed as a result of a survey of the evolution of geomorphological investigations in the tropics:

(a) Are there typical tropical landforms?
(b) How do climatic contrasts, especially in the amount and seasonal distribution of rainfall, affect geomorphic processes in low latitudes?
(c) What is the role of plate tectonics and other tectonic events, both associated with plate boundaries and otherwise, in tropical landform evolution?
(d) Is it possible to distinguish parallel sequences of landform evolution in the tropical areas of different continents?
(e) What has been the effect of sea-level changes?
(f) How closely are the dynamics and evolution of ecosystems related to landform development?

These questions have been addressed in differing forms in various chapters of this book; evidence has ranged from detailed investigations of small areas such as those in the Gunung Mulu National Park to widescale overviews particularly the studies of the tropical area of South America in Chapter 10, the studies of Pleistocene aridity in three continents to the eastern hemisphere tropics in Chapter 11 and the analysis of the tectonic background to long-term landform development in tropical Africa in Chapter 14. However, certain themes which recur across these various chapters provide some answers to the questions restated above.

17.1 Typical tropical landforms

The arrangement of surface drainage is often regarded as a simple yet sensitive morphometric measure of landscape organisation and evolution. In Chapter 5 Walsh adopted such an approach by the study of drainage density in the

Windward Islands of the Caribbean. However, he shows that drainage density is a function of lithology and clay mineralogy of soils and of the quantity and seasonal variability of rainfall. It does not seem to reflect the age of the landscape. Obviously such landscape created by volcanicity of varying ages is not necessarily the sort of landscape on which the landforms typical of particular climates would be expected to be found. The nature of the volcanic activity would in large measure determine the type of topography.

There is a greater likelihood that landscapes reflecting tropical climatic influences would be found on stable old shield areas. However, in his extensive review of the Brazilian and Guyana shields Tricart, in Chapter 10, points out that great caution is needed when trying to define the specific landforms of the present climatic zones. Only accurate studies may lead to a definition of that part of the landscape which is presently active and that which is inherited from former conditions which were usually quite different. In Chapter 16 Späth states that in the dry zone of Sri Lanka there are abundant signs that the landscape is largely inherited and that past climates thereby still influence present processes. Despite broad similarities in the large-scale geomorphology of the shield areas of the humid and seasonally wet tropics there are no clear-cut, typical tropical landforms that are universally recognised.

17.2 The effects of climatic contrasts on geomorphic processes

The data about geomorphic processes presented in Chapter 3 suggests that rainfall erosivity may be an important factor that should be taken into account. However, this somewhat begs the question about which precise attributes of climate are relevant to geomorphic processes. Each individual case study clearly demonstrates the role of edaphic conditions. The way in which the rainfall strikes the ground, the distribution of the water runoff as overland flow, subsurface flow or deep seepage to ground water greatly effects the efficiency of the erosion process. Avenard and Michel (Ch. 4) suggest that the crucial timing for the effectiveness of process is the seasonal transitions from the wet to the dry season and from the dry to the wet season. This sort of transition when the energy of the falling rain occurs at a time of lower protective efficiency by the vegetation finds its parallel in longer timescale considerations of transition periods. Evidence from enhanced sedimentation off the mouths of the Congo and Niger Rivers, from erosive phases in the Simen Mountains of Ethiopia (Hurni 1982) and from soil erosion periods in Nigeria (Smith 1982) points to the significance of such a transition period from about 12 000 to 10 500 BP. At this time increased rainfall coincided with the raising of high mountain snowlines and preceded the full response of vegetation to climatic change. Thus although this period was not of great hydrological significance it is likely to have been of important geomorphic impact because it was preceded by a long period of predominantly low runoff (see Street-Perrott *et al.*, Ch. 8).

The importance of the vegetation cover and the relationship between the vegetation cover and precipitation is crucial in both processes at the present time and in process change in the Quaternary. In Chapter 4 the geomorphic effects of degradation of vegetation in the Sahel are highlighted. In forested, humid and seasonally wet tropical environments, removal or destruction of the vegetation creates geomorphic process conditions akin to those of a semi-arid or arid environment. Thus in the phase of transition from aridity to humidity a greater frequency of rain events allied with a lack of ground cover creates a more rapid evolution of the landscape, a period of rhexistasy.

17.3 The role of plate tectonics and other tectonic events

Large areas of the tropics are in seismically active zones; 38% of the rainforests of the south-east Asian–Australasian area are in such zones, 14% of the American tropics and 1% of the African tropics. In the latter, for example, more or less continuous uplift, coincident with the development of intra-plate volcanism, has had a profound effect on landform development (Summerfield, Ch. 14). However, even outside these seismically active zones tectonic activity is important. In Chapter 10 the role of tectonic subsidence during the Tertiary and Quaternary in Amazonia is emphasised. The partially drowned delta island of Marajò is shown to be the result of the complicated interaction of sea-level changes, climatic fluctuations and tectonic movements. It has also been noted that there is a relationship between volcanic activity and climate. Rampino and Self (1984) have suggested that the development of the El Niño in 1982 (Ch. 2) could have been caused by ash from the El Chichon eruption warming the upper atmosphere in the tropics and so altering the thermal conditions in the eastern Pacific. Such an observation of the influence of tectonics ties in with recent considerations of frost rings in trees as records of major volcanic eruptions. La Marche and Hirschboeck (1984) have demonstrated that frost rings coincide well with four climatically effective northern hemisphere or equatorial eruptions in the past 100 years: Krakatoa, 1883, Pelée and Soufrière, 1902; Katmii, 1912; and Agung, 1963. This relationship between a tectonic event, a climatic change and an intensification of process as a result of the climatic change shows that tectonic activity causes erratic extreme events not only directly in the areas of tectonic activity but also elsewhere in the tropics as a result of short-term short-lived climatic shifts. At the other end of the scale, more fundamental and sustained trends are demonstrated by the mapping of the global distribution of duricrust (Petit, Ch. 13). This shows a relationship between ancient duricrusts and the breakup of Gondwanaland. The role of plate tectonics in tropical landform evolution, like that of climate and climatic change, is felt at number of differing but overlapping scales. The more detailed the examination of tectonics the more widespread the range of effects identified. The ideas set forth in Chapter 14 suggest that close examination of other tropical continents will reveal an even more closely linked pattern of the combined effects of tectonics and climate.

17.4 Parallel sequences of landform evolution in different continents

While environmental change in Africa appears to be better documented than that in other tropical areas, the chapters of this book reveal a remarkably close correspondence of environmental change in Africa, the Middle East, south Asia and the Americas. Evidence for this synchroneity is particulary well developed in Chapter 8 where lakes are shown to have developed in two periods from 27 000 to 21 000 BP and from 12 500 to 5000 BP separated by the arid interval of the last glacial maximum. From all continents lake-level fluctuations indicate the abruptness of climatic change and they reinforce the notion, developed originally in Africa and then separately in Australia, of phases of long-term stability and short-term landform disruption. The application of Büdel's (1982) concepts of relief generation to Sri Lanka shows a different scale of periodicity in the landscape. Both Bremer (Ch. 15) and Späth (Ch. 16) emphasise the point made in Chapter 10 that inheritance of landform is as important a feature of the tropics as it is of more closely studied better known areas such as Europe and North America.

There is a whole range of evidence yet to be deciphered in terms of palaeohydrology. In particular and as Chapter 6 (Smart *et al.*) shows, Quaternary drainage histories can be reconstructed where changing fluvial situations can be related to other kinds of evidence such as cave passage morphology and datable cave deposits. The inland delta of the Niger, the rivers draining into the Gulf of Carpentaria and the changes in the Amazon all suggest that in the continental deposits of such areas there awaits the reconstruction of an event-based stratigraphy which could reveal the significance of those climate shifts when the effectiveness of processes increased rapidly. Nevertheless, caution must be maintained in such interpretations. The palaeohydrologic record is notoriously difficult to interpret, especially because a single exceptional flood can account for a large accumulation of sediment. Unravelling the impact of individual catastrophic events from the effects of longer-term climatic shifts remains a vital task.

17.5 The effect of sea-level changes

As with climate, sea-level oscillations occur over different timespans but with the complication that eustatic fluctuations, consequent upon ice volume changes, have interacted with isostatic factors, including postglacial rebound, glacial margin sinking and hydro-isostatic loading (Bloom 1967, Walcott 1972, Chappell 1974), to produce regional sea-level curves (Clark *et al.* 1978). The key importance of the glacio-eustatic component is, however, well illustrated by two observations. First, there has been a widespread identification of the last interglacial (*c.* 125 000 ± 4000 BP) high sea stand at 1.5–9.0 m above present sea level in the Atlantic, Indian and Pacific Oceans. Secondly, there is a good agreement over the timing of high sea-level episodes from stepped sequences of dated coral terraces on the tectonically emergent islands of Barbados and New Guinea.

High sea stands in Barbados are observed during the penultimate interglacial at about 220 000, 200 000 and 180 000 BP and during the last interglacial at 125 000, 105 000 and 82 000 BP (Mesolella *et al.* 1969, Bender *et al.* 1979). In New Guinea and Timor relative high stands date from 220 000 and 180 000 BP and subsequently at 15 000–20 000 year intervals from 140 000 to the present (Bloom *et al.* 1974, Chappell & Veeh 1978, Aharon 1983). Elsewhere, on stable coasts, the more compressed record shows interglacial sea levels characterised by patch reef growth and shoal ridge development slightly above present sea level and glacial low stands indicated by palaeosol development and the deposition of speleothems in caves and 'blue holes' at elevations well below present sea level (e.g. Bahamas: Neumann & Moore 1975; Yucatan: Szabo *et al.* 1978, Bermuda: Harmon *et al.* 1983).

All these localities show that the last interglacial high sea stand, and the subsequent interstadials, were brief events, less than 10 000 years in duration, and bracketed by rates of sea-level rise and fall of 3.5–6.0 mm yr^{-1} (Harmon *et al.* 1983) and perhaps up to 14.0 mm yr^{-1} (Chappell & Veeh 1978). Such rates of sea-level rise would have been particularly catastrophic on low-angle continental shelves; on the Sahul Shelf, for example, transgression rates were 24–45 m yr^{-1} between 18 000 and 10 000 BP (Van Andel *et al.* 1967). Thus one needs to think of coastal transition periods in addition to, and at least as important as, those in the terrestrial record.

A sea-level lowering of at least 130 m at the last glacial maximum has been estimated by Chappell (1974). At Obidos, 700 km inland from the river mouth, the bed of the Amazon is cut down to 83 m below present sea level and thus during glacial low sea-level stands the river would have almost had an equilibrium profile. Modern tributaries in this area take on the character of fluvial rias and the present fluvial morphology is governed by the previous low sea levels (Tricart, Ch. 10): similar low sea levels would have made the Gulf of Carpentaria into an enclosed freshwater lake basin with river courses incising down to a low, temporary lake. The lake would probably have been the product of streamflow and ice melt from the New Guinea highlands as northern Australia would have been even more arid than at present. Similar incision of alluvial fans characterised the Melinau catchment, Sarawak (Smart *et al.*, Ch. 6) with enhanced river gradients across the then dry shelf.

17.6 Ecosystem dynamics and landform development

Throughout this book the theme of ecological variability recurs; Chapter 2 emphasises how short-term climatic and seismic events disrupt rainforest and reef ecosystems. As forests regenerate through the gap creation and regrowth process our data about forest dynamics must be carefully assessed to determine what stage of the forest regrowth process they represent. There is still no adequate knowledge of how the species composition of forests affects their nutrient cycling and how that nutrient cycling is related to the processes of erosion and denudation. A few studies have suggested that the chemical composition of water in rivers in tropical areas bears little relationship to the nutrient cycling in the vegetation. This difference may reflect the dry-season baseflow to rivers from deep circulation, delayed return flow and ground water which derives its chemical composition from its passage through the regolith and bedrock, whereas the water involved in nutrient cycling is essentially the water that is transpired through the plant community and therefore has little contact with the rivers. Such an approach would suggest that the partial source area for tropical runoff is a valid hypothesis and would corroborate the studies by Northcliff and Thornes (1981) and Gilmour *et al.* (1980) which emphasise the role of storm events and seasonal soil moisture conditions in runoff generation. If streamflow is generated from discrete areas within catchments and the whole catchment area contributes to streamflow only under extreme conditions, then two independent or semi-independent water circulation systems and associated nutrient losses within the forest can be postulated. Such a twofold system would link with the ideas of double planation surface development. It would suggest that the geomorphologically effective lowering is that at the basal surface of weathering and that the ecologically effective weathering system is at the vegetation root zone. This would seem to be the condition that would exist in biostasy.

However, abrupt change such as disruption of the vegetation, wind-throw or fire would lead to exposure of the soil, stripping of the soil and loss of material from the ground surface. This may occur linearly along the lines of gully development or also by massive surface wash of unprotected soil. In the drier, seasonally wet climates this may well happen in those crucial periods of transition between dry and wet seasons. In forested areas this condition would occur at times when the forest is disrupted by cyclones, landslides or similar events. Thus we can relate the concept of biostasy and rhexistasy to evolution of the land surface at the present time.

We can equally develop a relationship between land-forming processes and ground surface evolution in the times of the abrupt climate shifts that have been identified during the period since the last glaciation. The type of situation envisaged by Knox (1972) would seem to be prevalent in the recent geomorphological history of the tropics. The shift from one vegetation type to another is characterised by a long response time. Throughout the tropics examples of relict vegetation formations can be found and in edaphically favourable sites the remains of rainforest are situated well beyond the present limits defined by seasonal climates. Obviously such refuges or relict formations would have existed in places throughout the Quaternary. It is still not clear exactly which areas were refuges and certainly there is a lack of information about any geomorphological differentiation between the landforms of Pleistocene refuge areas and the landforms that underwent climatic change. Tricart's review (Ch. 10) of environmental change in tropical South America shows that while the sequence of events may have been similar the contrasts in vegetation type led to different ecological conditions and consequent morphological responses in different parts of the region. However, the potential of the dating techniques described in Chapter 9 for a better elucidation of the nature of these episodic events and their relationship to erosion processes clearly needs further exploration.

The relatively rapid temperature changes at the onset and end of glaciation were associated with changes in the chemistry of the atmosphere and the oceans. Whether chemical changes depend on the temperature changes or vice-versa is unclear, but such shifts could be associated with expansions and contractions of the terrestrial plant biomass as ice sheets and synchronous tropical arid zones expanded and contracted.

After 13 000 BP, towards the end of the last glaciation, carbon dioxide levels in the atmosphere increased by 30%, sufficient to cause temperatures to increase by $0.9 \pm 0.5^{\circ}C$ (Neftel *et al.* 1982). One hypothesis suggests that as sea levels rose, organic matter from the coastal ocean of glacial time was buried by river-derived silts deposited on the now drowned continental shelves. Such burial of organic matter would have reduced the carbon and phosphorus content of sea salt. The increased erosion of the immediate post-13 000 BP period, described earlier, would have made such burial relatively rapid.

Another hypothesis suggests that the change in atmospheric CO_2 was related to ecological shifts both on land and in the oceans. The expansion of tropical forests alone would have led to a decrease in CO_2, but if marine organisms incorporated more carbon in relation to phosphorus in their soft tissue during glacial periods and relatively less carbon after glaciation, the ocean's content of dissolved CO_2, and therefore the atmosphere's CO_2 content, would have indeed increased (Broecker 1983). In the whole of this hydrosphere–atmosphere–biosphere interrelationship, geomorphic processes and factors are intimately involved. Both sea-level change and accelerated erosion rates immediately after 13 000 BP play vital roles in the suggested global chemical changes.

Temperature shifts at the onset of glaciations would be equivalent to a cooling of $0.1^{\circ}C\ yr^{-1}$. In Chapter 11, Williams suggests that such events would last about 100 years and would occur once in every 1000 to 10 000 years. Similarly rapid changes in precipitation regime probably occurred at the same time in the tropics. The rapid rises in lake levels throughout Africa (e.g. Ethiopia: Gillespie *et al.* 1983) appear to have been associated with the return, and thereafter increasingly westerly influence, of the south-west monsoon (Kutzbach 1981).

The ability of the monsoon to recur seems to be crucial to the rapid change in environmental conditions. If such rapid changes are know to have occurred at least once in the postglacial they may well have occurred many times during the Quaternary. Thus studies of sediments associated with Quaternary events are likely to complicate the pictures of landform evolution derived from earlier less detailed studies.

Events in Sierra Leone described in Chapter 12 link these Quaternary events with longer-term landform evolution. Etchplanation or the development of double planation surfaces seems to have generally been produced by one or more distinct episodes of the stripping of relict saprolite materials. Certainly in recent times and in the Quaternary it would appear that such stripping was induced primarily by climatic change but in this chapter it is also associated with tectonic activity. Clearly there is a need for a closer examination of the possible links between local tectonic volcanic activity and global climatic change. In conclusion, the overwhelming emphasis at all timescales of the relationship between ecosystems and landform development is that periods of relative quiescence, or biostasy, are separated by periods of rapid change in environmental conditions, rapid adjustment of ecosystems and high rates of geomorphological activity, or rhexistasy.

17.7 Geomorphic events and scales of change

Four major sequences of change have emerged from the preceding discussion:

(1) Variation and ecological instability due to events such as tropical cyclones.
(2) Local disruption and world climatic effects due to volcanic eruptions such as those of El Chichon and Krakatoa and their effects on such phenomena as the El Niño.
(3) Phases of rapid climatic change and intense geomorphic activity such as the immediated postglacial events in the period 12 000 to 10 500 BP.
(4) The long-term events linking planation surface development to tectonic activity.

On the first of these variations much information has been presented in Chapter 2. It is clear that, if we are to have adequate measures of rates of geomorphic processes as influenced by climate, long periods of monitoring are necessary to establish the role that events like tropical cyclones play. However, such is the amount of variability that can occur that for any given period of monitoring we can only expect to achieve the rates of erosion due to the magnitude of events ocurring within such a period. Thus 25 years of observation ought to permit the recognition of the work done by a once in 25 year event but inevitably they might well contain a once in a 100 year or less frequent event.

The all-important links between climatic events and volcanoes are a subject for further investigation. The study of frost effects in tree rings provides some evidence of ecological disruption, but much more is needed on the geomorphic indicators of such extreme events. An event-based stratigraphy must be established to elucidate the links between volcanic activity and climate. The frost ring studies are not yet conclusive for there are many notable frost events which are not associated with volcanic eruptions. Some eruptions may result in a frost record at a certain locality while others may not, and quite clearly frost damage can be due to atmospheric variability from other causes. What does seem to be clear, however, is that air–sea interactions, like the El Niño, may in some instances be closely related to volcanic events.

The excellent evidence from several areas in Africa for rapid geomorphic changes in the 12 000 to 10 500 BP period suggests that a search for similar evidence from other tropical continents would be likely to be profitable.

Finally, an important task must be to relate the theories of King (1983) and Büdel (1982). Here are two great visions of the evolution of tropical areas and it is vital that the tectonic warping implied by King's cymatogeny be related to the etchplanation ideas developed by Büdel.

As described by Thomas and Thorp in Chapter 12, etchplanation essentially represents landscape evolution by the surface lowering of pre-weathered material, over long periods of time in terrain where resistant areas of unweathered rock may persist, or become exposed, as the surface is lowered. Büdel (1982) notes that the gaps between the resistant residuals, or inselbergs, are etchplain passes which could only have been produced by the relief-forming mechanism of double planation surface development (Fig. 1.1). However, despite the long periods of geological time required for the creation of such etchplains, relict etchplains often have surfaces which have been endogenously uplifted and peripherally dissected. Büdel (1982 p. 171) describes how etchplain stairways will develop where gentle arching of the surface occurs. He argues that such uplift can separate a higher and lower etchplain, the outer edge of the latter being part of the same original etchplain as the now higher surface. King (1983) envisages that cymatogenic, or domed, uplift would separate erosion surfaces so that in the main axis of uplift the oldest surface would be the most elevated, but at the margin, usually a coastline, it would dip beneath younger sedimentary formations. The six global planation cycles that King (1983) still recognises represent periods of etchplanation. Büdel (1982) would argue that those surfaces continue to evolve by 'traditional development', following a pattern of erosion made possible by the low relict and deep mantle of weathered rock. Both authors find that the geotectonic events to which geomorphic evolution is due are not continuous. Short episodes of uplift and deformation alternate with much longer periods of tectonic quiescence during which denudation and deposition are dominant. Evidence from both the land surface and the ocean floors supports this.

Tropical geomorphology, like all geomorphology, is event-dependent. This book has shown that, while episodes of as long a wavelength as that associated with cymatogenic uplift and etchplanation may dominate the Gondwanaland tropics, those occurring with much shorter mean frequencies determine the landscape changes measurable in a human lifespan. Such are the irregularities of the latter that it is difficult to separate the $0.1\,^{\circ}\mathrm{C}\ \mathrm{yr}^{-1}$ cooling at the onset of a glaciation from the irregular fluctuations about once every 10 to 20 years, due to such phenomena such as the El Niño and the Southern Oscillation.

As more is learnt about tropical landforms and their correlative sediments, the dimensions and consequences of environmental change become clearer, However, a sneaking suspicion remains that the true clarity is linked to the last 30 000 years and even in that period the effects of events of the magnitude of the 1970s Sahel drought or the 1926 and 1971 floods in peninsular Malaysia are not distinguished from longer-term trends. As suggested earlier, event-based stratigraphies may help resolve this dilemma. The fascination of the tropics is in many ways a consequence of the extremes and erosivity of its climates, extremes which not only complicate the pattern of environmental change but which also make the lessons of an historical geomorphology vitally necessary for application in the improvement of conditions for the people of low latitudes.

References and bibliography

Abrahams, A. D. 1972. Drainage densities and sediment yields in eastern Australia. *Aust. Geog. Stud.* **10**, 19–41.

Ackermann, E. 1936. Dambos in Nordrhodesien. *Wiss. Veröff. Dtsch. Mus. Länderkunde*, NF **4**, 148–57.

Adams, F. D. 1929. The geology of Ceylon. *Can. J. Res.* **1**, 425–511.

Adamson, D. A. 1982. The integrated Nile. In *A land between two Niles*, M. A. J. Williams and D. A. Adamson (eds), 221–34. Rotterdam: Balkema.

Adamson, D. A., F. Gasse, F. A. Street and M. A. J. Williams 1980. Late Quaternary history of the Nile. *Nature* **288**, 50–5.

Adamson, D. A., R. Gillespie and M. A. J. Williams 1982. Palaeogeography of the Gezira and of the lower Blue and White Nile valleys. In *A land between two Niles*, M. A. J. Williams and D. A. Adamson (eds), 165–219. Rotterdam: Balkema.

Adis, J., K. Furch and U. Irmler 1979. Litter production of a Central-Amazonian blackwater inundation forest. *Trop. Ecol.* **20**, 236–45.

Aharon, P. 1983. 140000-yr isotope climatic record from raised coral reefs in New Guinea. *Nature* **304**, 720–30.

Aleva, C. J. J., E. H. Bon, J. J. Nossin and W. J. Sluiter 1973. A contribution to the geology of part of the Indonesian tin belt: the sea areas between Singkep and Bangka Islands and around the Karimata Islands. *Geol Soc. Malay. Bull. 6*, 61–86.

Alexandre, J. 1967. L'action des animaux fournisseurs et des feux de brousse sur l'efficacité érosive du ruissellement dans une région de savane boisée. *Cong. Coll. Univ. Liège. Belg.* **40**, 43–9.

Allchin, B., A. S. Goudie and K. T. M. Hegde 1978. *The prehistory and palaeogeography of the great Indian desert*. London: Academic Press.

Appleby, P. G. and F. Oldfield 1978. The calculation of lead-210 dates assuming a constant rate of supply of unsupported ^{210}Pb to the sediment. *Catena* **5**, 1–8.

Appleby, P. G. and F. Oldfield 1978. The assessment of ^{210}Pb data from sites with varying sediment accumulation rates. *Hydrobiologia* **103** 29–35.

Arculus, R. J. 1976. Geology and geochemistry of the alkali basalt–andesite association of Grenada, Lesser Antilles island arc. *Bull. Geol Soc. Am.* **87**, 612–24.

Ashton, P. S. 1964. *Ecological studies in the mixed dipterocarp forests of Brunei State. Oxf.* Forest. Mem. 25.

Ashton, P. S. 1978. Crown characteristics of tropical trees. In *Tropical trees as living systems*, P. B. Tomlinson and M. H. Zimmerman (eds), 591–615. Cambridge: Cambridge University Press.

Asprey, G. F. and A. R. Loveless 1958. The dry evergreen formations of Jamaica. II. The raised coral beaches of the north coast. *J. Ecol.* **46**, 547–70.

Asseline, J. and C. Valentin 1977. *Construction et mise au point d'un infiltromètre à aspersion*. Abidjan: Centre ORSTOM d'Adiopodoumé.

Assemien, P., J. C. Fillerton, L. Martin and J. P. Tastet 1970. Le Quaternaire de la zone littorale de Côte d'Ivoire. *Bull. ASEQUA* **25**, 65–78.

Aubert, G. 1962. Les sols de la zone aride. Étude de leur formation, de leurs caractères, de leur utilisation et de leur conservation. (UNESCO, Recherches sur la zone aride.) *Actes Coll. Paris* **19**, 127–50.

Aubert, G. and R. Maignien 1949. L'érosion éolienne dans le Nord du Sénégal et du Soudan français. (C.R. Conf. Afr. Sols, GOMA.) *Bull. Agr. Congo Belge* **40**, 1309–16.

Aubréville, A. 1947. Érosion et bowal en Afrique Noire française. *Agron. trop. Fr.* **2**, 339–57.

Aubréville, A. 1949. *Climats, forêts et désertification de l'Afrique tropicale*. Paris: Société d'Éditions géographiques, maritimes et coloniales.

Aubréville, A. 1962. Savanasisation tropicale et glaciations quaternaires. *Adansonia* NS II **1**, 16–84.

Audley-Charles, M. G., J. R. Curray and G. Evans 1977. Location of major deltas. *Geology* **5**, 341–4.

Audley-Charles, M. G., J. R. Curray and G. Evans 1979. Significance and origin of big rivers: a discussion. *J. Geol.* **87**, 122–3.

Aufrère, L. 1932. La signification de la laterite dans l'évolution climatique de la Guinée. *Bull. Assoc. Géogrs Fr.* **60**, 95–7, 457–8.

Aufrère, L. 1936. La géographie de la laterite. *C.R. Soc. Biogéogr. Fr.* **13**, 3–11.

Avenard, J.-M. 1962. *La solifluxion ou quelques methodes de mécanique des sols appliquées au problème géomorphologique des versants*. Paris: Centre de Documentation Universitaire.

Avenard, J.-M. 1969. *Réflexions sur l'état de la recherche concernant les problèmes posés par les contacts forêts-savannes, essai de mise au point et de bibliographie*. ORSTOM Sér. Init. Doc. Tech. 14.

Avenard, J.-M. 1971. *La répartition des formations végétales en relation avec l'eau du sol dans la région de Man-Touba*. ORSTOM Sér. Trav. Doc. 12.

Avenard, J.-M. 1972. Rôle des régimes hydrologiques des sols dans l'Ouest de la Côte d'Ivoire. *Ann. Géog.* **81**, 421–50.

Avenard, J.-M. 1973a. Évolution géomorphologique au Quaternaire dans le Centre-Ouest de la Côte d'Ivoire. *Rev. Géomorph. Dyn.* **22**, 145–60.

Avenard, J.-M. 1973b. *Cartographie géomorphologique dans l'Ouest de la Côte d'Ivoire*. ORSTOM Sér. Notices Explic. 71.

Avenard, J.-M. 1982. *La dégradation du milieu à la périphérie de Ougadougou (Haute-Volta)*. Strasbourg: ULP.

Avenard, J.-M and E. J. Roose 1972. *Quelques aspects de la dynamique actuelle sur versants en Côte d'Ivoire*. Communication presented to 22nd Congr. Int. Geog., Canada, August 1972. Abidjan: Centre ORSTOM d'Adiopodoumé.

Avenard, J.-M., J. Bonvallot, M. Latham, M. Renard-Dugerdil and J. Richard 1972a. *Aspects du contact forêt–savane dans le centre de l'Ouest de La Côte d'Ivoire: étude descriptive*. ORSTOM, Sér. Trav. Doc. 35.

Avenard, J.-M., J. Bonvallot, M. Latham, D. Renard-Dugerdil and J. Richard 1972b. Le contact forêt savane et Côte d'Ivoire. *Ann. Géog.* **82**, 543–4.

Avenard, J.-M., M. Eldin, G. Girard, J. Sircoulon, P. Touchebeuf, J.-L. Guillaumet, E. Adjanohoun and A. Perraud 1971. *Le milieu naturel de Côte d'Ivoire*. ORSTOM Mém. 50.

Ayres, W. A. 1971. Radiocarbon dates from Easter Island. *J. Polynes. Soc.* **80**, 497–504.

Bailey, D. K. 1964. Crustal warping – a possible tectonic control of alkaline magmatism. *J. Geophys. Res.* **69**, 1103–11.

Bailey, D. K. 1974. Continental rifting and alkaline magmatism. In *The alkaline rocks* H. Sørenson (ed.), 148–59. New York: Wiley.

Bailey, D. K. 1977. Lithosphere control of continental rift magmatism. *J. Geol Soc. Lond.* **133**, 103–6.

Baines, G. B. K. and R. F. McLean 1976. Resurveys of 1972 hurricane rampart. on Funafuti Atoll. *Search* **7**, 36–7.

Baines, G. B. K., P. J. Beveridge and J. E. Maragos 1974, Storms and island building at Funafuti Atoll, Ellice Islands. *Proc. 2nd Int. Coral Reef Symp.*, Vol. 2, 485–96.

Bak, R. P. M. and B. E. Luckhurst 1980. Constancy and change in coral reef habitats along depth gradients in Curaçao. *Oecologia (Berl.)* **47**, 145–55.

Bak, R. P. M., R. M. Termaat and R. Dekker 1982. Complexity of coral interactions: influence of time, location of interaction and epifauna. *Marine Biol.* **69**, 215–22.

Baker, B. and J. Wohlenberg 1971. Structure and evolution of the Kenya Rift Valley. *Nature* **229**, 538–42.

Baker, H. G. 1970. Evolution in the tropics. *Biotropica* **2**, 101–11.

Baker, P. E., F. Buckley and J. G. Miller 1974. Petrology and geochemistry of Easter Island, *Contrib. Mineral. Petrol.* **44**, 85–100.

Baker, V. R. 1978. Adjustment of fluvial systems to climate and source terrain in tropical and subtropical environments. In *Fluvial sedimentology*, A. D. Miall (ed.), 211–30. Can. Soc. Petrol. Geol. Mem. 5.

Bakker, J. P. 1957. Quelques aspects du problème de sediments correlatifs en climat tropical humide. *Z. Geomorph.* NF **1**, 1–34.

Bakker, J. P. and H. J. Muller 1957. Zweiphasige Flussablagerungen und Zweiphasen-Verwitterung in den Tropen unter besonderer Berücksichtigung von Surinam. *Stuttgarter Geogr. Stud.* **69**, 365–97.

Baldy, C. 1977. *Introduction à l'agrométéorologie*. Project PNUD/OMM.

Ball, M. M., E. A. Shinn and K. W. Stockman 1967. The geologic effects of Hurricane Donna in South Florida. *J. Geol.* **75**, 583–97.

Barbetti, M. and H. Allen 1972. Prehistoric man at Lake Mungo, Australia, by 32 000 years BP. *Nature* **240**, 46–8.

Bardossy, G. 1973. Bauxite formation and plate tectonics. *Acta Geol. Acad. Sci. Hung.* 141–54.

Barnes, D. J. 1973. Growth in colonial scleractinians. *Bull. Marine Sci.* **23**, 280–98.

Batchelor, B. C. 1979. Discontinuously rising late Cainozoic eustatic sea-levels, with special reference to Sundaland, Southeast Asia. *Geol. Mijnb.* **58**, 1–20.

Baynes, J. 1938. *Les sols d'Afrique centrale, spécialement du Congo Belge*. Bruxelles: INEAC.

Baynes, A., D. Merrilees and J. K. Porter 1975. Mammal remains from the upper levels of a late Pleistocene deposit in Devil's Lair, Western Australia. *J. R. Soc. W. Aust.* **58**, 97–126.

Bazzaz, F. A. and S. T. A. Pickett 1980. The physiological ecology of plant succession. *Ann. Rev. Ecol. Syst.* **11**, 287–310.

Beard, J. S. 1945–6a. The progress of plant succession on the Soufrière of St Vincent. *J. Ecol.* **33**, 1–9.

Beard, J. S. 1945–6b. The Mona forests of Trinidad, British West Indies. *J. Ecol.* **33**, 173–92.

Beard, J. S. 1949. *The natural vegetation of the Windward and Leeward Islands*. Oxf. Forest. Mem. 21.

Beaumont, C. and J. F. Sweeney 1978. Graben generation of major sedimentary basins. *Tectonophysics* **50**, T19–23.

Behrmann, W. 1917. Der Sepik (Kaiserin-August-Fluss) und sein Strömgebeit. *Mitt. Dtsch. Schutzgebieten* **12**, 1–100.

Behrmann, W. 1921. Die Oberflächenformen in den feuchtwarmen tropen. *Z. Ges. Erdkd. Berlin* **56**, 44–60.

Behrmann, W. 1924. Das westliche Kaiser-Wilhelms-Land in New Guinea. *Z. Ges. Erdkd. Berlin Enganzungsheft* **1**, 1–72.

Behrmann, W. 1928. Die Insel Neu Guinea: Grundzuge ihrer Oberflächengestaltung nach dem gegenwörtigen Stande der Forschung. *Sonderb. Z. Ges. Erdkd. Berlin, Hundertjahrfeier 1927–1928.*

Bellouard, P. 1948. Érosion des sols du Sénégal oriental, du Soudan occidental, du Fouta Djalon. *Bull. Agr. Congo Belge* **40**, 1299–308.

Bender, M. L., R. G. Fairbanks, F. W. Taylor, R. K. Matthews, J. G. Goddard and W. S. Broecker 1979. U-series dating of Pleistocene reef tracts of Barbados, West Indies. *Bull. Geol Soc. Am.* **90**, 577–94.

Benson, W. W. 1982. Alternative models for infrageneric diversification in the humid tropics: tests with passion vine butterflies. In *Biological diversification in the tropics*, G. T. Prance (ed.), 608–40. New York: Columbia University Press.

Berggren, W. A., L. H. Burckle, M. B. Cita, H. B. S. Cooke, B. M. Funnell, S. Gartner, J. D.

Hays, J. P. Kennett, N. D. Opdyke, L. Pastouret, N. J. Shackleton and Y. Takayanagi 1980. Towards a Quaternary time scale. *Quat. Res.* **13**, 277–302.

Bernard, F. 1970. Étude de la litiere et de sa contribution au cycle des éléments minéraux en forêt ombrophile de Côte d'Ivoire. *Oecologia Plantarum* **5**, 247–66.

Bernhard-Reversat, F. 1975. Nutrients in throughfall and their quantitative importance in rain forest mineral cycles. In *Tropical ecological systems*, F. B. Golley and E. Medina (eds), 153–9. Berlin: Springer.

Bernard-Reversat, F., C. Huttel and G. Lemée 1978. Structure and functioning of evergreen rain-forest ecosystems of the Ivory Coast. In *Tropical forest ecosystems*, 557–74. Paris: UNESCO.

Bertrand, R. 1967a. L'érosion hydrique. Nature et évolution des matériaux enlevés. Relation et conséquences sur le sol érodé (Station de Bouaké). *Coll. sur la Fertilité des Sols Tropicaux*, Tananarive, November 1967, 107, 1296–301.

Bertrand, R. 1967b. Étude de l'érosion hydrique et de la conservation des eaux et du sol en pays Baoulé. *Coll. sur la Fertilité des Sols* Tropicaux, Tananarive, November 1967, 106, 1281–95.

Bibus, E. 1982. Boden-und Reliefentwicklung in ausgewählten Gabieten Mittelbrasiliens. *Tagungsber. Wiss. Abh. Dtsch. Geographentag Mannheim 1981*. Wiesbaden: Steiner.

Bigarella, J. J. 1975. International Symposium of the Quaternary. *Bol. Par. Geosci.* **33**.

Bigarella, J. J. and G. O. de Andrade 1965. Contributions to the study of the Brazilian Quaternary. In *International studies on the Quaternary*, A. E. Wright and D. G. Frey (eds), Geol Soc. Am. Spec. Pap. 84, 433–51.

Birks, H. J. B. 1981. The use of pollen analysis in the reconstruction of past climate. In *Climate and history*, T. M. L. Wigley, M. J. Ingram and G. Farmer (eds), 111–38. Cambridge: Cambridge University Press.

Birks, H. J. B. and H. H. Birks 1980. *Quaternary palaeoecology*. London: Arnold.

Birot, P. 1958. Les domes crystallines. *CNRS Mem. Doc.* 6, 8–34.

Birot, P. 1960. *Géographie physique générale de la zone intertropicale*. Paris: Centre de Documentation Universitaire.

Birot, P. 1966. *General physical geography*. London: Harrap.

Birot, P. 1970. *L'Influence du climat sur la sédimentation continentale*. Paris: Centre du Documentation Universitaire.

Birot, Y., J. Galabert, E. Roose and J. Arrivets 1968. *Deuxième campagne d'observations sur la station de mesure de l'érosion de Gampela: 1968*. Rapport multigr., CTFT.

Bishop, W. W. 1966. Stratigraphic geomorphology: a review of some East African landforms. In *Essays in Geomorphology*, G. H. Dury (ed.), 139–76. London: Heinemann.

Bishop, W. W. and A. F. Trendall 1967. Erosion-surfaces, tectonics and volcanic activity in Uganda. *Q. J. Geol Soc. Lond.* **122**, 385–420.

Biswas, B. 1973. Quaternary changes in sea-level in the South China Sea. *Geol. Soc. Malay. Bull.* **6**, 229–56.

Blackwelder, E. 1925. Exfoliation as a phase of rock weathering. *J. Geol.* **33**, 793–806.

Bloemendal, J., F. Oldfield and R. Thompson 1979. Magnetic measurements used to assess sediment influx at Llyn Goddionduon. *Nature* **200**, 50–3.

Blondel, F. 1929a. Les altérations des roches en Indochine française. *Proc. 4th Congr. Sci. Pacifique*. Bull. Serv. Geol. Indochine 18(3).

Blondel, F. 1929b Les phénomènes karstiques en Indochine Française. *Proc. 4th Congr. Sci. Pacifique*, IIB.

Blong, R. J. 1982. *The time of darkness: legend and reality*. Canberra: ANU Press.

Bloom, A. L. 1967. Pleistocene shorelines: a new test of isostasy. *Bull. Geol Soc. Am.* **78**, 1477–94.

Bloom, A. L., W. S. Broecker, J. M. A. Chappell, R. K. Matthews and K. J. Mesolella 1974. Quaternary sea level fluctuations on a tectonic coast: New ^{230}Th/^{234}U dates from the Huon Peninsula, New Guinea. *Quat. Res.* **4**, 184–205.

Bonatti, E. and S. Gartner, Jr 1973. Caribbean climate during Pleistocene Ice Ages. *Nature* **244**, 563–5.

Bond, G. 1978. Evidence for Late Tertiary uplift of Africa relative to North America, South America, Australia and Europe. *J. Geol.* **86**, 47–65.

Bonvallot, J. 1972. Utilisation des courbes granulométriques pour la cartographie des phénomènes de dynamique actuelle. *Cah. ORSTOM Ser. Sci. Hum.* **9**(2).

Bornhardt, W. 1900. *Zur Oberflächen-gestaltung and Geologie, Deutsch-Ostafrika.* Berlin: Collection Deutsch-Ostafrika.

Bougère, J. 1978. Saison sèche, saison humide: approche methodologique pour les régions à longue saison sèche. *Bull. Écol.* **9**, 301–5.

Bowler, J. M. 1975. Deglacial events in southern Australia: their age, nature and palaeoclimatic significance. In *Quaternary studies*, R. P. Suggate and M. M. Cresswell (eds), 75–82. Wellington: Royal Society, NZ.

Bowler, J. M. 1976. Aridity in Australia: origins and expression in aeolian landforms and sediments. *Earth Sci. Rev.* **12**, 279–310.

Bowler, J. M. 1978. Glacial age aeolian events at high and low latitudes: a Southern Hemisphere perspective. In *Antarctic glacial history and world palaeoenvironments*, E. M. Van Zinderen Bakker (ed.), 149–72. Rotterdam: Balkema.

Bowler, J. M. 1981. Australian salt lakes: a palaeohydrologic approach. *Hydrobiologia* **81/82**, 431–44.

Bowler, J. M., G. S. Hope, J. N. Jennings, G. Singh and D. Walker 1976. Late Quaternary climates of Australia and New Guinea. *Quat. Res.* **6**, 359–94.

Bowler, J. M., A. G. Thorne and H. A. Polach. 1972. Pleistocene man in Australia: age and significance of the Mungo skeleton. *Nature*, **420**, 48–50.

Bowles, F. A. 1975. Palaeoclimatic significance of quartz/illite variations in cores from the eastern equatorial North Atlantic. *Quat. Res.* **5**, 225–35.

Brabben, T. 1978. *Reservoir sedimentation study, Selorejo, East Java, Indonesia: Reservoir survey and field data.* Wallingford Hydraul. Res. Stn Rep. OD 15.

Brabben, T. 1979. *Reservoir sedimentation study, Karangkates, East Java, Indonesia.* Wallingford Hydraul. Res. Stn Rep. OD 22.

Bradbury, J. P., B. Leyden, M. Salgado-Labouriau, W. M. Lewis, C. Schubert, M. W. Binford, D. G. Frey and D. R. Whitehead undated. *Late Pleistocene and Holocene environmental history of Lake Valencia, Venezuela: the effects of man and climate on a low elevation tropical lake.* Unpubl. ms.

Bradbury, J. P., B. Leyden, M. L. Solgado-Labouriau, W. M. Lewis, C. Schubert, M. W. Binford, D. G. Frey, D. R. Whitehead and F. M. Weibezahn 1981. Late Quaternary environmental history of Lake Valencia, Venezuela. *Science* **214**, 1299–305.

Brakenbridge, G. R. 1978. Evidence for a cold, dry full-glacial climate in the American Southwest. *Quat. Res.* **9**, 22–40.

Branner, J. C. 1896. Decomposition of rock in Brazil. *Bull. Geol Soc. Am.* **7**, 255–314.

Brasell, H. M., G. L. Unwin and G. C. Stocker 1980. The quantity, temporal distribution and mineral-element content of litterfall in two forest types at two sites in tropical Australia. *J. Ecol.* **68**, 123–39.

Bremer, H. 1965. Ayers Rock, ein Beispiel für klimagenetische Morphologie. *Z. Geomorph.* NF **9**, 249–84.

Bremer, H. 1971. Flüsse, Flächen-und Steufenbildung in den feuchten Tropen. *Würzburger Geog. Arb.* **35**.

Bremer, H. 1979. Relief und Böden in den Tropen. *Z. Geomorph.* NF Suppl. **33**, 25–37.

Bremer, H. 1980. Landform development in the humid tropics, German geomorphological research. *Z. Geomorph.* NF Suppl. **36**, 162–75.

Bremer, H. 1981a. Reliefformen und reliefbildende Prozesse in Sri Lanka. *Relief, Boden, Paläoklima* **1**, 7–184.

Bremer, H. 1981b. Inselberge – Beispiele für eine ökologische Geomorphologie. *Geogr. Z.* **69**, 199–216.

Bremer, H. and J. N. Jennings (eds) 1978. Inselbergs/Inselberge. *Z. Geomorph.* NF Suppl. **31**.

Bremer, H., A. Schnütgen and H. Späth (eds) 1981. Zur Morphogenese in den feuchten Tropen. Verwitterung und Reliefbildung am Beispiel von Sri Lanka. *Relief, Boden, Paläoklima* **1**.

Briden, J. C. and I. G. Gass 1974. Plate movement and continental magnetism. *Nature* **248**, 650–3.

Briden, J. C., D. C. Rex, A. M. Faller and J. F. Tomblin 1979. K–Ar geochronology and palaeomagnetism of volcanic rocks in the Lesser Antilles island arc. *Phil Trans R. Soc. Lond.* A **291**, 485–528.

Brinkman, R., A. G. Jongmans, R. Miedema and P. Maskaut 1973. Clay decomposition in seasonally wet, acid soils: micromorphological, chemical and mineralogical evidence from individual argillans. *Geoderma* **10**, 259–70.

Brinkmann, W. L. F. 1983. Nutrient balance of a central Amazonian rainforest: comparison of natural and man-managed systems. *Assoc. Int. Hydrol. Sci. Publn.* **140**, 153–63.

Brockway, L. 1979. *Science and colonial expansion: the role of the British Royal Botanic Gardens*. New York: Academic Press.

Broecker, W. S. 1983. The ocean. *Sci. Am.* **249**(3), 100–12.

Broecker, W. S. and J. Van Donk 1970. Insolation changes, ice volumes, and the ^{18}O record in deep-sea cores. *Rev. Geophys Space Phys* **8**, 169–98.

Broecker, W. S., D. L. Thurber, J. Goddard, T.-L. Ku, R. K. Matthews and J. K. Mesolella 1968. Milankovitch hypothesis is supported by precise dating of coral reefs and deep-sea sediments. *Science* **159**, 297–300.

Bruckner, D. 1955. The mantle rock (laterite) of the Gold Coast and its origin. *Geol. Rundsch.* **43**, 307–27.

Bruijnzeel, L. A. 1982. *Hydrological and biogeochemical aspects of man-made forests in south-central Java, Indonesia*. Amsterdam: Academish Proefschrift, Vrije Universiteit te Amsterdam.

Bruijnzeel, L. A. 1983. Evolution of runoff sources in a forested basin in a wet monsoon environment: a combined hydrological and hydrochemical approach. *Assoc. Int. Hydrol. Sci. Publn.* **140**, 165–74.

Brunet-Moret, Y. 1963. *Étude générale des averses exceptionnelles en Afrique Occidentale: République de Haute-Volta*. ORSTOM, Comité Inter-États d'études Hydrauliques.

Brunner, C. A. 1982. Paleoceanography of surface waters in the Gulf of Mexico during the Late Quaternary. *Quat. Res.* **17**, 105–19.

Brunsden, D. and J. B. Thornes 1979. Landscape sensitivity and change. *Trans. Inst. Br. Geogrs* NS **4**, 463–84.

Bryant, E. 1983. Regional sea level, Southern Oscillation and beach change, New South Wales, Australia. *Nature* **305**, 213–16.

Bryson, R. A. and A. M. Swain 1981. Holocene variations of monsoon rainfall in Rajasthan. *Quat. Res.* **16**, 135–45.

Buchanan, F. 1807. *A journey from Madras through the countries of Mysore, Canara and Malabar, performed under the orders of the Most Noble the Marquis Wellesley, Governor-General of India, for the express purpose of investigating the state of agriculture, arts and commerce; the religions, manners and customs; the history natural and civil and antiquity in the Dominions of the Rajah of Mysore, and the countries acquired by the Honourable East India Company*. London: The Author.

Buddemeier, R. W. and R. A. Kinzie 1976. Coral growth. *Oceanogr. Marine Biol. Ann. Rev.* **14**, 183–225.

Büdel, J. 1957. Die Doppelten Einebnungsflächen in den feuchten Tropen. *Z. Geomorph.* NF **1**, 201–88.

Büdel, J. 1965. Die relief typen der Flächenspül-zone Sud-Indiens am Ostabfall Dekans gegen Madras. *Coll. Geog. Bonn* **8**.

Büdel, J. 1970. Pedimente, Rumpfflächen und Rückland Steilhänge; deren aktive und passive Rückverlegung in verschiedenen Klimaten. *Z. Geomorph.* NF **14**, 1–57.

Büdel, J. 1977. *Klima-Geomorphologie*. Berlin and Stuttgart: Borntraeger.

Büdel, J. 1980. Climatic and climatomorphic geomorphology. *Z. Geomorph.* NF Suppl. **36**, 1–8.

Büdel, J. 1982. *Climatic geomorphology*. Engl. trans. by L. Fischer and D. Busche. Princeton: Princeton University Press.

Bull, P. A. 1981. Some fine-grained sedimentation phenomena in caves. *Earth Surf. Processes* **6**, 11–22.

Burke, K. 1976. The Chad Basin: an active intra-continental basin. *Tectonophysics* **36**, 197–206.

Burke, K. and J. F. Dewey 1973. Plume-generated triple junctions: key indicators in applying plate tectonics to old rocks. *J. Geol.* **81**, 406–33.

Burke, K. and B. Durotoye 1971. Geomorphology and superficial deposits related to large Quaternary climatic variation in south western Nigeria. *Z. Geomorph.* NF **15**, 430–44.

Burke, K. and A. J. Whiteman 1973. Uplift, rifting and the break-up of Africa. In *Implications of continental drift for the earth sciences*, D. H. Tarling and S. K. Runcorn (eds), 735–55. London: Academic Press.

Burke, K. and J. T. Wilson 1972. Is the African plate stationary? *Nature* **239**, 387–9.

Burke, K. and J. T. Wilson 1976. Hot spots on the Earth's surface. *Sci. Am.* **235**(2), 46–57.

Burke, K., T. F. J. Dessauvagie and A. J. Whiteman 1971a. Opening of the Gulf of Guinea and geological history of the Benue Depression and Niger Delta. *Nature Phys. Sci.* **223**, 51–5.

Burke, K., A. B. Durotoye and A. J. Whiteman, 1971b. A dry phase south of the Sahara 20 000 years ago. *W. Afr. J. Archaeol.* **1**, 1–8.

Burke, K., W. S. F. Kidd and J. T. Wilson 1973. Relative and latitudinal motion of Atlantic hotspots. *Nature* **245**, 133–7.

Burns, R. V. 1947. Soil erosion in Ceylon with particular reference to floods. *Trop. Agr. (Ceylon)* **103**, 240–5.

Busche, D., J. Grünert, E. Schulz and A. Skowronek 1979. Erste Radiokarbondaten aus dem vorland des Messak Mellet und plateau du Mangueni, Zentral-Sahara. *Würzburger Geog. Arb.* **49**, 183–98.

Butler, B. E. 1959. *Periodic phenomena in landscapes as a basis for soil studies*. CSIRO Aust. Soil. Publn. 14.

Butzer, K. W. 1980a. The Holocene lake plain of North Rudolf, East Africa. *Phys. Geog.* **1**, 42–58.

Butzer, K. W. 1980b. Pleistocene history of the Nile Valley in Egypt and Lower Nubia. In *The Sahara and the Nile*, M. A. J. Williams and H. Faure (eds), 253–80. Rotterdam: Balkema.

Butzer, K. W. 1980c. Holocene alluvial sequences: problems of dating and correlations. In *Timescales in geomorphology*, R. A. Cullingford, D. A. Davidson and J. Lewin (eds), 133–41. London: Wiley.

Butzer, K. W. and C. L. Hansen 1968. *Desert and river in Nubia*. Madison: University of Wisconsin Press.

Butzer, K. W., G. L. Isaac, J. L. Richardson and C. Washbourn-Kamau 1972. Radiocarbon dating of East African lake levels. *Science* **175**, 1069–76.

Byers, D. S. (ed.) 1967. *The prehistory of the Tehuacan Valley*, Vol. 1. Austin: University of Texas Press.

Cain, S. A., G. M. de Oliveira Castro, J. Mura Pires and N. Tomas de Silva 1956. Application of some phytosociological techniques to Brazilian rainforest. *Am. J. Bot.* **33**, 911–41.

Caratini, C. and P. Giresse 1979. Contribution palynologique à la connaissance des environments continentaux et marins du Congo à la fin du Quaternaire. *C.R. Acad. Sci. Paris D* **288**, 379–82.

Carbonnel, J. P. 1965. Sur les cycles de mise en solution du fer et de la silice en milieu tropical. *C.R. Acad. Sci. Paris* **260**, 4035–8.

Carbonnel, J. P. and J. Guiscafré 1965. *Grand lac du Cambodge: sédimentologie et hydrologie 1962–63*. Ministère des Affaires Étrangères, Paris.

Chamard, P. C. 1973. Monographic d'une sebkha continentale du Sud ouest saharien: la sebkha de Chemchane (Adrar de Mauritanie). *Bull. Inst. Fond. Afr. Noire* **35A**, 207–43.

Chappell, J. 1973. Astronomical theory of climatic change: status and problem. *Quat. Res.* **3**, 221–36.

Chappell, J. 1974. Late Quaternary glacio- and hydro-isostasy on a layered earth. *Quat. Res.* **4**, 429–40.

Chappell, J. 1976. Aspects of late Quaternary palaeo-geography of the Australian–East Indonesia region. In *The origin of the Australians*, R. L. Kirk and A. G. Thorne (eds), 11–22. Canberra: Aust. Inst. Aborig. Studies.

Chappell, J. 1978. Theories of Upper Quaternary ice ages. In *Climatic change and variability: a southern perspective*, A. B. Pittock *et al.* (eds), 211–25. Cambridge: Cambridge University Press.

Chappell, J. 1983a. Aspects of sea levels, tectonics and isostasy since the Cretaceous. In *Mega-geomorphology*, R. Gardner and H. Scoging (eds), 56–72. Oxford: Clarendon Press.

Chappell, J. 1938b. Thresholds and lags in geomorphic changes. *Aust. Geogr.* **15**, 357–66.

Chappell, J. and H. H. Veeh. 1978. Late Quaternary tectonic movements and sea level changes at Timor and Atauro Islands. *Geol. Soc. Am. Bull.* **89**, 356–68.

Charreau, C. 1969. Influence des techniques culturales sur le développement du ruissellement et de l'érosion en Casmance. Bambey. *VIIe Cong. Int. du Génie Rural*, CNRA.

Charreau, C. and R. Nicou, 1971. L'amélioration du profil cultural dans les sols sableux et sablo-argileux de la zone tropicale sèche ouest-africaine et ses incidences agronomiques. *Agron. Trop.* **26**, 903–78, 1183–247.

Charreau C. and L. Seguy 1969. Mesure de l'érosion et du ruissellement à Séfa en 1968. *Agron. Trop. Fr.* **24**, 1055–97.

Chase, C. G. 1979. Subduction, the geoid, and lower mantle convection. *Nature* **282**, 464–8.

Cheke, A. S., Wecrachal Nanakorn and Chusec Yankoses 1979. Dormancy and dispersal of seeds of secondary forest species under the canopy of a primary tropical rain forest in northern Thailand. *Biotropica* **11**, 88–95.

Chia, L. S. 1968. An analysis of rainfall patterns in Selangor. *J. Trop. Geog.* **27**, 1–18.

Chia, L. S. and K. K. Chang 1971. The record floods of 10th December 1969 in Singapore. *J. Trop. Geog.* **33**, 9–19.

Chinn, S. S., G. A. Tateishi and J. S. J. Yee, 1983. Water resources data. *Hawaii and other Pacific Areas Water Year 1982*, Vol. 1: *Hawaii*. U.S. Geol. Surv. Water-Data Rep. HI-82-1.

Chorley, R. J., A. J. Dunn and R. P. Beckinsale 1964. *The history of the study of landforms or the development of geomorphology*. vol. 1: *Geomorphology before Davis*. London: Methuen.

Church, M. 1980. Records of recent geomorphological events. In *Timescales in geomorphology*, R. A. Cullingford, D. A. Davidson and J. Lewin (eds), 13–29. Chichester: Wiley.

Churchill, D. M., R. W. Galloway and G. Singh 1978. Closed lakes and the palaeoclimatic record. In *Climatic change and variability: a southern perspective*, A. B. Pittock *et al.* (eds), 97–108. Cambridge: Cambridge University Press.

Clark, J. and J. Dymond 1974. Age, chemistry and tectonic significance of Easter Island. *Am. Geophys. Union, 55th Ann. Meet. Sec. Oceanogr.* **300** (paper 0111).

Clark, J. A., W. E. Farrell and W. R. Peltier 1978. Global changes in post-glacial sea level: a numerical calculation, *Quat. Res.* **9**, 265–87.

Clark, J. C., M. A. J. Williams and A. B. Smith 1973. The geomorphology and archaeology of Adrar Bous, central Sahara: a preliminary report. *Quaternaria* **17**, 245–97.

CLIMAP Project Members 1976. The surface of the ice-age earth. *Science* **191**, 1131–7.

Cloos, H. 1939. Hebung-Spatlung-Vulcanismus. *Geol. Rundsch.* **30**, 405–527.

Cointepas, J.-P. 1956. Premiers résultats des mesures de l'érosion en Moyenne Casamance. *Congr. Int. Sci. Sol, Paris D*, 569–76.

Colebrook, J. M. 1976. Trends in the climate of the North Atlantic Ocean over the past century. *Nature* **263**, 576–7.

Coleman, J. M. 1980. Recent increased aridity in Peninsular Florida. *Quat. Res.* **13**, 75–9.

Coleman, J. M. 1982. Recent seasonal rainfall and temperature relationships in Peninsular Florida. *Quat. Res.* **18**, 144–51.

Colinvaux, P. A. 1972. Climate and the Galapagos Islands. *Nature* 240, 17–20.

Colinvaux, P. A. and E. K. Schofield 1976a. Historical ecology in the Galapagos Islands I. A historical pollen record from El Junco Lake, Isla San Cristobal. *J. Ecol.* **64**, 989–1012.

Colinvaux, P. A. and E. K. Schofield 1976b. Historical ecology in the Galapagos Islands II. A Holocene spore record from El Junco Lake, Isla San Cristobal. *J. Ecol.* **64**, 1013–28.

Collinet, J. and A. Lafforgue 1978. *Mesures du ruissellement et d'érosion sous pluies simulées pour quelques types de sols de Haute-Volta*. Centres ORSTOM d'Adipodoumé, Abidjian.

Connell, J. H. 1973. Population ecology of reef building corals. In *Biology and Geology of Coral Reefs*. Vol. II: *Biology 1*, O. A. Jones and R. Endean (eds), 271–324. Amsterdam: Elsevier.

Connell, J. H. 1978. Diversity in tropical rain forests and coral reefs. *Science* **199**, 1302–10.

Conrad, G. 1969. *L'évolution continentale post-hercynienne du Sahara algerien*. Paris: CNRS.

Cooke, H. B. S. 1958. *Observations relating to Quaternary environments in East and Southern Africa*. Alex du Toit Mem. Lect. No. 5, Annex to vol. LX, Geol. Soc. S. Afr. Bull.

Coombe, D. E. 1960. An analysis of the growth of *Trema guineensis*. *J. Ecol.* **48**, 219–31.

Coombe, D. E. and W. Hadfield 1962. An analysis of the growth of *Musanga cecropioides*. *J. Ecol.* **50**, 221-34.

Cooper, M. 1966. Destruction of marine fauna and flora on Fiji caused by the hurricane of February 1965. *Pacif. Sci.* **20**, 137–41.

Cooray, P. G. 1963. The Erunvala Gravel and the probable significance of its ferricrete Cap. *Ceylon Geog.* **17**, 39–48.

Cooray, P. G. 1967. *An introduction to the geology of Ceylon*. Colombo: Nat. Mus. Ceylon.

Cooray, P. G. 1968. The geomorphology of part of the northwestern coastal plain of Ceylon. *Z. Geomorph.* NF, Suppl. **7**, 95–113.

Corbel, J. 1957. *Les karsts du nord-ouest de l'Europe*. Mem. Inst. Études Rhodaniennes 12.

Corbel, J. 1959. Vitesse de l'érosion. *Z. Geomorph.* NF **3**, 1–28.

Corbel, J. 1971. Les karsts des régions chaudes (des deserts aux zones tropicales humides). *Stud. Geom. Carpatho-Balcania* **5**, 49–74.

Corbel, J. and R. Muxart 1970. Karsts des zones tropicales humides. *Z. Geomorph.* NF **14**, 411–74.

Cornet, J. 1896. Les depôts superficiels et l'érosion continentale dans le bassin du Congo. *Bull. Soc. Belge Géol.* **10**, 44–116.

Cornforth, I. S. 1970. Leaf fall in a tropical rain forest. *J. Appl. Ecol.* **7**, 609–15.

Corvinus, G. 1969. Stratigraphy and geological background of an Acheulian site at Chirki-on-Pravara, India. *Anthropos* **63/64** 431–40.

Cotton, C. A. 1942. *Climatic accidents in landscape making*. Wellington: Whitcombe and Tombs.

Cotton, C. A. 1958. Fine-textured erosional relief in New Zealand. *Z. Geomorph.* NF **2**, 187–210.

Cotton, C. A. 1961. Theory of savanna planation. *Geography* **46**, 89–96.

Coventry, R. J. and D. Hopley 1980. The Quaternary of Northeastern Australia: Introduction. In *The geology and geophysics of northeastern Australia*, R. A. Henderson and P. J. Stephenson (eds), 375–82. Brisbane: Geological Society of Australia.

Cox, R. E., M. A. Mazurek and B. R. T. Simoneit 1982. Lipids in Harmattan aerosols of Nigeria. *Nature* **296**, 848–9.

Crickmay, C. H. 1974. *The work of the river*. London: Macmillan.

Crough, S. T. 1979. Hot spot epeirogeny. *Tectonophysics* **61**, 321–33.

Crough, S. T. 1981. Mesozoic hotspot epeirogeny in eastern North America. *Geology* **9**, 1–6.

Crough, S. T. and D. M. Jurdy 1980. Subducted lithosphere, hot spots and the geoid. *Earth Planet. Sci. Lett.* **48**, 15–22.

Crow, T. R. 1980. A rain forest chronicle: a 30 year record of change in structure and composition at El Verde, Puerto Rico. *Biotropica* 12, 42–55.

Cry, G. W. 1967. *Effects of tropical cyclone rainfall on the distribution of precipitation over the Eastern and Southern United States*. US Dept Commerce, ESSA Prof. Pap. 1.

CTFT 1971. *Défense et restauration des sols. Station de Gampela. Rapport annuel 1971*. Haute-Volta, Ministere de l'Agriculture, de l'Elevage et des Eaux et Forêts, CTFT.

CTFT/HV 1973. *Rapport de synthèse 1972*. CTFT Ministère Agric. de Haute-Volta, Ouagadougou.

Cuisinier, L. 1929. Régions calcaires de l'Indochine. *Ann. Géog.* **38**, 266–73.

Cunha, S. B., M. B. Machado and M. R. Mousinho de Meis 1975. Drainage basin morphometry on deeply weathered bedrocks. *Z. Geomorph*. NF **19**, 125–39.

Curray, J. R., F. J. Emmal, D. G. Moore and R. W. Raitt 1982. Structure, tectonics and geological history of the northeastern Indian Ocean. In *The ocean basins and margins*. Vol. 6: *The Indian Ocean*, A. E. M. Nairn and F. G. Stehli (eds), 399–450. New York: Plenum.

Da Dapola, E. 1980. *Contribution à l'étude géographique des paysages voltaiques (Monographie de la région de Gaoua)*. Mém. Maîtrise, Univ. Ouagadougou, ESLSH, Géog.

Daly, D. D. 1882. Surveys and explorations in the native states of the Malayan Peninsula, 1875–82. *Proc. R. Geogr. Soc*. **4**, 393–412.

Daly, R. A. 1910. Pleistocene glaciation and the coral reef problem. *Am. J. Sci*. **30**, 297–303.

Daly, R. A. 1934. *The changing world of the Ice Age*. New Haven: Yale University Press.

Damuth, J. E. 1977. Late Quaternary sedimentation in the western equatorial Atlantic. *Bull. Geol Soc. Am*. **88**, 695–710.

Damuth, J. E. and R. W. Fairbridge 1970. Equatorial Atlantic deep-sea arkosic sands and Ice-age aridity in tropical South America. *Bull. Geol Soc. Am*. **81**, 189–206.

Dana, J. D. 1850. On denudation in the Pacific. *Am. J. Sci. (Ser. 2)* **9**, 48–62. Reprinted 1972 in *River morphology*, S. A. Schumm (ed.), 24–39. Stroudsberg: Dowden, Hutchinson and Ross.

Dansgaard, W. and H. Tauber 1969. Glacier oxygen-18 content and Pleistocene ocean temperatures. *Science* **166**, 499–502.

Darwin, C. 1890. *On the structure and distribution of coral reefs: also geological observations on the volcanic islands and parts of South America*. London: Ward Lock.

Davis, G. E. 1982. A century of natural change in coral distribution at the Dry Tortugas: a comparison of reef maps from 1881 and 1976. *Bull. Marine Sci*. **32**, 608–23.

David, M. B. 1976. Erosion rates and land-use history in Southern Michigan. *Env. Cons*. **3**, 139–48.

Davis, M. B. 1981. Quaternary history and the stability of forest communities. In *Forest succession: concepts and applications*, D. C. West, H. H. Shugart and D. B. Botkin (eds), 132–53, New York: Springer Verlag.

Davis, W. M. 1899. The geographical cycle. *Geogr. J*. **14**, 481–504.

Davis, W. M. 1909. *Geographical essays*. Boston: Dover.

Davis, W. M. 1928. *The coral reef problem*. Am. Geogr. Soc. Spec. Publn.

Day, D. G. 1980. Lithologic controls of drainage density: a study of six small rural catchments in New England, NSW. *Catena* **7**, 339–51.

Day, M. J. 1980. Landslides in the Gunung Mulu National Park. *Geogr. J*. **146**, 7–13.

De Alwis, K. A. 1980. *The origin of the Red Earth Formation of Sri Lanka: evidence from a pedologic study*. Colombo (unpubl.).

De Alwis, K. A. and C. R. Panabokke 1973. *Handbook of the soils of Sri Lanka* Dept. Irrigation, Colombo.

Dearing, J. A. 1979. *The application of magnetic measurements to studies of particulate flux in lake-watershed ecosystems*. PhD thesis, University of Liverpool.

Dearing, J. A. 1982. *Core correlation and total sediment influx*. IGCP Project, B. E. Berglund (ed.), Section 10.16.

Dearing, J. A. 1984. Changing patterns of sedimentation in a small lake in Scania, S. Sweden. *Hydrobiologia* **103**, 59–64.

Dearing, J. A., J. K. Elner and C. M. Happey-Wood 1981. An examination of total sediment flux and erosional processes in a Welsh upland lake-catchment based on magnetic susceptibility measurements of the recent lake sediments. *Quat. Res*. **16**, 356–72.

De Dapper, M. 1978. Couvertures limono-sableuse, stone-line, indurations ferrugineuses et action des termites sur le plateau de la Manika (Kolwezi, Shaba, Zaire). *Géo. Éco. Trop*. **2**, 256–78.

Degens, E. T. and R. E. Hecky 1974. Palaeoclimatic reconstruction of Late Pleistocene and Holocene based on biogenic sediments from the Black Sea and a tropical African lake. In *Les méthodes quantitatives d'étude des variations du climat au cours du Pléistocène*, Coll. Int. CNRS 219, 13–37.

De Heinzelin, J. 1952. *Sols, paléosols et désertifications anciennes dans le secteur oriental du bassin du Congo*. Publ. INEAC Coll. 4.

Delacourt, F. and J. P. Troy 1971. *Carte des sols de l'Inde*. Inst. fr. Pondicherry.

Delvigne, J. and G. Grandin 1969. Études des cycles morphogenetiques et tentative de chronologie paleoclimatique dans la région granitique de Toumodi en Côte d'Ivoire. *C. R. Acad. Sci., Paris* **296**, 1327–9.

Delwaulle, J.-C. 1973. Résultats de six années d'observations sur l'érosion au Niger. *Bois Forêts Trop.* **150**, 15–37.

Demangeot, J. 1975. Recherches géomorphologiques en Indie du Sud. *Z. Geomorph.* NF **19**, 229–72.

De Martonne, E. 1940. Problèmes morphologiques du Brésil tropical atlantique. *Ann. Géog.* **49**, 1–27, 106–29.

De Martonne, E. 1946. Géographie zonale. Le zone tropicale. *Ann Géog.* **55**, 1–8.

De Martonne, E. 1951. *Traité de géographie physique*. Vol. II: *Le relief du sol*, 9th edn. Paris: Colin.

De Martonne, E. and P. Birot 1944. Sur l'évolution des versants en climat tropical humide. *C.R. Acad. Sci. Paris* **218**, 529–32.

De Meis, M. R. M. and A. M. F. Monteiro 1979. Upper Quaternary 'ramps': Doce river valley, southeastern Brazil plateau. *Z. Geomorph.* NF **23**, 132–51.

Denslow, J. S. 1980. Gap partitioning among tropical rainforest trees. *Biotropica* **12**, 47–55.

Denton, G. H. and T. J. Hughes (eds) 1981. *The last great ice sheets*. New York: Wiley Interscience.

De Ploey, J. 1963. Quelques indices sur l'évolution morphologique et paléoclimatique des environs du Stanley-Pool (Congo). *Stud. Univ. Lovanium Fac. Sci.* **17**, 1–16.

De Ploey, J. 1965. Position géomorphologique, genèse et chronologie de certains dépôts superficiels au Congo occidental. *Quaternaria* **7**, 131–54.

Deraniyagala, P. E. P. 1958. *The Pleistocene of Ceylon*. Colombo: Nat. Mus. Ceylon.

Deraniyagala, P. E. P. 1973. The geomorphology and pedology of three sedimentary formations containing a Mesolithic industrie in the lowlands of the Dry Zone of Sri Lanka. Symp. *72 Annu. Meet. Am. Anthrop. Assoc.*, New Orleans.

Descamps, M. and Alii 1978. Étude des écosystemès guyanais, II. Données biogéographiques sur la partie orientale des Guyanes. *C.R. Soc. Biogéogr. Fr.* **461**(9), 55–82.

De Swardt, A. M. J. and A. F. Trendall 1969. The physiographic development of Uganda. *Over. Geol. Min. Res.* **10**, 241–88.

Deuser, W. G., E. H. Ross and L. S. Waterman 1976. Glacial and pluvial periods: their relationship revealed by Pleistocene sediments of the Red Sea and Gulf of Aden. *Science* **191**, 1168–70.

De Vera, M. R. 1981. Assessment of sediment yield using the universal soil loss equation. *Publns Int. Assoc. Hydrol Sci.* **132**, 600–14.

D'Hoore, J. 1963. *Soil map of Africa scale 1 to 5 000 000*. Explanatory monograph. CCTA Publn 93.

Dias de Avila-Pires, F. 1974. Caracterizacao zoogeografica de Provincia Amazonica. *An. Acad. Brasil. Ciencias* **46**(1), 133–81.

Dickinson, W. R. 1974. Plate tectonics and sedimentation. In *Tectonics and sedimentation*, W. R. Dickinson (ed.), 1–27. Soc. Econ. Paleont. Mineral. Tulsa, Spec. Publ. 22.

Diester-Haass, L. 1976. Late Quaternary climatic variations in Northwest Africa deduced from East Atlantic sediment cores. *Quat. Res.* **6**, 299–314.

Dingle, R. V. 1973. Mesozoic palaeogeography of the southern Cape, South Africa. *Palaeogeog. Palaeoclim. Palaeoecol.* **13**, 203–13.

Dingle, R. V. and R. A. Scrutton 1974. Continental breakup and the development of post-Palaeozoic sedimentary basins around southern Africa. *Bull. Geol Soc. Am.* **85**, 1467–74.

Dixey, F. 1922a. Notes on lateritisation in Sierra Leone. *Geol Mag*, **57**, 211–20.

Dixey, F. 1922b. The physiography of Sierra Leone. *Geogr. J.* **60**, 41–65.

Dixey, F. 1938. Some observations on the physiographical development of Central and Southern Africa. *Trans. Geol. Soc. S. Afr.* **41**, 113–70.

Dixey, F. 1942. Erosion cycles in central and southern Africa. *Trans. Geol Soc. S. Afr.* **45**, 151–81.

Dixey, F. 1943. The morphology of the Congo–Zambesi watershed. *S. Afr. Geog. J.* **25**, 20–41.

Dixey, F. 1946. Erosion and tectonics in the East African Rift System. *Q. J. Geol Soc. Lond.* **102**, 339–88.

Dollar, S. J. 1982. Wave stress and coral community structure in Hawaii. *Coral Reefs*, **1**, 71–81.

Domrös, M. 1971. 'Wet Zone' and 'Dry Zone' – Möglichkeiten einer klimaökologischen Raumgliederung der Insel Ceylon. *Erdkdl. Wiss.* **27**, 205–32.

Domrös, M. 1976. *Sri Lanka. Die Tropeninsel Ceylon*. Darmstadt: Wiss. Buchges.

Doornkamp, J. C. and C. A. M. King 1971. *Numerical analysis in geomorphology: an introduction*. London: Edward Arnold.

Douglas, I. 1967. Erosion of granite terrains under tropical rain forest in Australia, Malaysia and Singapore. *Publns Assoc. Int. Hydrol. Sci.* **75**, 31–9.

Douglas, I. 1969. The efficiency of humid tropical denudation systems. *Trans. Inst. Br. Geogrs* **46**, 1–16.

Douglas, I. 1970. Measurements of river erosion in West Malaysia. *Malay. Nat. J.* **23**, 78–83.

Douglas, I. 1971. *Aspects of the water balance of catchments in the main range near Kuala Lumpur*. Univ. Hull Dept. Geog. Misc. Ser. 11, 23–48.

Douglas, I. 1972. The geographical interpretation of river water quality data. *Progr. Geog.* **4**, 1-81.

Douglas, I. 1973. *Rates of denudation in selected small catchments in Eastern Australia*. Univ. Hull. Occ. Pap. Geog. 21.

Douglas, I. 1976. Erosion rates and climate: geomorphological implications. In *Geomorphology and climate*. E. Derbyshire (ed.), 269–87. London: Wiley.

Douglas, I. 1977. *Humid landforms*. Cambridge, Mass: MIT Press.

Douglas, I. 1978a. The impact of urbanisation on fluvial geomorphology in the humid tropics. *Géo. Éco. Trop* **2**, 229–42.

Douglas, I. 1978b. Tropical geomorphology: present problems and future prospects. In *Geomorphology: present problems and future prospects*, C. Embleton, D. Brunsden and D. K. C. Jones (eds), 162–84. Oxford: Oxford University Press.

Douglas, I. 1980a. Geomorphic processes during the Quaternary. In *The geology and geophysics of Northeastern Australia*, R. A. Henderson and P. J. Stephenson (eds), 393–5. Brisbane: Geological Society of Australia.

Douglas, I. 1980b. Climatic geomorphology. Present-day processes and landform evolution. Problems of interpretation. *Z. Geomorph.* NF Suppl. **36**, 27–47.

Douglas, I. 1981. Soil conservation measures. In *River basin planning: theory and practice*, S. V. Sama and C. J. Barrow (eds), 49–73. Chichester: Wiley.

Dresch, J. and G. Rougerie. 1960. Observations morphologiques sur le Sahel du Niger. *Rev. Géom. Dyn. Fr.* **11**, 4–6, 49–58.

Drew, F. 1875. *The Jummoo and Kashmir Territories*. London.

Dunne, T. 1979. Sediment yield and land use in tropical catchments. *J. Hydrol.* **42**, 281–300.

Duplessy, J. C. 1982. Glacial to interglacial contrasts in the Northern Indian Ocean. *Nature* **295**, 494–8.

Dury, G. H. 1973. Palaeohydrologic implications of some pluvial lakes in northwestern New South Wales, Australia. *Bull. Geol Soc. Am.* **84**, 3663–76.

Dustan, P. 1975. Growth and form of the reef-building coral *Montastrea annularis*. *Marine Biol.* **3**, 101–7.

Eavis, A. J. 1981. *Caves of Mulu '80*. London: Royal Geographical Society.

Eden, M. J. 1971. Some aspects of weathering and landforms in Guyana. *Z. Geomorph.* NF **15**, 181–98.

Edmunds, W. M. and E. P. Wright 1979. Groundwater recharge and palaeoclimate in the Sirte and Kufra basins, Libya. *J. Hydrol.* **40**, 215–41.

Edwards, K. A. 1979. Regional contrasts in rates of soil erosion and their significance with respect to agricultural development in Kenya. In *Soil physical properties and crop production in the tropics*, R. Lal and D. J. Greenland (eds), 441–54. Chichester: Wiley.

Edwards, P. J. 1982. Studies of mineral cycling in a montane rain forest in New Guinea v. rates of cycling in throughfall and litter fall. *J. Ecol.* **70**, 807–27.

Emiliani, C. 1966. Palaeotemperature analysis of Caribbean cores P6303-8 and P6304-9 and a generalized temperature curve for the past 425 000 years. *J. Geol.* **74**, 109–26.

Emiliani, C. 1971. Pleistocene climatic cycles at low latitudes. In *Late Cenozoic glacial ages*, K. K. Turekian (ed.), 183–97. Yale University Press.

Emiliani, C. 1972. Quaternary palaeotemperatures and the duration of the high-temperature intervals. *Science* **178**, 398–401.

Endean, R. 1973. Population explosions of *Acanthaster planci*, and associated destruction of hermatypic corals in the Indo-West Pacific region. In *Biology and geology of coral reefs*. Vol. II: *Biology 1*, O. A. Jones and R. Endean (eds), 389–428. New York: Academic Press

Endler, J. A. 1982. Pleistocene forest refuges: fact or fancy? In *Biological diversification in the tropics*, G. T. Prance (ed.), 608–40. New York: Columbia University Press.

Erhart, H. 1955. Biostasie et Rhexistasie: Esquisse d'une theorie sur le rôle de la pedogenèse en tant que phénomène géologique. *C.R. Acad. Sci. Paris* **241**, 1218–20.

Erhart, H. 1956. *Le genèse des sols, en tant que phénomène geologique. Esquisse d'une théorie géologique et géoghimique. Biostasie et rhexistasie*. Paris: Masson.

Erhart, H. 1961. Sur la genèse de certaines gîtes miniers sedimentaires, en rapport avec le phénomène de bio-rhexistasie et avec des mouvements tectoniques de faible amplitude. *C.R. Acad. Sci. Paris* **252**, 2904–6.

Erhart, H. 1966. Bio-rhexistasie, biostasies évolutives, hétérostasie. Importance de ces notions en gîtologie minière exogène. *C.R. Acad. Sci. Paris* **263**, 1048–51.

Ewel, J. J. 1976. Litter fall and leaf decomposition in a tropical forest succession in eastern Guatemala. *J. Ecol.* **64**, 293–308.

Ewers, R. O. 1972. *A model for the development of sub-surface drainage routes along bedding planes*. MSc Thesis, University of Cincinnati.

Eyles, N. 1979. Facies of supra-glacial sedimentation on Icelandic and Alpine temperate glaciers. *Can. J. Earth Sci*, **16**, 1341–61.

Eyles, R. J. 1966. Stream representation on Malayan maps. *J. Trop. Geog*, **22**, 1–9.

Faegri, K. 1950. On the value of palaeoclimatological evidence. *Centenary Proc. R. Meteor. Soc. 1950*, 188–94.

Fairbridge, R. W. 1970. World paleoclimatology of the Quaternary. *Rev. Géogr. Phys. Géol. Dyn.* **12**(2), 97–104.

Fairbridge, R. W. 1976. Effects of Holocene climatic change on some tropical geomorphic processes. *Quat. Res.* **6**, 529–57.

Fairbridge, R. W. 1979. Vertical crustal movements and the rifting of continents. *Geol. Mijnb.* **58**, 273–6.

Fairbridge, R. W. and C. W. Finkl Jr. 1978. Geomorphic analysis of the rifted cratonic margins of Western Australia. *Z. Geomorph.* NF **22**, 369–89.

Fairbridge, R. W. and C. W. Finkl Jr 1980. Cratonic erosional unconformities and peneplains. *J. Geol.* **88**, 69–86.

Fairhead, J. D. and N. B. Henderson 1977. The seismicity of southern Africa and incipient rifting. *Tectonophysics*, **41**, T19–26.

Fairhead, J. D. and C. V. Reeves 1977. Teleseismic delay times, Bouguer anomalies and inferred thickness of the African lithosphere. *Earth Planet. Sci. Lett.* **36**, 63–76.

Falconer, J. D. 1911. *The geology and geography of Northern Nigeria*. London: Macmillan.

Falconer, J. D. 1912. The origin of kopjes and inselbergs. *Br. Assoc. Adv. Sci. Trans Sec. C*, 476.

FAO–UNESCO 1973. *World Soil Map*.

Fauck, R. 1954. Les facteurs et les intensités de l'érosion en moyenne Casamance. *Congr. Int. Sci. Sol*, Léopoldville, Vol. 3, 753–93.

Faure, H. 1966. Évolution des grand lacs sahariens à l'Holocene. *Quaternaria* **8**, 167–75.

Faure, H. 1971. Relations dynamiques entre la croûte et le manteau d'après l'étude de l'évolution paléogeographique des bassins sedimentaires. *C.R. Acad. Sci. Paris* **272D**, 3239–42.

Ferguson, R. I. 1978. Drainage density–basin area relationship. Comment. *Area* **10**, 350–2.

Ferhi, A. and R. Letolle 1977a. Variation de la composition isotopique de l'oxygene organique de quelques plantes on fonction de leur milieu de vie. *C.R. Acad. Sci. Paris* **284D**, 1887–9.

Ferhi, A. and R. Letolle 1977b. Transpiration and evaporation as the principal factors in oxygen isotope variations of organic matter in land plants. *Phys. Veg.* **15**, 363–70.

Fieldes, M. and K. W. Perrott 1966. A rapid test for allophane. *NZ J. Sci.* **9**, 623–9.

Fileux, M. and H. Stommel 1975. Preliminary look at feasibility of using marine reports of sea temperature for documenting climatic change in the western North Atlantic. *J. Marine Res.* **33**(Suppl.), 83–95.

Fink, J. and E. J. Kukla 1977. Pleistocene climates in Central Europe: at least 17 interglacials after the Olduvai Event. *Quat. Res.* **7**, 363–71.

Finkl, C. W. Jr 1979 Stripped (etched) landsurfaces in southern western Australia. *Aust Geogr. Stud.* **17**(1) 33–52

Finkl, C. W. Jr 1982. On the geomorphic stability of cratonic planation surfaces. *Z. Geomorph.* NF **26**, 137–90.

Finkl, C. W. Jr and H. M. Churchward 1973. The etched landsurfaces of south western Australia. *J. Geol Soc. Aust.* **20**, 295–307.

Firing, E., R. Lukas, J. Sadler and K. Wyrkti 1983. Equatorial undercurrent disappears during 1982–83. El Niño. *Science* **222**, 1121–3.

Fittkau, E. J. 1973–4. Esboço de uma divisa ecolôgica da Região Amazônica. *A. Amazonia Brasil. em Fôco* **9**, 17–23.

Flageollet, J. C. 1981. Aspects morphoscopiques et exoscopiques des quartz dans quelques sols ferrallitiques de la région de Cechi (Côte d'Ivoire). *Cah. ORSTOM Sér. Pedol.* **18**(2), 111–22.

Flenley, J. R. 1972. Evidence of Quaternary vegetational change in New Guinea. In *The Quaternary era in Malesia*, P. & M. Ashton (eds), 99–109. *Trans. 2nd Aberdeen–Hull Symp. on Malesian Ecology*. Univ. of Hull Dept of Geog., Misc. Ser. 13.

Flenley, J. R. 1979. *The equatorial rain forest, a geological history*. London: Butterworths.

Flenley, J. R. and S. M. King 1984. Late Quaternary pollen records from Easter Island. *Nature* **307**, 47–50.

Flint, R. F. 1959a. On the basis of Pleistocene correlation in East Africa. *Geol Mag.* **96**, 265–84.

Flint, R. F. 1959b. Pleistocene climates in eastern and southern Africa. *Bull. Geol Soc. Am.* **70**, 343–74.

Flint, R. F. 1971. *Glacial and Quaternary geology*. New York: Wiley.

Flint, R. F. 1976. Physical evidence of Quaternary climatic change. *Quat. Res.* **6**, 519–28.

Flohn, H. 1953. Studien uber die atmosphärische Zirkulation in der letzten Eiszeit. *Erdkunde* **7**, 266–75.

Flohn, H. 1979. On time scales and causes of abrupt palaeoclimatic events. *Quat. Res.* **12**, 135–49.

Flohn, H. and S. Nicholson 1980. Climatic fluctuations in the arid belt of the 'Old World' since the last glacial maximum: possible causes and future implications. *Palaeoecol. Afr.* **12**, 3–21.

Folk, R. L. and R. Robles 1964. Carbonate sands of Isla Perez, Alacran Reef complex, Yucatan. *J. Geol.* **72**, 255–92.

Fölster, H. 1969. Slope development in SW Nigeria during the Late Pleistocene and Holocene. *Geiss. Geogr. Schr.* **20**, 3–56.

Fölster, H. and G. de las Salas 1976. Litterfall and mineralization in three tropical evergreen forest stands, Colombia. *Acta Cientia Venez.* **27**, 196–202.

Fölster, H., G. de las Salas and P. Khanna 1976. A tropical evergreen forest site with perched water table, Magdalena valley, Colombia. Biomass and bioelement inventory of primary and secondary vegetation. *Oecologia Plantarum* **11**, 297–320.

Fosberg, R. R. 1962. The physical background of the humid tropics substratum. In *Symposium*

on the impact of man on humid tropics vegetation, 35–7. Administration of the Territory of Papua and New Guinea, Port Moresby.
Fournier, F. 1954. *Les parcelles expérimentales. Méthode d'étude expérimentale de la conservation du sol, de l'érosion, du ruissellement*. Rapp. ORSTOM, Paris, 1623.
Fournier, F. 1955. Les facteurs de l'érosion du sol par l'eau en Afrique Occidentale Française. *C.R. Acad. Agr. Fr.* 660–5.
Fournier, F. 1956. Les formes et types d'érosion du sol par l'eau en Afrique Occidentale Française. *C.R. Acad. Agr. Fr.* **42**, 215–21.
Fournier, F. 1958. *Étude de la relation entre l'érosion du sol par l'eau et les précipitations atmosphériques*. Thèse Lettres, Paris. Paris: Presses Univ. France.
Fournier, F. 1960. *Climat et érosion*. Paris: Presses Univ. France.
Fournier, F. 1962. *Carte du danger d'érosion en Afrique au Sud du Sahara (fondé sur l'aggressivité climatique et la topographie)*. CEE/CCTA, Bur. Interafr. Sols.
Fournier, F. 1967. La recherche en érosion et conservation des sols sur le continent Africain. *Sols Afr.* **12**, 5–53.
Freeth, S. J. 1978. Tectonic activity in West Africa and the Gulf of Guinea since Jurassic times – an explanation based on membrane tectonics. *Earth Planet. Sci. Lett.* **38**, 298–300.
Freise, F. W. 1930. Beobachtungen über den Schweb einiger Flüsse des brasilianischen Staates Rio de Janeiro. *Z. Geomorph.* **5**, 214.
Freise, F. W. 1932. Beobachtungen über Erosion am Urwaldsgebirgsflüssen des brasilianischen Staates Rio de Janeiro. *Z. Geomorph.* **7**, 1–9.
Freise, F. W. 1934. Erscheinungen des Erdfliessens in Tropenurwalde. *Z. Geomorph.* **9**, 88–98.
Freise, F. W. 1936. Das Binnenklima von Urwäldern im subtropischen Brasilien. *Petermanns Geogr. Mitt.* **82**, 301–4.
Freise, F. W. 1938. Inselberge und inselberg-landschaften im granit und gneissgebiete Brasiliens. *Z. Geomorph.* **10**, 137–68.
Freson, R., G. Goffinet and F. Malaisse 1974. Ecological effects of the regressive succession muhulu–miombo–savannah in Upper-Shaba (Zaire). *Proc. 1st Int. Congr. Ecol.*, 365–71. The Hague: Pudoc.
Frey, D. G. 1979. Palaeolimnology of Lake Valencia, Venezuela. *IPPCCE Newslett.* **2**, 37–9.
Fryberger, S. G. 1980. Dune forms and wind regime, Mauritania, West Africa: implications for past climate. *Palaeoecol. Afr.* **12**, 79–96.
Furtado, J. I., S. Verghese, K. S. Liew and T. H. Lee 1980. Litter production in a freshwater swamp forest Tasek Bera, Malaysia. In *Tropical Ecology and Development Proc. Int. Symp. Tropical Ecology*, J. I. Furtado (ed.), 815–22 International Society of Tropical Ecology, Kuala Lumpur.
Fyfe, W. S. and O. H. Leonardos Jr 1977. Speculations on the causes of crustal rifting and subduction, with applications to the Atlantic margin of Brazil. *Tectonophysics*, **42**, 29–36.

Gac, J. Y. and M. Pinta 1973. Bilan de l'érosion et de l'altération en climat tropical humide. Estimation de la vitesse d'approfondissement des profiles. Étude du bassin versane de l'Ouham, Rep. Centre Africaine. *Cah. ORSTOM Sér Géol.* **5**, 83–96.
Galabert, J. and E. Millogo 1972. *Indice d'érosion par la pluie en Haute-Volta*. CTFT, Ministère de l'Agriculture, de l'Élevage et des Eaux et Forêts.
Galloway, R. W. 1965a. A note on world precipitation during the last glaciation. *Eiszeit. Gegenwart* **16**, 76–7.
Galloway, R. W. 1965b. Late Quaternary climates in Australia. *J. Geol*, **73**, 603–18.
Galloway, R. W. 1970. The full-glacial climate in the southwestern United States. *Ann. Assoc. Am. Geogrs* **60**, 245–56.
Galloway, R. W. and E. Löffler 1972. Aspects of geomorphology and soils in the Torres Strait Region. In *Bridge and barrier: the natural and cultural history of Torres Strait*. ANU Publn BG/3, 11–28.
Gardner, R. A. 1981. Reddening of dune sand – evidence from Southeast India. *Earth Surf. Processes Landforms.* **6**, 459–68.

Gardner, R. and K. Pye 1981. Nature, origin and palaeoenvironmental significance of red coastal and desert dune sands. *Progr. Phys. Geog.* **5**, 514–34.

Garner, H. F. 1974. *The origin of landscapes*. New York: Oxford University Press.

Garwood, N. C., D. P. Janos and N. Brokaw 1979. Earthquake-caused landslides: a major disturbance to tropical forests. *Science* **205**, 997–9.

Gaskin, A. R. C. 1975. Investigation of the residual iron ores of Tonkilili district, Sierra Leone. *Trans. Inst. Min. Metal. Sec. B. Appl. Earth Sci.* B98–119.

Gass, I. G. 1977. The age and extent of the Red Sea oceanic crust. *Nature*, **265**, 722–4.

Gass, I. G., D. S. Chapman, H. N. Pollack and R. S. Thorpe 1978. Geological and geophysical parameters of mid-plate volcanism. *Phil Trans R. Soc. Lond.* A. **288**, 581–96.

Gasse, F. 1975. *L'évolution des lacs de l'Afar Central (Ethiope et TFAI) du Plio-Pléistocène à l'actuel: reconstitution des paléomilieux lacustres à partir de l'étude des diatomées*. DSc Thesis, Univ. Paris VI.

Gasse, F. and G. Delibrias 1977. Les lacs de l'Afar Central (Ethiope et TFAI) au Pléistocène supérieur. In *Palaeolimnology of Lake Biwa and the Japanese Pleistocene*, S. Horie (ed.), 529–75.

Gasse, F. and F. A. Street 1978. Late Quaternary lake-level fluctuations and environments of the northern Rift Valley and Afar region (Ethiopa and Djibouti). *Palaeogeog. Palaeoclim. Palaeoecol.* **24**, 279–325.

Gasse, F., J. C. Fontes and P. Rognon 1974. Variations hydrologiques et extension des lacs holocènes du désert danakil. *Palaeogeog. Palaeoclim. Palaeoecol.* **15**, 109–48.

Gasse, F., P. Rognon and F. A. Street, 1980. Quaternary history of the Afar and Ethiopian Rift lakes. In *The Sahara and the Nile*, M. A. J. Williams and H. Faure (eds), 361–400. Rotterdam: Balkema.

Gaur, J. P. and H. N. Pandey 1978. Litter production in two tropical deciduous forest communities at Varanasi, India. *Oikos* **24**, 430–5.

Gerstenhauer, A. 1960. Der tropische Kegelkarst in Tabasco (Mexico). *Z. Geomorph.* NF, Suppl. **2**, 22–48.

Ghose, B., S. Pandey, S. Singh and G. Lal 1967. Quantitative geomorphology of the drainage basins in the central Luni basin in western Rajasthan. *Z. Geomorph.* NF **11**, 146–60.

Gibbs, R. J. 1967. The geochemistry of the Amazon River system: Part I. The factors that control the salinity and the composition and concentration of suspended solids. *Bull. Geol Soc. Am.* **78**, 1203–32.

Gibbs, R. J. 1972. Water chemistry of the Amazon River. *Geochim. Cosmochim. Acta* **36**, 1061–6.

Gidskehaug, A., K. M. Creer and J. G. Mitchell 1975. Palaeomagnetism and K–Ar ages of the south west African basalts and their bearing on the time of initial rifting of the South Atlantic ocean. *Geophys. J. R. Astr. Soc.* **42**, 1–20.

Gill, A. E. and A. M. Rasmusson 1983. The 1982–83 climatic anomaly in the equatorial Pacific. *Nature* **306**, 229–34.

Gillespie, R., F. A. Street-Perrott and R. Switsur 1983. Post-glacial arid episodes in Ethiopia have implications for climate prediction. *Nature* **306**, 680–3.

Gilmore, M. D. and B. R. Hall 1976. Life history, growth habits and constructional roles of *Acropora cervicornis* in the patch reef environment. *J. Sed. Petrol.* **46**, 519–22.

Gilmour, D. A., M. Bonell and D. F. Sinclair 1980. *An investigation of storm drainage processes in a tropical rainforest catchment*. Aust. Water Res. Counc. Tech. Pap. 56.

Giresse, P. 1978. Le contrôle climatique de la sédimentation marine et continentale en Afrique centrale atlantique à la fin du Quaternaire – problèmes de corrélation. *Palaeogeog. Palaeoclim. Palaeoecol.* **23**, 57–77.

Giresse, P. and L. Le Ribault 1981. Contribution de l'étude exoscopique des quartz à la reconstitution paléogéographique des derniers épisodes du Quaternaire littoral du Congo. *Quat. Res.* **15**, 86–100.

Giresse, P. and G. Moguedet 1980. Chronosequences fluvio-marines de l'Holocene de l'estuaire du Kouilou et des colmatages côtiers voisins du Congo. *Trav. Doc. Géog. Trop. CEGET* **39**, 21–46.

Giresse, P., G. Bongo-Passi, G. Delibrias and J.-C. Duplessy 1982. La lithostratigraphie des sédiments hémipelagiques du delta profond du fleuve Congo et ses indications sur les paléoclimats de la fin du Quaternaire. *Bull. Soc. Géol. Fr. Sér. 7*, **24**, 803–15.

Gjessing, J. 1967. Norway's Paleic surface. *Norsk. Geog. Tidsskr.* **21**, 69–132.

Glynn, P. W. 1968. Mass mortalities of echinoids and other reef flat organisms coincident with midday low water exposures in Puerto Rico. *Marine Biol.* **1**, 226–43.

Glynn, P. W., L. R. Almodovar and J. G. Gonzales 1964. Effects of Hurricane Edith on marine life at La Parguera, Puerto Rico. *Carib. J. Sci.* **4**, 335–45.

Glynn, P. W., G. M. Wellington and C. Birkeland 1979. Coral reef growth in the Galapagos: limitation by sea urchins. *Science* **203**, 47–9.

Godefroy, J., M. Muller and E. Roose 1970. Estimation des pertes par lixivation des éléments fertilisants dans un sol de bananeraie de basse Côte d'Ivoire. *Fruits* **25**(6), 403–23.

Golley, F. B., J. J. McGinnis, R. G. Clements, G. I. Child and M. J. Deuver 1975. *Mineral cycling in a tropical moist forest ecosystem*. Athens, Georgia: University of Georgia Press.

Golson, J. 1977. The making of the New Guinea highlands. In *The Melanesian environment*, J. H. Winslow (ed), 45–56. Canberra: ANU Press.

Goodbody, I. 1961. Mass mortality of tropical fauna after rains. *Ecology*, **42**, 150–5.

Goreau, T. F. 1964. Mass expulsion of zooxanthellae from Jamaican reef communities after Hurricane Flora. *Science* **145**, 383–6.

Goose, W. C. 1874. *Report and diary of Mr W. C. Gosse's central and western exploring expedition, 1873*. S. Aust. Parliament. Pap. 48.

Goudie, A. S. 1983a. The arid earth. In *Mega-geomorphology*, R. Gardner and H. Scoging (eds), 152–71. Oxford: Clarendon Press.

Goudie, A. S. 1983b. *Environmental change*, 2nd edn. Oxford: Clarendon Press.

Goudie, A. S., B. Allchin and K. T. M. Hegde 1973. The former extensions of the Great Indian Sand Desert. *Geogr. J.* **134**, 243–57.

Grant, P. J. 1981. Recently increased tropical cyclone activity and inferences concerning coastal erosion and inland hydrological regimes in New Zealand and Eastern Australia. *Clim. Change* **3**, 317–32.

Gray, W. M. 1968. Global view of the origin of tropical disturbances and storms. *Mon. Weather Rev.* **96**, 669–700.

Green, D. G. 1981. Time series and postglacial forest ecology. *Quat. Res.* **15**, 265–77.

Gregory, J. W. 1894a. Contributions to the physical geography of British East Africa. *Geogr. J.* **4**, 408–24

Gregory, J. W. 1894b. The glacial geology of Mount Kenya. *Q. J. Geol Soc. Lond.* 1.

Gregory, K. J. 1976. Drainage networks and climate. In *Geomorphology and Climate*, E. Derbyshire (ed.), 289–315. London: Wiley.

Grigg, R. W. and J. E. Maragos 1974. Recolonization of hermatypic corals on submerged lava flows in Hawaii. *Ecology* **55**, 387–95.

Grimes, K. G. and H. F. Doutch 1978. The late Cainozoic evolution of the Carpentaria Plains, North Queensland. *BMR J. Aust. Geol. Geophys.* **3**, 101–12.

Grove, A. T. 1982. Discussion. In Whitmore, F. C., J. R. Flenley and D. R. Harris. The tropics as the norm in biogeography. *Geogr. J.* **148**, 19.

Grove, A. T. and A. Warren 1968, Quaternary landforms and climate on the south side of the Sahara. *Geogr. J.* **134**, 194–208.

Grove, J. M. 1966. The Little Ice Age in the massif of Mont Blanc. *Trans Inst. Br. Geogrs* **40**, 129–34.

Grove, J. M. 1972. The incidence of landslides, avalanches and floods in Western Norway during the Little Ice Age. *Arctic Alpine Res.* **4**, 131–8.

Grubb, P. J. 1974. Factors controlling the distribution of forest-types on tropical mountains: new facts and a new perspective. In *Altitudinal zonation in Malesia*, J. R. Flenley (ed.), 13–46.

Trans. 3rd Aberdeen–Hull Symp. on Malesian Ecology, Hull 1973. Univ. Hull Dep. Geog. Misc. Ser. 16.

Grund, A. 1914. Der geographische Zyklus im Karst. *Z. Ges. Erdk. Berlin* **52**, 621–40.

Guillemain, C. 1908. Ergebnisse geologischer Forschung im Schutzgebiet Kamerun. *Mitt. Dtsch. Schutzgebieten* **21**, 15–35.

Guillemain, C. 1914. Geomorphologische Forschungen aus Kamerun. *Petermanns Geog. Mitt.* **60**(2), 131–5, 183–6.

Hack, J. T. 1960. Interpretation of erosional topography in humid temperate regions. *Am. J. Sci.* **258A**, 80–97.

Hack, J. T. 1975. Dynamic equilibrium and landscape evolution. In *Theories of landform development*, W. N. Melhorn and R. C. Flemal (eds). London: George Allen and Unwin.

Haile, N. S. 1969. Quaternary shorelines in West Malaysia and adjacent parts of the Sunda Shelf. *Quaternaria* **15**, 333–43.

Haile, N. S. and M. Ayob 1968. Note on radiometric age determination of samples of peat and wood from tin-bearing Quaternary deposits at Sungai Besi Tin Mines, Kuala Lumpur, Selangor. *Malay. Geol Mag.* **105**, 519–20.

Haile, N. S. and N. D. Watkins 1972. The use of palaeomagnetic reversals in Pleistocene geochronology in Southeast Asia. *Geol Soc. Malay. Newslett.* **4**, 17–18.

Hall, P. K. 1969. *The diamond fields of Sierra Leone*. Bull. Geol Surv. Sierra Leone 5.

Hallé, F. 1978. Architectural variation at the specific level in tropical trees. In *Tropical trees as living systems*, P. B. Tomlinson and M. H. Zimmerman (eds), 209–21. Cambridge: Cambridge University Press.

Halley, E. 1715. On the causes of saltness of the ocean and of the several lakes that emit no rivers; with a proposal, by help thereof, to discover the age of the world. *Phil Trans R. Soc. Lond.* 29.

Halpern, D., S. P. Hayes, A. Lectmaa, D. V. Hansen and S. G. H. Philander 1983. Oceanographic observations of the 1982 warning of the tropical eastern Pacific. *Science* **221**, 1173–5.

Hamilton, A. C. 1976. The significance of patterns of distribution shown by forest plants and animals in tropical Africa for the reconstruction of Upper Pleistocene palaeoenvironments: a reivew. *Palaeoecol. Afr.* **9**, 63–97.

Hamilton, A. C. 1982. *Environmental history of East Africa: a study of the Quaternary*. London: Academic Press.

Hamilton, N. and A. I. Rees 1970. The use of magnetic fabric in palaeocurrent estimation. In *Palaeogeophysics*, S. K. Runcorn (ed.). London: Academic Press.

Hamilton, R. 1964. Microscopic studies of laterite formations. In *Soil micromorphology*, A. Jungerius (ed.), 269–76. Amsterdam: Elsevier.

Hamilton, W. 1979. *Tectonics of the Indonesia region*. US Geol Surv. Prof. Pap. 1078.

Harding, S. T. 1965. *Recent variation in the water supply of the western Great Basin*. Calif. Water Res. Center Arch. Ser. Rep. 16.

Hardy, F. 1939. Soil erosion in St Vincent, BWI. *Trop. Agr.* **16**, 58–65.

Harlan, J. R. and J. M. J. de Wet 1973. On the quality of evidence for origin and dispersal of cultivated plants. *Curr. Anthropol.* **14**, 51–5.

Harmon, R. S., P. Thompson, H. P. Schwarcz and D. C. Ford 1975. Uranium-series dating of speleothems. *Bull. Nat. Speleol. Soc.* 1–35.

Harmon, R. S., R. M. Mitterer, N. Krjausakul, L. S. Land, H. P. Schwarz, P. Garrett, G. J. Lawson, H. L. Vacher and M. Rowe 1983. U-series and amino-acid racemization geochronology of Bermuda. *Palaeogeog. Palaeoclim. Palaeoecol.* **44**, 41–70.

Harrison, J. B. 1933. *The katamorphism of igneous rocks under humid tropical conditions*. Harpenden: Imperial Bureau of Soil Science.

Hartshorn, G. S. 1978. Tree falls and tropical forest dynamics. In *Tropical trees as living systems*, P. B. Tomlinson and M. H. Zimmerman (eds), 617–38. Cambridge: Cambridge University Press.

Hartshorn, G. S. 1980. Neotropical forest dynamics. *Biotropica* **12**(Suppl.), 23–30.
Harvey, T. J. 1976. *The palaeolimnology of Lake Mobutu Sese Seko, Uganda-Zaire: the last 28 000 years*. PhD Thesis, Duke University, North Carolina.
Haxby, W. F., G. D. Karner, J. L. La Breque and J. K. Weissel 1983. Digital images of combined oceanic and continental data sets and their use in tectonic studies. *EOS* **64**, 995–1004.
Hay, R. L. 1959a. Origin and weathering of late Pleistocene ash deposits on St Vincent, BWI. *J. Geol.* **67**, 65–87.
Hay, R. L. 1959b. Formation of the crystal-rich glowing avalanche deposits of St Vincent, BWI. *J. Geol.* **67**, 540–62.
Hayes, C. W. 1899. Physiography and geology of regions adjacent to the Nicaragua canal route. *Bull. Geol. Soc. Am.* **10**, 285–348.
Haynes, C. V., P. J. Mehringer and E. L. A. Zaghloul 1979. Pluvial lakes of north-western Sudan. *Geogr. J.* **145**, 437–45.
Hays, J. D., J. Imbrie and N. J. Shackleton 1976. Variations in the earth's orbit: pacemaker of the ice ages. *Science* **194**, 1121–32.
Herbertson, A. J. 1913. The higher units. *Scientia* **41**, 199–212.
Hernandez-Avila, M. L., H. H. Roberts and L. J. Rouse 1977. Hurricane-generated waves and coastal boulder rampart formation. *Proc. 3rd Int. Coral Reef Symp.*, Miami, Vol. 2, 71–8.
Heron, A. M. 1938. The physiography of Rajputna. *Proc. 25th Indian Sci. Congr.* Vol 2, 421–2.
Herrera, R. 1979. *Nutrient distribution and cycling in an Amazon coatinga forest on spodosols in southern Venezuela*. PhD thesis, University of Reading.
Heusser, C. J. 1981. Palynology of the last interglacial–glacial cycle in midlatitudes of southern Chile. *Quat. Res.* **16**, 293–321.
Heusser, C. J. and C. S. Streeter 1980. A temperature and precipitation record of the past 160 000 years in southern Chile. *Science* **210**, 1345–7.
Heyerdahl, T. and E. N. Ferdon Jr (eds) 1961. *Reports of the Norwegian Archaeological Expedition to Easter Island and the East Pacific*. Vol. 1: *Archaeology of Easter Island*. London: George Allen & Unwin.
Highsmith, R. C. 1982. Reproduction by fragmentation in corals. *Marine Ecol. Prog. Ser.* **7**, 207–26.
Highsmith, R. C., A. C. Riggs and C. M. d'Antonio 1980. Survival of hurricane-generated coral fragments and a disturbance model of reef calcification/growth rates. *Oecologia* **46**, 322–9.
Hodgkin, E. P. 1959. Catastrophic destruction of the littoral fauna and flora near Fremantle, January 1959. *W. Aust. Naturalist* **7**, 6–11.
Hoffman, R. S. and J. K. Jones 1970. Influence of Late Glacial and post-glacial events on the distribution of recent mammals on the northern Great Plains. In *Pleistocene and Recent environments of the Central Great Plains*, W. Dort and J. K. Jones (eds), 355–94. Kansas: Kansas University Press.
Hollis, G. E. 1978. The falling levels of the Caspian and Aral Seas. *Geogr. J.* **144**, 62–80.
Holmes, A. 1965. *Principles of physical geology*, 2nd edn. London: Nelson.
Hopkins, B. 1965. Vegetation of the Olokemeji Forest Reserve, Nigeria. III. The microclimates with special reference to their seasonal changes. *J. Ecol.* **53**, 125–38.
Hopkins, W. 1844. On the transport of erratic blocks. *Trans. Camb. Phil Soc.* **8**, 220–40.
Horton, R. E. 1932. Drainage basin characteristics. *Trans Am. Geophys. Union* **13**, 350–61.
Horton, R. E. 1945. Erosional development of streams and their drainage basins: hydrophysical approach to quantitative morphology. *Bull. Geol Soc. Am.* **56**, 275–370.
Hubbard, B. 1923. The geology of the Lores District, Porto Rico. *NY Acad. Sci. Scient. Surv. Porto Rico and the Virgin Islands*, Vol. 2, 1–115.
Hubbard, F. H. 1983. The Phanerozoic cover sequences preserved as xenoliths in the Kimberlite of eastern Sierra Leone. *Geol Mag.* **120**(1), 67–71.
Hudson, J. H. 1981. Growth rates in *Montastrea annularis*: a record of environmental change in Key Largo coral reef marine sanctuary, Florida. *Bull. Marine Sci.* **31**, 444–59.
Hurni, H. 1982. Klima und Dynamik der Höenstutung von der letzten Kaltzeit bis zur Gegenwart: Hochgebirge von Semien-Äthiopien Vol. II. *Geog. Bernensia* **13**, 1–191.

Hutchinson, G. E. 1959. Homage to Santa Rosalia or Why are there so many kinds of animals? *Am. Nat.* **93**, 145–59.

IITA 1975. *1975 annual report*. Int. Inst. Trop. Agric., Ibadan, Nigeria.

Imbrie, J. and W. S. Broecker 1970. Wisconsin climates recorded in Atlantic deep-sea cores (abs.). *Geol Soc. Am., Abstr. with Progs.* **2**, 584.

Inman, D. L. and C. E. Nordstrom 1971. On the tectonic and morphologic classification of coasts. *J. Geol.* **79**, 1–21.

International atlas of West Africa 1971.

Irion, G. 1976. Die Entwicklung des zentral- und oberamazonischen Tieflands im Spät-Pleistozän und im Holozän. *Amazoniana* **6**(1), 67–79.

Jackson, I. J. 1977. *Climate water and agriculture in the tropics*. London: Longman.

Jackson, J. 1833. *Observations on lakes being an attempt to explain the laws of nature regarding them; the cause of their formation and gradual diminution; the different phenomena they exhibit, etc., with a view of the advancement of useful science*. London: Bossange, Barthes and Lovell.

Jaeger, P. 1969. Vers une destruction accélérée de la savane soudanaise. *Nature Fr.* **3300**, 155–7.

Jäkel, E. 1979. Runoff and fluvial formation processes in the Tibesti Mountains as indicators of climatic history in the Central Sahara during the Late Pleistocene and Holocene. *Palaeoecol. Afr.* **11**, 13–39.

Janzen, D. H. 1974. Tropical blackwater rivers; animals and mast fruiting by Dipterocarpaceae. *Biotropica* **6**, 69–103.

Janzen, D. H. 1978. Seeding patterns of tropical trees. In *Tropical trees as living systems*, P. B. Tomlinson and M. H. Zimmerman (eds), 83–128. Cambridge: Cambridge University Press.

Jeje, L. K. 1980. A review of geomorphic evidence for climatic change since the Late Pleistocene in the rainforest areas of southern Nigeria. *Palaeogeog. Palaeoclim. Paelaeoecol.* **1**, 63–86.

Jennings, J. N. 1971. Sea level changes and land links. In *Aboriginal man and environment in Australia*, D. J. Mulvaney and J. Golson (eds), 1–25. Canberra: ANU Press.

Jennings, J. N. 1975. Desert dunes and estuarine fill in the Fitzroy Estuary, northwestern Australia. *Catena* **2**, 215–62.

Jessen, O. 1936. *Reisen und Forschungen im Angola*. Berlin.

Joachim, A. W. R., S. Kandiah and D. G. Pandithesekera 1938. Studies of Ceylon soils. *Trop. Agr.* **90**, 136–41.

Johannes, R. E. and L. Tepley 1974. *In situ* examination of the reef coral, *Porites lobata*, using time lapse photography. *Proc. 2nd Int. Coral Reef Symp.* Vol. 1, 127–31.

Johnston, A. R. 1844. Note on the island of Hong Kong. *J. R. Geog. Soc.* **14**, 112–7.

Jones, G. A. and W. F. Ruddiman 1982. Assessing the global meltwater spike. *Quat. Res.* **17**, 148–72.

Jones, S. B. 1938. Geomorphology of the Hawaiian Islands: a review. *J. Geomorph.* **1**, 55–61.

Jordan, C. F. 1970. A progress report on studies of mineral cycles at El Verde. In *A tropical rain forest*, H. T. Odum and R. F. Pidgeon (eds), H217–9. Oak Ridge, Tennessee: US Atomic Energy Commission.

Jordan, E. 1981. Die rezenten Dünengebiete Boliviens und ihre regional-genetische Differenzierung. *Würzburger Geog. Arb.* **53**, 59–94.

Joshua, W. D. 1977. *Soil erosive power of rainfall in the different climatic zones of Sri Lanka*. IAHS/AISM Publ. 122, 51–61.

Journaux, A. 1975a. Géomorphologie des bordures des l'Amazonie brésilienne: le modelé des versants, essai d'évolution paléoclimatique. *Bull. Assoc. Géog. Fr.* **422/3**, 5–18.

Journaux, A. 1975b. Recherches géomorphologiques en Amazonie brésilienne. *Bull. Centre Geom.* Caen: CNRS.

Jutson, J. T. 1934. Physiography (geomorphology) of Western Australia. *Bull. Geol. Surv. W. Aust.* **95**, 254–6.

Kalms, J.-M. 1975. Influence des techniques culturales sur l'érosion et le ruissellement en région centre de Côte d'Ivoire. IRAT, Bouaké. *Coll. sur la conservation et l'aménagement du sol dans les tropiques humides*, Iabadan, 30 June–4 July 1975.

Katz, M. B. 1971. The Precambrian metamorphic rocks of Ceylon. *Geol. Rdsch.* **60**, 1523–85.

Kellman, M. C. 1969. Some environmental components of shifting cultivation in upland Mindanao. *J. Trop. Geog.* **28**, 40–56.

Kendall, R. L. 1969. An ecological history of the Lake Victoria Basin. *Ecol. Monogr.* **39**, 121–76.

Kent, P. E. 1980. Vertical tectonics associated with rifting and spreading. *Phil Trans R. Soc. Lond. A* **294**, 125–35.

Kenworthy, J. B. 1971. *Water and nutrient cycling in a tropical rain forest*. Univ. Hull Dept. Geog. Misc. Ser. 11, 49–65.

Kerr, R. A. 1981. Milankovitch climate cycles: old and unsteady. *Science* **213**, 1095–6.

Kershaw, A. P. 1970. A pollen diagram from Lake Euramoo, north-east Queensland, Australia. *New Phytol.* **69**, 785–805.

Kershaw, A. P. 1971. A pollen diagram from Quincan Crater, north-east Queensland, Australia. *New Phytol.* **70**, 669–81.

Kershaw, A. P. 1975. Late Quaternary vegetation and climate in northeastern Australia. *Bull. R. Soc. NZ* **13**, 181–7.

Kershaw, A. P. 1976. A Late Pleistocene and Holocene pollen diagram from Lynch's Crater, northeastern Queensland, Australia. *New Phytol.* **77**, 469–98.

Kershaw, A. P. 1978. Record of last interglacial-glacial cycle from northeastern Queensland. *Nature* **272**, 159–61.

Kershaw, A. P. 1980. Evidence for vegetation and climatic change in the Quaternary. In *The geology and geophysics of northeastern Australia*, R. A. Henderson and P. J. Stephenson (eds), 398–402. Brisbane: Geological Society of Australia.

Kershaw, A. P. 1981. Quaternary vegetation and environments. In *Ecological biogeography of Australia*, A. Keast (ed.), 83–101. The Hague: Junk.

Kes, A. S. 1979. The causes of water level changes of the Aral Sea in the Holocene. *Sov. Geog.* **20**, 104–13.

Kesel, R. H. 1977. Slope runoff and denudation in the Rupunumi savanna, Guyana. *J. Trop. Geog.* **44**, 33–42.

Killigrew, L. P. and R. J. Gilkes 1974. Development of playa lakes in SW Australia. *Nature* **274**, 454–5.

King, B. C. 1978. Structural and volcanic evolution of the Gregory Rift Valley. In *Geological background to fossil man*, W. W. Bishop (ed.), 29. Edinburgh: Scottish University Press.

King, L. C. 1950. The study of the world's plainlands. *Q. J. Geol Soc. Lond.* **106**, 101–31.

King, L. C. 1953. Canons of landscape evolution. *Bull. Geol Soc. Am.* **64**, 721–52.

King, L. C. 1962. *The morphology of the earth: a study and synthesis of world scenery*. Edinburgh: Oliver and Boyd.

King, L. C. 1976. Planation remnants upon high lands. *Z. Geomorph.* NF **20**, 133–48.

King, L. C. 1983. *Wandering continent and spreading sea floors on an expanding earth*. Chichester: Wiley.

Kinsman, D. J. J. 1975a. Rift valley basins and sedimentary history of trailing continental margins. In *Petroleum and global tectonics*, A. G. Fischer and S. Judson (eds), 83–126. New Jersey: Princeton University Press.

Kinsman, D. J. J. 1975b. Salt floors to geosynclines. *Nature* **255**, 375–8.

Kira, T. 1978. Community architecture and organic matter dynamics in tropical lowland rain forests of Southeast Asia with special reference to Pasoh forest, West Malaysia. In *Tropical trees as living systems*, P. B. Tomlinson and M. H. Zimmermann (eds), 561–90. Cambridge: Cambridge University Press.

Klammer, G. 1981. Landforms, cyclic erosion and deposition, and Late Cenozoic changes in climate in southern Brazil. *Z. Geomorph.* NF **25**, 146–65.

Klammer, G. 1982. Die Paläowüste des Pantanal von Mato Grosso und die pleistozäne Klimazeschichte der brasilianischan Randtropen. *Z. Geomorph.* NF **26**, 393–416.

Klinge, H. 1977. Fine litter production and nutrient return to the soil in three natural forest stands of Eastern Amazonia. *Géo.-Éco.-Trop.* **1**, 159–67.

Klinge, H. and W. A. Rodrigues 1968. Litter production in an area of Amazonian *terra firme* forest. *Amazoniana* **1**, 287–310.

Klinge, H., K. Furche, E. Harms and J. Revilla 1983. Foliar nutrient levels of native tree species from Central Amazonia I. Inundation forests. *Amazoniana* **8**, 19–45.

Klinge, H., K. Furche, U. Irmler and W. Junk 1981. Fundamental ecological parameters in Amazonia, in relation to the potential development of the region. In *Tropical agricultural hydrology*, R. Lal and E. W. Russell (eds), 37–57. Chichester: Wiley.

Knowlton, N., J. C. Lang, M. C. Rooney and P. Clifford 1981. Evidence for delayed mortality in hurricane-damaged Jamaican staghorn corals. *Nature* **294**, 251–2.

Knox, J. C. 1972. Valley alluviation in southwestern Wisconsin. *Ann. Assoc. Am. Geogrs* **62**, 401–10.

Kolla, V., P. E. Biscaye and A. F. Hanley 1979. Distribution of quartz in Late Quaternary Atlantic sediments in relation to climate. *Quat. Res.* **11**, 261–77.

Komanda, A. 1978. Le rôle des Termites dans la mise en place des sols de plateau dans le Shaba méridional. In *Géomorphologie dynamique dans les régions intertropicales*, J. Alexandre (ed.), 81–93. Zaire: Presse Universitaire du Zaire.

Koopmans, B. N. and P. H. Stauffer 1968. Glacial phenomena on Mt Kinabalu, Sabah. *Geol. Surv. Malay. Borneo Region Bull.* **8**, 25–35.

Krynine, P. D. 1936. Geomorphology and sedimentation in the humid tropics. *Am. J. Sci.* **232**, 297–306.

Kukla G. J. 1975. Missing link between Milankovitch and climate. *Nature* **253**, 600–3.

Kukla, G. J. 1980. End of the last interglacial: a predictive model of the future. *Palaeoecol. Afr.* **12**, 395–408.

Kukla, G. J., J. K. Angell, J. Korshover, H. Dronia, M. Hoshiai, J. Namias, M. Rodewald, R. Yamamoto and T. Iwashima 1977. New data on climatic trends. *Nature* **270**, 573–80.

Kutzbach, J. E. 1980. Estimates of past climate at Palaeolake Chad, North Africa, based on a hydrological and energy-balance model. *Quat. Res.* **14**, 210–23.

Kutzbach, J. E. 1981. Monsoon climate of the early Holocene: climate experiment with the Earth's orbital parameters for 9000 years ago. *Science* **214**, 59–61.

La Brecque, J. L., D. V. Kent and S. C. Cande 1977. Revised magnetic polarity time scale for Late Cretaceous and Cenozoic time. *Geology*, **5**, 330–5.

Ladd, W. L., Dickson and W. C. Pitman III 1973. The age of the south Atlantic. In *The ocean basins and margins*, Vol. 1: *The South Atlantic*, A. E. M. Nairn and F. G. Stehli (eds), 555–73. New York: Plenum.

Lafforgue, A. 1977. Inventaire et examen des processus élémentaires de ruissellement et d'infiltration sur parcelles. Application à une exploitation méthodique dans données obtenues sous pluies simulées. *Cah. ORSTOM Hydrol.* **14**(4), 299–344.

Lafforgue, A. and E. Naah 1976. Exemple d'analyse des facteurs de ruissellement sous pluies simulées. *Cah. ORSTOM Hydrol.* **13**, 3.

La Marche, V. C. and K. K. Hirschboeck 1984. Frost rings in trees as records of major volcanic eruption. *Nature* **307**, 321–6.

Lamb, H. H. 1969. Climatic fluctuations. In *General climatology, World survey of climatology*, Vol. 2, H. E. Landsberg (ed.), 173–249.

Lamb, H. H. and A. I. Johnson 1961. Climatic variation and observed changes in the general circulation. *Geog. Ann.* **43**, 363–400.

Lambert, J. D. H., J. T. Arnason and J. L. Gale 1980. Leaf-litter and changing nutrient levels in a seasonally dry tropical hardwood forest, Belize, C.A. *Plant Soil* **55**, 429–43.

Lamotte, M. and G. Rougerie 1962. Les apports allochtones dans la genèse des cuirasses ferrugineuses. *Rév. Géomorph. Dyn.* **13**, 145–60.

Landsberg, H. E. 1960. Do tropical storms play a role in the water balance of the Northern Hemisphere? *J. Geophys. Res.* **65**, 1305–7.

Lang, D. M. 1967. *Soil and Land-use Surveys No 21 Dominica*. Regional Research Centre of the British Caribbean at the University of the West Indies, Imperial College of Tropical Agriculture, Trinidad, WI.

Lang, H. D. 1975. Secondary rutile deposition in Sierra Leone. *Natural Res. Dev.* **1**, 59–68.

Lang, J. 1973. Interspecific aggression by scleractinian corals 2. Why the race is not only to the swift. *Bull. Marine Sci.* **23**, 260–79.

Langbein, W. B. and S. A. Schumm 1958. Yield of sediment in relation to mean annual precipitation. *Trans Am. Geophys. Union* **39**, 1076–1084.

Lapperent, J. de 1939. La décomposition laterique du granite dans la région de Macenta (Guinée française). L'arènisation prétropicale et prédesertique en A.O.F. *C. R. Acad. Sci., Paris* **208**, 1767–9 and **209**, 9.

Larson, R. L. and J. W. Ladd 1973. Evidence for the opening of the South Atlantic in the early Cretaceous. *Nature* **246**, 209–12.

Laudelot, M. and J. Meyer 1954. Les cycles d'éléments minéraux et de matière organique en forêt équatoriale congolaise. *Trans 5th Int. Congr. Soil Sci.* Vol. 2, 267–72.

Lauer, W. and P. Frankenberg 1980. Modelling of climate and plant cover in the Sahara for 5500 BP and 1800 BP. *Palaeoecol. Afr.* **12**, 307–14.

Laurie, A. 1983. An ill wind for iguanas. *New Scient.* **100**, 108.

Laverty, M. 1982. Cave minerals of the Gunung Mulu National Park, Sarawak. *Trans Br. Cave Res. Assoc.* **35**, 17–25.

Le Bas, M. J. 1971. Per-alkaline volcanism, crustal swelling, and rifting. *Nature Phys. Sci.* **230**, 85–7.

Le Bas, M. J. 1980. Alkaline magmatism and uplift of continental crust. *Proc. Geol Assoc.* **91**, 33–8.

Le Buanec B. 1972. Dix ans de culture motorisée sur un bassin-versant du centre Côte d'Ivoire. Évolution de la fertilité et de la production. *Agron. Trop.* **27**(11), 1191–211.

Ledger, D. C. 1969. Dry season flow characteristics of West African rivers. In *Environmental and land use in Africa*, M. F. Thomas and G. Wittington (eds). London: Methuen.

Lehmann, H. 1954. Das Karstphänomen in der verschiedenen Klimazonen. *Erdkunde* **8**, 114–39.

Lehmann, H. 1964. States and tasks of research on karst phenomena. *Erdkunde* **16**, 81–3.

Lehmann, H. 1970. Kelgelkarst und Tropengrenze. *Tübinger Geogr. Stud.* **34**, 107–12.

Leichti, P., F. W. Roe and N. S. Haile 1960. *The geology of Sarawak, Brunei, and the western part of North Borneo*. Bull. Geol. Surv. Dept. Br. Borneo 3.

Leigh, C. H. 1978. Slope hydrology and denudation in the Pasoh Forest Reserve. II. Throughflow: experimental techniques and some preliminary results. *Malay. Nature J.*, **30**, 199–210.

Leigh, C. H. 1982. Sediment transport by surface wash and throughflow at the Pasoh Forest Reserve, Negri Sembilan, Peninsular Malaysia. *Geogr. Ann.*, **64A**, 171–80.

Leopold, L. B., M. G. Wolman and J. P. Miller 1964. *Fluvial processes in geomorphology*. San Francisco and London: Freeman.

Lerman, J. T. 1975. Reconstructing the rate of accumulation of lake sediment: the effect of sediment focusing. *Quat. Res.* **5**, 541–50.

Lewis, W. M. and F. H. Weibezahn 1981. Chemistry of a 7.5 m sediment core from the Lake Valencia, Venezuela. *Limnol. Oceanogr.* **26**, 907–24.

Lézine, A. 1981. *Le lac Abiyaba (Ethiopie) – Palynologie et paléoclimatologie du Quaternaire recent*. Thèse, Univ. Bordeaux I.

Libby, W. F. 1952. *Radiocarbon dating*. Chicago: University of Chicago Press.

Lidmar-Bergström K. 1982. Pre-Quaternary geomorphological evolution in southern Fennoscandia. *Sv. Geol. Undersokning, Ser. C.* **785**, 202.

Liew, T. C. 1974. A note on soil erosion study at Tawau Hills Forest Reserve. *Malay. Nature J.* **27**, 20–6.

Liew, T. C. and F. O. Wong 1973. Density, recruitment, mortality and growth of dipterocarp seedlings in virgin and logged-over forests in Sabah. *Malay. For.* **36**, 3–15.

Lim, C. L. 1969. Storm rainfall study of Sungei Kelang catchment, Malaysia. MSc thesis, University of Hull.

Lim, M. T. 1978. Litterfall and mineral-nutrient content of litter in Pasoh Forest Reserve. *Malay. Nature J.* **30**, 375–80.

Limbrey, S. 1976. Tlapacoya: problems of interpretation of lake margin sediments at an early occupation site in the basin of Mexico. In *Geoarchaeology*, D. A. Davidson and M. L. Shackley (eds), 213–26. London: Duckworth.

Livingstone, D. A. 1975. Late Quaternary climatic change in Africa. *Ann. Rev. Ecol. Syst.* **6**, 249–80.

Livingstone, D. A. 1980. Environmental changes in the Nile headwaters. In *The Sahara and the Nile*, M. A. J. Williams and H. Faure (eds), 339–59. Rotterdam: Balkema.

Livingstone, D. A. and W. D. Clayton 1980. An altitudinal cline in tropical African grass floras and its palaeoecological significance. *Quat. Res.* **13**, 392–402.

Lobeck, A. K. 1922. The physiography of Porto Rico. *NY Acad. Sci., Scient. Surv. of Porto Rico and the Virgin Islands* Vol. 1, 301–79.

Lockwood, J. G. 1980. Some problems of humid equatorial climates. *Malay. J. Trop. Geog.* **1**, 12–20.

Löffler, E. 1977. *Geomorphology of Papua New Guinea*. Canberra: ANU Press.

Louchet, A. 1981. Sur les basses surfaces de l'île de Ceylon. *Bull. Assoc. Géogrs Fr.* **58**, 475–6.

Louis, H. 1961. *Allegemeine Geomorphologie*. Berlin: de Gruyter.

Louis, H. 1964. Über Rumpfflachen- und Talbildung in den wechselfeuchten Tropen, besonders nach Studien in Tanganyika. *Z. Geomorph.* NF **8**, (Sonderheft), 43–70.

Louis, H. 1979. *Allegemeine Geomorphologie* Vol. 4: *Aufl. unter Mitarbeit von K. Fischer*. New York: De Gruyter.

Lourenz, R. S. 1977. *Tropical cyclones in the Australian region, July 1969 to June 1975*. Canberra Bureau of Meteorology.

Loveless, A. R. and G. F. Asprey 1957. The dry evergreen formations of Jamaica. I. The limestone hills of the south coast. *J. Ecol.* **45**, 799–822.

Loya, Y. 1976. Recolonization of Red Sea corals affected by natural catastrophes and man-made perturbations. *Ecology* **57**, 278–89.

Lundgren, B. 1978. *Soil conditions and nutrient cycling under natural and plantation forests in Tanzanian Highlands*. Swedish Univ. Agr. Sci., Dept Forest Soils, Rep. Forest Ecol. Forest Soils 31.

Lutz, T. M. and K. A. Foland 1978. Meridional pattern of the oceanic rift system. *Geology* **6**, 179–83.

Lynts, G. W., J. B. Judd and C. F. Steman 1973. Late Pleistocene history of the Tongue of the Ocean, Bahamas. *Bull. Geol Soc. Am.* **84**, 2665–84.

Mabbutt, J. A. 1961. A stripped landsurface in Western Australia. *Trans Inst. Br. Geogrs 29*, 101–14.

Mabbutt, J. A. 1965. The weathered land surface in central Australia. *Z. Geomorph.* NF **9**, 82–114.

MacArthur, R. H. 1955. Fluctuations of animal populations and a measure of community stability. *Ecology* **36**, 533–6.

McClure, H. A. 1976. Radiocarbon chronology of Late Quaternary lakes in the Arabian desert. *Nature* **263**, 755–6.

McConnell, R. B. 1972. Geological development of the Rift System of Eastern Africa. *Bull. Geol Soc. Am.* **83**, 2549–72.

McConnell, R. B. 1977. East African Rift system dynamics in view of Mesozoic apparent polar wandering. *J. Geol Soc. Lond.* **134**, 33–9.

McFarlane, A., M. J. Crow, J. W. Arthurs, A. F. Wilkinson and J. W. Aucott 1981. *The geology and mineral resources of northern Sierra Leone*. Inst. Geol. Sci. Overseas Mem. 7. London: HMSO.

McFarlane, M. J. 1976. *Laterite and landscape*. London: Academic Press.

Machatschek, F. 1955. *Das Relief der Erde: Versuch einer regionalen Morphologia der Erdoberflach*. II Band, 2nd Edn. Berlin: Borntraeger.

McIntyre, A. and CLIMAP Project Members 1976. The surface of the Ice-Age Earth. *Science* **191**, 1131–44.

Mäckel, R. 1974. Dambos: a study of morphodynamic activity on the plateau regions of Zambia. *Catena* **1**, 327–65.

Mäckel, R. 1975. *Untersuchungen zur Reliefentwicklung des Sambesi-Eskarpmentlandes und des Zentralplateaus von Sambia*. Gießener Geog. Schr. 36.

Mäckel, R. 1979. Zur Entstehungund geoökologischem Stellung der Bolis in Sierra Leone/Westafrika. *Ber. Naturf. Ges. Freiburg i. Br.* **69**, 547–71.

McKenzie, D. P., J. M. Roberts and N. O. Weiss 1974. Convection in the Earth's mantle: towards a numerical simulation. *J. Fluid Mech.* **62**, 458–65.

McKenzie, D. P. and F. Richter 1976. Convection currents in the Earth's mantle. *Sci. Am.* **235**(5), 72–89.

McKenzie, D. P., A. Watts, B. Parsons and M. Roufosse 1980. Planform of mantle convection beneath the Pacific Ocean. *Nature* **288**, 442–6.

Mackereth, F. J. H. 1966. Some chemical observations on post glacial lake sediments. *Phil Trans R. Soc. Lond. B* **250**, 165–213.

Mackereth, F. J. H. 1969. A short core sampler for subaqueous deposits. *Limnol. Oceanogr.* **14**, 154–61.

Madduma Bandara, C. M. 1974a. Drainage density and effective precipitation. *J. Hydrol.*, **21**, 187–90.

Madduma Bandara, C. M. 1974b. The orientation of straight slope forms on the Hatton Plateau of central Sri Lanka. *J. Trop. Geog.* **38**, 37–44.

Maguire, P. K. H. and M. A. Khan 1980. The deep structure and dynamics of the East African Plateau, the Kenya Dome and the Gregory Rift. *Proc. Geol Assoc.* **91**, 25–31.

Mainguet, M. 1983. Tentative mega-morphological study of the Sahara. In *Mega-geomorphology*, R. Gardner and H. Scoging (eds), 113–33. Oxford: Clarendon Press.

Mainguet, M. and L. Canon 1976. Vents et paléovents du Sahara: tentative d'approche paléoclimatique. *Rev. Géog. Phys. Géol. Dyn.* **18**, 241–50.

Mainguet, M., L. Cossus and A.-M. Chapelle 1980. Utilisation des images satellites pour préciser les trajectoires éoliennes au sol, au Sahara et sur les marges sohéliennes. Interprétations des documents Météosat du 28 mai 1978 ou 9 février 1979. *Bull. Soc. Fr. Photogram.* **78**, 1–15.

Makin, M. J., T. J. Kingham, A. E. Waddams, C. J. Birchall, and R. Teferra 1975. *Development prospects in the southern Rift Valley, Ethiopia*. Land Res. Study 21, Land. Res. Div., Min. Overseas Dev.

Maley, J. 1976. Essai sur le rôle de la zone tropicale dans les changements climatiques; l'exemple Africain. *C.R. Acad. Sci. Paris* **283D**, 337–40.

Maley, J. 1977. Paleoclimates of central Sahara during the early Holocene. *Nature* **269**, 573–7.

Maley, J. 1981. *Études palynologiques dans le bassin du Tchad et paléoclimatologie de l'Afrique nord-tropicale de 30 000 ans à l'époque actuelle*. Trav. Doc. ORSTOM 129.

Maloney, B. K. 1976. *Man's influence on the vegetation of North Sumatra: a palynological study*. PhD thesis, University of Hull.

Maloney, B. K. 1980. Pollen analytical evidence for early forest clearance in North Sumatra. *Nature* **287**, 324–6.

Mamedov, V. I., V. A. Bronevoi, I. O. Makstenek, V. A. Ivanov and V. V. Pokrovskii 1983. Subsurface water regime – the main controlling factor of the mineralogic and geochemical zonality of the crust of weathering on the Literian shield. *Lithol. Mineral Res.*, **18**, 1–8.

Manabe, S. and D. G. Hahn 1977. Simulation of the tropical climate of an Ice Age. *J. Geophys. Res.* **82**, 3889–911.

Manokaran, N. 1979. Stemflow, throughfall and rainfall interception in a lowland tropical rain forest in peninsular Malaysia. *Malay. Forest.* **42**, 174–201.

Manokaran, N. 1980. The nutrient contents of precipitation, throughfall and stemflow in a lowland tropical rainforest in Peninsular Malaysia. *Malay. Forest.* **43**, 266–89.

Maragos, J. E., G. B. K. Baines and P. J. Beveridge 1973. Tropical cyclone Bebe creates a new land formation on Funafuti Atoll. *Science* **181**, 1161–4.

Mathieu, Ph. 1971. Érosion et transport solide sur un bassin versant forestier tropical (Bassin de l'Amitoro, Côte d'Ivoire). *Cah. ORSTOM Sér. Géol. III* **2**, 115–44.

Maurer, H. 1928. Das Schwankungsmass der jährlichen Niederschlagsmengen. *Met. Z.* **45**, 166–74.

Medway, Lord 1972. The Quaternary mammals of Malesia, a review. *Trans 2nd Hull-Aberdeen Symp. on Malesian Ecology* Univ. Hull Dept. Geog. Misc. Ser. 13, 63–98.

Melhorn, W. M. and D. E. Edgar 1975. The case for episodic, continental-scale erosion surfaces: a tentative geodynamic model. In *Theories of landform development*, W. M. Melhorn and R. C. Flemal (eds), 243–76. Binghamton: State University of New York.

Melton, M. A. 1957. *An analysis of the relations among elements of climate, surface properties and geomorphology*. Off. Naval Res. Geog. Branch, Project NR 389–042, Tech. Rep. 11. Columbia University.

Mendes, J. C. 1957. Notas sôbre a bacia sedimentar amazônica. *Bol. Paulista Geog.* **26**, 3–37.

Mensching, H. 1970. Geomorphologische Beobachtungen in der Inselberglandschaft südlich des Victoria-Sees (Tanzania). *Abh. 1. Geog. Inst. FU Berlin* NF **13**, 111–24.

Mesolella, K. J., R. K. Matthews, W. S. Broecker and D. S. Thurber 1969. The astronomical theory of climatic change. Barbados data. *J. Geol.* **77**, 250–74.

Messerli, B. and M. Winiger 1980. The Saharan and East African uplands during the Quaternary. In *The Sahara and the Nile*, M. A. J. Williams and H. Faure (eds), 87–118. Rotterdam: Balkema.

Meybeck, M. 1976. Total mineral dissolved transport by world major rivers. *Int. Assoc. Sci. Hydrol. Bull.* **21**, 265–84.

Meybeck, M. 1979. Concentration des eaux fluviales en éléments majeures et apports en solution aux océans. *Rev. Géol. Dyn. Géog. Phys.* **21**, 215–46.

Meyerhoff, H. A. 1927. Tertiary physiographic development of Porto Rico and the Virgin Islands. *Bull. Geol Soc. Am.* **38**, 557–75.

Meyerhoff, H. A. 1933. *Geology of Puerto Rico*. Puerto Rico Univ. Monogr. Ser. B 1.

Meyerhoff, H. A. 1937. The texture of the karst topography in Cuba and Puerto Rico. *J. Geomorph.* **1**, 279–95.

Michel, P. 1959. *Rapport de mission au Soudan occidental et dans le Sud-Est du Sénégal*, Vol. 2: *Dépôts alluviaux et dynamique fluviale*. Arch. Bur. Rech. Géol. Min., Dakar.

Michel, P. 1960. *Note sur l'évolution morphologique des vallées de la Kolimbiné, du Karakoro et du Sénégal dans la région de Kayes*. Rapp. Bur. Rech. Géol. Min, Dakar.

Michel, P. 1962. *Observations sur la géomorphologie et les dépôts alluviaux des cours moyens du Bafing et du Bakoy (Rep. du Mali)*. Rapp. Bur. Rech. Géol. Min., Dakar.

Michel, P. 1966. *Les applications de recherches géomorphologique en Afrique occidentale*. Rev. Géog. Afr. Occid. Dakar 3.

Michel, P. 1968a. Morphogenèse et pedogenèse. Exemples d'Afrique occidentale. *Sols Afr.* 13.

Michel, P. 1968b. Le façonnement actuel de la vallée du Sénégal et de ses bordures, de Bakel à Richard-Toll. *Com. Trav. Hist. Sci., Paris, Bull. Sect. Géog.*, **80**, *Hydrol. continentale*, 447–84.

Michel, P. 1969. Les grandes étapes de la morphogenèse dans les bassins des fleuves Sénégal et Gambie pendant le Quaternaire. *Comm. 6e Congr. Panafr. Préhist. Quat.*, Dakar 1967. Bull. IFAN 31 (2).

Michel, P. 1973. *Les bassins des fleuves Sénégal et Gambie. Étude géomorphologique*. Mem. ORSTOM 63.

Michel, P. 1978. La dynamique actuelle de la géomorphologie dans le domaine soudanien de l'Ouest africain: l'exemple du Mali occidentale et du Sénégal orientalé. In *Géomorphologie dynamique les régions intertropicales. Coll.* Lubumbashi, 1975. *Géo. Eco. Trop.* **2**(1).

Michel, P. 1980. The southwestern Sahara margin: sediments and climatic changes during the recent Quaternary. *Palaeoecol. Afr.* **12**, 297–314.

Mietton, M. 1980. *Recherches géomorphologiques au Sud de la Haute-Volta. La dynamique actuelle dans la région de Pô-Tiébélé.* Univ. Grenoble 1, UER de Géographie, Thèse 3ème cycle.

Mifflin, M. D. and M. M. Wheat 1979. *Pluvial lakes and estimated pluvial climates of Nevada.* Nevada Bur. Mines Geol. Bull. 94.

Milliman, J. D. and R. H. Meade 1983. World-wide delivery of river sediment to the oceans. *J. Geol.* **91**, 1–21.

Millot, G. 1970. *Geology of clays.* Paris: Masson.

Milne, G. 1935. Some suggested units of classifications and mapping particularly for East African soils. *Bodenkundl. Forsch.* **4**, 183–98.

Milton, D. 1974. Some observations of global trends in tropical cyclone frequencies. *Weather* **29**, 267–70.

Minster, J. and T. H. Jordan 1978. Present-day plate motions. *J. Geophys. Res.* **83**, 5331–54.

Molengraaff G. A. F. and M. Weber 1920. On the relation between the Pleistocene glacial period and the origin of the Sunda Sea (Java and South China Sea), and its influence on the distribution of coral reefs and on the land and fresh water fauna. *Proc. R. Acad. Amsterdam* **23**, 395–439.

Molnar, P. and T. Atwater 1973. Relative motion of hotspots in the mantle. *Nature* **246**, 288–9.

Molnar, P. and J. Francheteau 1975. The relative motion of 'hot spots' in the Atlantic and Indian Oceans during the Cenozoic. *Geophys. J. R. Astr. Soc.* **43**, 763–74.

Monnier, Y. 1968. *Les effets des feux de brousse sur un savane préforestière de Côte d'Ivoire.* Études Eburnéennes, no. IX, Abidjan.

Moran, J. 1975. Return of the ice age drought in peninsular Florida? *Geology* **3**, 695–6.

Moreau, R. E. 1963. Vicissitudes of the African biomes in the Late Pleistocene. *Proc. Zool Soc. Lond.* **141**, 395–421.

Morgan, R. P. C. 1976. The role of climate in the denudation system: a case study from West Malaysia. In *Geomorphology and Climate*, E. Derbyshire (ed.). London: Wiley.

Morisawa, M. E. 1964. Development of drainage systems on an upraised lake floor. *Am. J. Sci.* **262**, 340–54.

Morley, R. J. 1976. *Vegetation change in West Malesia during the Late Quaternary period: a palynological study of selected lowland and lower mountain sites.* PhD thesis, University of Hull.

Morley, R. J. 1982. A palaeoecological interpretation of a 10 000 year pollen record from Danau Padang, Central Sumatra, Indonesia. *J. Biogeog.* **9**, 151–90.

Moss, R. P. 1965. Slope develpment and soil morphology in a part of south-west Nigeria. *J. Soil Sci.* **16**, 192–209.

Mullins, C. E. 1977. Magnetic susceptibility of the soil and its significance in soil science. *J. Soil Sci.* **28**, 223–46.

Nagle, F., J. J. Stipp and D. E. Fisher 1976. K–Ar geochronology of the Limestone Caribees and Martinique, Lesser Antilles, West Indies. *Earth Planet Sci. Lett.* **29**, 401–12.

National Energy Authority of Thailand 1965. *Hydrologic data.* Bangkok: Ministry of National Development.

Neftel, A., H. Oeschger, J. Swander, B. Stauffer and R. Zumbrunn 1982. Ice core sample measurements give atmospheric CO_2 content during the post 40 000 yr. *Nature* **295**, 220–3.

Neill, W. M. 1973. Possible continental rifting in Brazil and Angola related to the opening of the South Atlantic. *Nature Phys. Sci.* **145**, 104–7.

Netterberg, F. 1978. Dating and correlation of calcretes and other pedocretes. *Trans Geol Soc. S. Afr.* **81**, 379–91.

Neumann, A. C. and W. S. Moore 1975. Sea level events and Pleistocene coral ages in the northern Bahamas. *Quat. Res.* **5**, 215–24.

Neumann, C. J., G. W. Cry, E. D. Caso and B. R. Jarvinen 1978. *Tropical Cyclones of the North Atlantic Ocean, 1871–1977*. National Climatic Center, Asheville, North Carolina, NOAA. Washington DC: US Govt Printing Office.

Newson, M. D. 1978. Drainage basin characteristics, their selection, derivation and analysis for a flood study of the British Isles. *Earth Surf. Processes* **3**, 277–93.

Ng, F. S. P. 1978. Strategies of establishment in Malayan forest trees. In *Tropical trees as living systems*, P. B. Tomlinson and M. H. Zimmerman (eds), 129–62. Cambridge: Cambridge University Press.

Nicholson, D. I. 1965. A review of natural regeneration in the dipterocarp forest of Sabah. *Malay. Forest.* **28**, 4–25.

Nicholson, S. E. and H. Flohn 1980. African environmental and climatic changes and the general atmospheric circulation in Late Pleistocene and Holocene. *Clim. Change* **2**, 313–48.

Niederberger, C. 1979. Early sedentary economy in the basin of Mexico. *Science* **203**, 131–42.

Nilsson, E. 1931. Quaternary glaciations and pluvial lakes in British East Africa. *Geogr. Ann. A* **13**, 249–349.

Nilsson, E. 1940. Ancient changes of climate in British East Africa and Abyssinia. A study of ancient lakes and glaciers. *Geogr. Ann. A* **22**, 1–79.

Ninkovitch, D., N. J. Shackleton, A. A. Abdel-Monem, J. D. Obradovich and G. Izett 1978. K–Ar age of the late Pleistocene eruption of Toba, north Sumatra. *Nature* **276**, 574–7.

Noel, M. 1982. Palaemognetic study of the Upper Cricket Mud, Clearwater Cave, Sarawak. *Trans. Br. Cave Res. Assoc.* **35**, 37–42.

Noltimier, H. C. 1974. The geophysics of the North Atlantic Basin. In *The ocean basins and margins*, Vol. 2: *The North Atlantic*, A. E. M. Nairn and F. G. Stehli (eds), 539–88. New York: Plenum.

Nortcliff, S. and Thornes, J. B. 1978. Water and cation movement in a tropical rainforest environment. 1. Objectives, experimental design and preliminary results. *Acta Amazonica* **8**, 245–58.

Nortcliff, S. and Thornes, J. B. 1981. Seasonal variations in the hydrology of a small forested catchment near Manaus Amazonas, and the implications for its management. In *Tropical agricultural hydrology*, R. Lal and E. W. Russell (eds), 37–57. Chichester: Wiley.

Nouvelot, J. F. 1969. Mésure et étude des transports solides en suspension au Cameroun. *Cah. ORSTOM Sér. Hydrol.* **6**(4), 43–86.

Nye, P. H. 1961. Organic matter and nutrient cycles under moist tropical forest. *Plant Soil* **13**, 333–46.

Nye, P. H. and D. J. Greenland 1960. *The soil under shifting cultivation*. Tech. Commun. Commonw. Bur. Soils 51.

Obst, E. 1915. Das abflusslose Rumpfschollenland im nordostlichen Deutsch-Ostafrika. *Mitt. Geogr. Ges. Hamburg* **29**, 3–105.

Odum, E. P. 1963. *Ecology*. New York: Holt, Rinehart and Winston.

Odum, H. T. 1970. Summary: an emerging view of the ecological system at El Verde. In *A tropical rain forest*, H. T. Odum and R. F. Pidgeon (eds), I-199–289. Oak Ridge, Tennessee: US Atomic Energy Commission.

Ogden, J. 1981. Dendrochronological studies and the determination of tree ages in the Australian tropics. *J. Biogeog.* **8**, 405–20.

Oldeman, R. A. A. 1978. Architecture and energy exchange of dicotyledonous trees in the forest. In *Tropical trees as living systems*, R. B. Tomlinson and M. H. Zimmerman (eds), 535–60. Cambridge: Cambridge University Press.

Oldfield, F. 1977. Lake sediments and human activity in prehistoric Papua New Guinea. In *The Melanesian environment*, J. H. Winslow (ed.). Canberra: ANU Press.

Oldfield, F. 1981. Peats and lake sediments. *Geomorphological techniques*, A. Goudie (ed.), London: George Allen & Unwin.

Oldfield, F. and P. G. Appleby 1984. Empirical testing of ^{210}Pb dating models for lake sediments.

Oldfield, F., P. G. Appleby, and R. Thompson 1980. Paleocological studies of lakes in the Highlands of Papua New Guinea. I. The chronology of sedimentation. *J. Ecol.* **68**, 457–78.

Oldfield, F., R. Thompson and D. P. E. Dickinson 1981. Artificial magnetic enhancement of stream bedload: a hydrological application of supramagnetism. In *Rock mechanics of five particles*, E. R. Deutsch (ed.). *Phys. Earth Planet. Int.* **26**, 107–24.

Oldfield, F., T. Rummery, R. Thompson and D. E. Walling 1979. Identification of suspended sediment sources by means of magnetic measurements: some preliminary results. *Wat. Res. Res.* **15**, 211–17.

Oliver, J. 1980. Considerations affecting palaeoclimatic models of northeastern Australia. In *The geology and geophysics of northeastern Australia*. R. A. Henderson and P. J. Stephenson (eds), 395–8.

Ollier, C. D. 1965. Some features of granite weathering in Australia. *Z. Geomorphol.* NF **9**, 285–304.

Ollier, C. D. 1969. *Weathering*. Edinburgh: Oliver and Boyd.

Ollier, C. D. 1979. Evolutionary geomorphology of Australia and Papua New Guinea. *Trans. Inst. Br. Geogrs* NS **4**, 516–39.

Ollier, C. D. 1981. *Tectonics and landforms*. London: Longman.

Opp, C. 1983. Eine Diskussion zum Catena-Begriff. *Hall. Jb. Geowiss.* **8**, 75–82.

Orians, G. *et al.* 1974. Tropical population biology. In *Fragile ecosystems*, E. G. Farnworth and F. B. Golley (eds), 5–65. New York: Springer-Verlag.

Osmaston, E. A. and M. M. Sweeting 1982. The geomorphology of the Gunung Mulu National Park. *Sarawak Mus. J. NS* 30, 75–93.

Oxburgh, E. R. 1978. Rifting in East Africa and large-scale tectonic processes. In *Geological background to fossil man*, W. W. Bishop (ed.), 7–18. Edinburgh: Scottish University Press.

Oxburgh, E. R. and D. L. Turcotte 1974. Membrane tectonics and the East African Rift. *Earth Planet. Sci. Lett.* **22**, 133–40.

Oyebande, L. 1981. Sediment transport and river basin management in Nigeria. In *Tropical agricultural hydrology* R. Lal and E. W. Russell (eds), 201–25. Chichester: Wiley.

Pachur, H. J. and G. Braun 1980. The paleoclimate of the Central Sahara, Libya and the Libyan Desert. *Palaeoecol. Afr.* **12**, 351–64.

Paddaya, K. 1978. New research designs and field techniques in the Palaeolithic archaeology of India. *World Archaeol.* **10**, 94–110.

Pain, C. F. and J. M. Bowler 1973. Denudation following the November 1970 earthquake at Madang, Papua-New Guinea. *Z. Geomorph.* Suppl. **18**, 92–104.

Palerm, A. 1973. *Obras hidraulicas prehispanicas en el sistema lacustre del Valle de Mexico*. Mexico City: Inst. Nac. Antrop. Hist.

Pallier, G. 1978. *Géographie générale de la Haute-Volta*. Limoges: Université de Limoges.

Palmer, C. E. 1951. Tropical meteorology. In *Compendium of meteorology,* T. F. Malone (ed.), 859–80. Boston: Am. Met. Soc.

Palmer, H. S. 1927. Lapies in Hawaiian basalts. *Geogr. Rev.* **17**, 627–31.

Panabokke, C. R. 1959. A study of some soils in the dry zone of Ceylon. *Soil. Sci.* **87**, 67–74.

Pannekoek, A. U. 1941. Einige karstterreinen in Nederlandsch-Indie. *Nederlands indische geografische Mededelingen* 1, 16–19.

Parkin, D. W. 1974. Trade-winds during the glacial cycles. *Proc. R. Soc. Lond. A* **337**, 73–100.

Parkin, D. W. and N. Shackleton 1973. Trade-winds and temperature correlations down a deep-sea core off the Saharan coast. *Nature* **245**, 455–7.

Parmenter, C. and D. W. Folger 1974. Eolian biogenic detritus in deep sea sediments: a possible index of equatorial Ice Age aridity. *Science* **185**, 695–8.

Passarge, S. 1903. Bericht über eine Reise im venezolanischen Guyana. *Z. Ges. Erdk. Berlin* **37**, 5–43.

Passarge, S. 1904. Rumpfflächen und Inselberge. *Z. Dtsch. Geol. Ges.* **56**, 193–215.

Passarge, S. 1910. Geomorphologische Probleme aus Kamerun. *Z. Ges. Erdkd. Berlin* **44**, 448–65.

Passarge, S. 1923. Inselberglandschaft der Massaisteppe. *Petermanns Geogr. Mitt.* **69**, 205–9.

Pastouret, L., H. Charnley, G. Delibrias, J. C. Duplessy and J. Thiede 1978. Late Quaternary climatic changes in western tropical Africa deduced from deep-sea sedimentation off the Niger Delta. *Oceanol. Acta* **1**, 217–32.

Paterson, T. T. and H. J. H. Drummond 1962. *Soan: The Palaeolithic of Pakistan*. Karachi: Dept of Archaeology, Govt of Pakistan.

Partridge, G. and S. Woodruff 1981. Changes in global surface temperature from 1880 to 1977 derived from historical records of sea surface temperature. *Mon. Weather Rev.* **109**, 2427–34.

Peh, C. H. 1978. *Rates of sediment transport by surface work in three forested areas of peninsular Malaysia*. Dept Geog., Univ. Malaya, Occas. Pap. 3.

Penck, A. and E. Brückner 1909. *Die Alpen in Eiszeitalter*. Leipzig.

Penck, W. 1924. *Die Morphologische Analyse*. Stuttgart.

Pendleton, R. L. 1940. Soil erosion as related to land utilization in the humid tropics. *Proc. Sixth Pacific Sci. Congr.* Vol. 4, 905–20.

Pereira, H. C. 1973. *Land use and water resources*. Cambridge: Cambridge University Press.

Perkins, R. D. and P. Enos. 1968. Hurricane Betsy in the Florida–Bahama area – geologic effects and comparison with Hurricane Donna. *J. Geol.* **76**, 710–16.

Perry, A. H. 1974. The downward trend of air and sea surface temperatures over the North Atlantic. *Weather* **29**, 451–5.

Peterson, G. M., T. Webb, J. E. Kutzbach, T. van der Hammen, T. A. Wijmstra and F. A. Street 1979. The continental record of environmental conditions at 18 000 yr BP: an initial evaluation. *Quat. Res.* **12**, 47–82.

Petit, J. R., M. Briat and A. Royer 1981. Ice age aerosol content from East Antarctic ice core samples and past wind strength. *Nature* **293**, 391–4.

Philander, S. G. H. 1983a. El Niño Southern Oscillation phenomena *Nature* **302**, 295–301.

Philander, S. G. H. 1983b. Anomalous El Niño of 1982–83. *Nature* **305**, 16.

Pickup, G., R. J. Higgins and R. F. Warner 1981. Erosion and sediment yield in the Fly River drainage basins, Papua New Guinea. *Publns Assoc. Int. Hydrol. Sci.* **132**, 438–56.

Pinkayan, S. and T. Ketratanaborvorn 1975. Measurement and analysis of very short duration rainfall. *Hydrol. Sci. Bull.* **20**, 87–92.

Polach, H. A. 1984. Pyroclastic deposits and eruptive sequences of Long Island, Part II: Radiocarbon dating of Long Island and Tibito tephras. In *Cooke-Ravian Volume of Volcanological Papers*, R. W. Johnson (ed.). Geol. Surv., Papua New Guinea Mem, 14–21.

Pollack, H. M. and D. S. Chapman 1977. On the regional variation of heat flow, geotherms and lithospheric thickness. *Tectonophysics*, **38**, 279–96.

Pollack, H. N., I. G. Gass, R. S. Thorpe and D. S. Chapman. 1981. On the vulnerability of lithospheric plates to mid-plate volcanism: reply to comments by P. R. Vogt. *J. Geophys. Res.* **86** B, 961–6.

Poore, M. E. D. 1968. Studies in Malaysian rainforest: I. The forest on Triassic sediments in Jengka Forest Reserve. *J. Ecol.* **56**, 143–96.

Porter, J. W. 1974. Community structure of coral reefs on opposite sides of the Isthmus of Panama. *Science* **186**, 543–5.

Porter, J. W. 1976. Autotrophy, heterotrophy and resources partitioning in Caribbean reef-building corals. *Am. Nat.* **110**, 731–42.

Porter, J. W., J. F. Battey and G. J. Smith 1982. Perturbation and change in coral reef communities. *Proc. Natl Acad. Sci. USA* **79**, 1678–81.

Porter, J. W., J. D. Woodley, G. J. Smith, J. E. Neigel, J. F. Battey and D. G. Dallmeyer 1981. Population trends among Jamaican reef corals. *Nature* **294**, 249–50.

Potter, P. E. 1978. Significance and origin of big rivers. *J. Geol.* **86**, 13–33.

Poulain, J. P. and J. Arrivets 1971. *Effets des principaux éléments fertilisants autres que l'azote sur les rendements des cultures vivrières de base (sorgho, mil, mais) au Sénégal et en Haute-Volta*. Séminaire CSTR/Hv.

Pouquet, J. 1956a. Le plateau de Labé (Guinée). Remarques sur le caractère dramatique des phénomènes d'érosion des sols et sur les remèdes proposés. *Bull. Inst. Fr. Afr. Noire, A, Sénégal*, **17**(1), 1–16.

Pouquet, J. 1956b. Aspects morphologiques du Fouta Djalon, régions de Kindia et de Labé, Guinée. Caractères alarmants des phénomènes d'érosion des sols déclenchés par les activités humaines. *Rev. Géog. Alp. Fr.* **2**, 231–45.

Prell, W. L. 1978. Upper Quaternary sediments of the Columbia Basin: spatial and stratigraphic variation. *Bull. Geol Soc. Am.* **89**, 1241–55.

Prell, W. L. and J. D. Hays 1976. Late Pleistocene faunal and temperature patterns of the Columbian basin, Caribbean Sea. In *Investigation of Late Quaternary palaeoceanography and palaeoclimatology*, R. M. Cline and J. D. Hays (eds), 201–20. Geol. Soc. Am. Mem. 145.

Prell, W. L., W. H. Hutson, D. F. Williams, A. W. H. Be, K. Geitzenauer and B. Molfino 1980. Surface circulation of the Indian Ocean during the last glacial maximum, approximately 18 000 yr BP. *Quat. Res.* **14**, 309–36.

Prescott, J. A. and Pendleton, R. L. 1952. *Laterite and lateritic soils*. Tech. Commun. Commonw. Bur. Soils 47.

Proctor, J. 1983. Mineral nutrients in tropical forests. *Progr. Phys. Geog.* **7**, 422–31.

Proctor, J., J. M. Anderson and H. W. Vallack 1982. Ecological studies in four forest types. *Sarawak Mus. J. NS* 30, 93–121.

Proctor, J., J. M. Anderson, S. C. L. Fogden and H. W. Vallack 1983. Ecological studies in four contrasting lowland rainforests in Gunung Mulu National Park, Sarawak. II. Litterfall, litter standing crop and preliminary observations on herbivary. *J. Ecol.* **71**, 261–84.

Prospero, J. M., R. A. Glaccum and R. T. Nees 1981. Atmospheric transport of soil dust from Africa to south America. *Nature* **289**, 570–2.

Quinn, W. H. 1974. Monitoring and predicting El Niño invasions. *J. Appl. Meteor.* **13**, 825–30.

Radambrasil 1976a. *Folha SA.21 Santarem*. Vol. 10. Rio de Janeiro: Min. Minas e En., DNPM.

Radambrasil 1976b. *Folha NA.19, Pico da Neblina*. Vol. 11. Rio de Janeiro: Min. Minas e En., DNPM.

Rai, B. and A. K. Srivastava 1982. Litter production in a tropical dry mixed deciduous forest stand. *Acta Oecol. Oecol. Appl.* **3**, 169–76.

Rai, S. N. 1981. Productivity of tropical rainforests of Karnatka. PhD thesis, University of Bombay.

Rampino, M. R. and P. Self 1984. The atmospheric effects of El Chichon. *Sci. Am.* **250**(1), 34–43.

Ranalli, G. and E. I. Tanczyk 1975. Meridional orientation of grabens and its bearing on geodynamics. *J. Geol.* **83**, 526–31.

Randall, R. H. and L. G. Eldredge 1977. Effects of typhoon Pamela on the coral reefs of Guam. *Proc. Third Int. Coral Reef Symp.* Vol. 2, 526–31.

Rapp, A., D. H. Murray-Rust, C. Christiansson and L. Berry 1972. Soil erosion and sedimentation in four catchments near Dodoma, Tanzania. *Geogr. Ann.* **54A**, 255–318.

Rasmussen, E. M. and J. M. Hall 1983. El Niño: the great equatorial Pacific Ocean warming event of 1982–83. *Weatherwise* **36**, 166–75.

Reyment, R. A. and E. A. Tait 1972. Faunal evidence for the origin of the South Atlantic. *Proc. 24th Int. Geol. Congr.* Montreal, Sect. 7, 316–23.

Richards, P. W. 1952. *The tropical rain forest*. Cambridge: Cambridge University Press.

Richards, P. W. 1969. Speciation in the tropical rain forest and the concept of the niche. *Biol. J. Linnean Soc.* **1**, 149–54.

Richardson, J. A. 1947. An outline of the geomorphological evolution of British Malaya. *Geol Mag.* **84**, 129–44.

Richter, F. M. and B. Parsons 1975. On the interaction of two scales of convection in the mantle. *J. Geophys. Res.* **80**, 2529–41.

Rigby, J. K. and H. H. Roberts 1976. *Geology, reefs and marine communities of Grand Cayman Island, British West Indies*. Spec. Publ. 4, Geol Stud. Brigham Young Univ., Provo, Utah, 1–95.

Robert, M. 1927. *Le Katanga physique*. Bruxelles: Lamertin.

Roberts, H. H., L. J. Rouse Jr, N. D. Walker and J. H. Hudson 1982. Cold-water stress in Florida Bay and Northern Bahamas: a product of winter cold-air outbreaks. *J. Sed. Petrol.* **52**, 145–55.

Roberts, N., A. Street-Perrott and A. Perrott 1981. The 'late-glacial' in the tropics. *Quat. Newslett.* **35**, 1–5.

Robson, G. R. 1965. Field trip guide: St Vincent. *Trans. 4th Caribb. Geol. Conf.*, Trinidad, 454–7.

Rodier, J. and C. Auvray 1965. *Estimation des débits de crues décennales pour les bassins-versants de superficie inférieure à 200 km*^{2} en Afrique occidentale. Rapp. ORSTOM-CIEH.

Rodin, L. E. and N. I. Bazilevich 1967. *Production and mineral cycling in terrestrial vegetation.* Edinburgh: Oliver and Boyd.

Rogers, C. S., T. H. Suchanek and F. A. Pecora 1982. Effects of hurricanes David and Frederic (1979) on shallow *Acropora palmata* reef communities: St Croix, US Virgin Islands. *Bull. Marine Sci.* **32**, 532–48.

Rognon, P. and M. A. J. Williams 1977. Late Quaternary climatic changes in Australia and North Africa: a preliminary interpretation. *Palaeogeog., Palaeoclim., Palaeoecol.* **21**, 285–327.

Rohdenburg, H. 1969. Hangpedimentation und Klimawechsel als wichtigste Faktoren der Flächen und Stufenbildung in den wechselfeuchten Tropen an Beispielen aus Westafrika, besondersaus dem Schichtstufenland Sudost-Niegerien. *Gottinger Bodenkundl. Ber.* **10**, 57–152.

Rohdenburg, H. 1970. Hangpedimentation und klimawechsel als wichtigste Faktoren der Flächen und Stufenbildung in den wechsel feuchten Tropen. *Z. Geomorph.* NF **14**, 58–78.

Rollinson, H. R. 1978. Zonation of supracrustal relics in the Archaean of Sierra Leone, Liberia, Guinea and Ivory Coast. *Nature* **272**, 440–2.

Roose, E. J. 1967. Dix années de mesure de l'érosion et du ruissellement au Sénégal. *Agron. Trop.* **22**(2), 123–52.

Roose, E. J. 1968. *Mesure de l'érodibilité d'un sol (facteur K) sur la parcelle de référence de Wischmeier. Deuxième projet du protocole standard et sa discussion.* Note multigr., ORSTOM, Abidjan.

Roose, E. J. 1971. *Influence des modifications du milieu naturel sur l'érosion, le ruissellement, le bilan hydrique et chimique, suite à la mise en culture sous climat tropical. Synthèse des observations en Côte d'Ivoire et en Haute-Volta.* Rapp. multigr., ORSTOM, Abidjan, 22. Commun. Sem. Sols Trop., Ibadan, May 1972, 22 multigr.

Roose, E. J. 1972a. *Contribution a l'étude de l'appauvrissement de quelques sols ferrallitiques et ferrugineux tropicaux situés entre Abidjan and Ouagadougou par l'utilisation de méthodes éxperimentales de terrain.* ORSTOM Bull. liaison, thème A, no. 1, 19–41. Commun. J. Ped. ORSTOM, 28 Sept. 1971.

Roose, E. J. 1972b. *Comparison des causes de l'érosion et des principes de lutte antiérosive en région tropicale humide, tropicale sèche et méditerranéenne.* Commun. J. Étude Génie Rural Florence 12–16 July 1972, 417–41.

Roose, E. J. 1973. *Dix-sept années de mesures éxperimentales de l'érosion et du ruissellement sur un sol ferrallitique sableux de basse Côte d'Ivoire. Contribution à l'étude de l'érosion hydrique en milieu inter-tropical.* Thèse, Abidjan.

Roose, E. J. 1975. *Application de l'équation de prévision de l'érosion de Wischmeier et Smith en Afrique de l'Ouest.* ORSTOM, Abidjan, 22.

Roose, E. J. 1977. *Érosion et ruissellement en Afrique de l'Ouest. Vingt ans de mesures en petites parcelles éxperimentales.* Trav. Doc. ORSTOM 78, 108.

Roose, E. J. 1980. Dynamique actuelle de quelques types de sols en Afrique de l'Ouest. *Z. Geomorph* NF Suppl. **35**, 32–9.

Roose, E. J. and R. Bertrand 1971. Contribution a l'étude de la methode des bandes d'arrêt pour lutter control l'érosion hydrique en Afrique de l'Ouest. Résultats éxperimentaux et observations sur le terrain. *Agron. Trop.* **26**(11), 1270–83.

Roose, E. and Y. Birot 1970. *Mésure de l'érosion et du lessivage oblique et vertical sous une savane arborée du plateau mossi (Gonse-Haute-Volta). 1. Résultats des campagnes 1968-69.* Rapp. multigr., CTFT, ORSTOM, Abidjan.

Roose, E. J., J. Arrivets and J. F. Poulain 1974. *Étude du ruissellement, du drainage et de l'érosion sur deux sols ferrugineux de la région Centre Haute-Volta. Bilan de trois années d'observation à la station de Saria.* Rapp. ORSTOM/Abidjan-IRAT/Haute-Volta, 83 multigr.

Rose, J. 1982. The Melinau River and its terraces. *Trans. Br. Cave Res. Assoc.* **9**, 113–27.

Rosen, B. R. 1981. The tropical high diversity enigma – the corals'-eye view. In *The evolving biosphere*, P. L. Forey (ed.), 103–29. Cambridge: British Museum of Natural History and Cambridge University Press.

Rossignol-Strick, M. and D. Duzer 1979. A Late Quaternary continuous climatic record from palynology of three marine cores off Senegal. *Palaeoecol. Afr.* **11**, 185–8.

Rossignol-Strick, M., W. Nesteroff, P. Olive and C. Vergnaud-Grazzini 1982. After the deluge: Mediterranean stagnation and sapropel formation. *Nature* **295**, 105–10.

Rougerie, G. 1954. Methode d'étude éxperimentale des phénomènes d'érosion en milieu naturel. *Rev. Géomorph. Dyn.* **5**, 220–7.

Rougerie, G. 1960. *Le façonnement actuel des modelés en Côte d'Ivoire forestière.* Mem. Inst. Afr. Noire. 58.

Ruhe, R. V. 1952. Topographic discontinuities of the Des Moines Lobe. *Am. J. Sci.* **250**, 46–56.

Rutten, L. M. R. 1928. *Science in the Netherlands East Indies.* Amsterdam: Koningklijke Akademie van Wetenschappen.

Rymer, L. 1978. The use of uniformitarianism and analogy in palaeoecology, particularly pollen analysis. In *Biology and Quaternary environments*, D. Walker and J. C. Guppy (eds). Canberra: Australian Academy of Science.

Sa-ard Boonkird 1968. Effects of forest fire on runoff and erosion. In *Conservation in tropical SE Asia*, L. M. Talbot and M. H. Talbot (eds), 103–10. Morges: IUCN.

Sabiels, B. E. 1966. Climatic variations in the tropical Pacific as evidenced by trace elements analysis of soils. In *Pleistocene and post-Pleistocene climatic variations in the Pacific Area.* D. J. Blumenstock (ed.), 131–51. Honolulu: Bishop Museum Press.

Sahagian, D. 1980. Sublithospheric upwelling distribution. *Nature* **287**, 217–18.

Sahni, A. 1982. The structure, sedimentation and evolution of Indian continental margins. In *The ocean basins and margins*, Vol. 6: *The Indian Ocean*, A. E. M. Nairn and F. G. Stehli (eds), 353–98. New York: Plenum.

Sall, M. 1971. Dynamique et morphogenèse actuelles (Contribution à l'étude géomorphologique du Sénégal occidental). Thèse 3ème cycle, Univ. Dakar.

Sall, M. 1978. Le processus géomorphologiques actuels dans l'environment des campagnes de la presqu'île du Cap-Vert (Sénégal). *Géo.-Éco.-Trop.* **2**, 21–30.

Sall, M. 1979. Hydrologie et géomorphologie du Delta du Sénégal et de ses bordures (Aftout es Saheli et Ferlo nord-occidental) d'après les images Landsat du 30.9.72. *Photointerprétation* **79**, 5, numéro special Sénégal, 29–37.

Sall, M. 1982. Dynamique et morphogenèse actuelles au Sénégal. Thèse d'État, Strasbourg.

Sammarco, P. W. 1980. *Diadema* and its relationship to coral spat mortality: grazing, competition, and biological disturbance. *J. Exp. Marine Biol. Ecol.* **45**, 245–72.

Sancetta, C., J. Imbrie, N. G. Kipp, A. McIntyre and W. F. Ruddiman 1972. Climatic record in North Atlantic deep-sea core V23-82: comparison of the Last and Present Interglacials based on quantitative time series. *Quat. Res.* **2**, 363–7.

Sanchez, W. A. and J. E. Kutzbach 1974. Climate of the American tropics and subtropics in the 1960s and possible comparisons with climatic variations of the last millenium. *Quat. Res.* **4**, 128–35.

Sankalia, H. D. 1974. *Prehistory and protohistory of India and Pakistan.* Poona: Deccan College Postgraduate and Research Institute.

Sankalia, H. D. 1978. The Early Palaeolithic of India and Pakistan. In *Early Palaeolithic in South and East Asia*, F. Ikawa-Smith (ed.), 97–127. The Hague: Mouton.

Sanou, D. 1981. *Étude comparative entre une parcelle pourvue de bourrelets anti-érosifs et des parcelles traditionnelles à Sirgui (Kaya): introduction aux problèmes de dynamique érosive.* Mem. Maitrise, Univ. Ouagadougou, ESLS, Géographie.

Sapper, K. 1909. Neu Mecklenburg (New Ireland). *Geogr. Z.* **15**, 425–50.

Sapper, K. 1910a. Beitrag zur Kenntris Neu Pommerns und des Kaiser Wilhelms Landes III. Zur Kenntris des Kaiser Wilhelms Landes und einiger Neu Guinea und Neu Pommern vorgelagerten Inseln. *Petermanns Geogr. Mitt.* **56**, 189–93, 255–6.

Sapper, K. 1910b. Aus den Schutzgebieten der Sudsee-Eine Durchquerung von Bougainville, *Mitt. Dtsch. Schutzgebieten* **23**, 206–17.

Sapper, K. 1910c. Wissenschaftliche Ergenbnisse einer amtlichen Forschungsreise nach dem Bismarck Archipel im Jarh 1908: 1. Beitrag zur Landeskunde von New-Mecklenburg und seiner Nachbarinseln. *Mitt. Schutzgebieten Erganzungshefte* **3**, 1–130.

Sapper, K. 1914. Uber Abtragungsvorgänge in den regenfeuchten Tropen und ihre morphologischen Wirkungen. *Geogr. Z.* **20**, 5–16, 81–92.

Sapper, K. 1917a. Beiträge zur Geographie der tätigen Vulkane. *Z. Vulkanol.* **3**, 65–197.

Sapper, K. 1917b. Melanesien-Melanesische Vulkanzone. In *Katalog der Geschichtlichen Vulkanaubruche* K. Sapper (ed.). *Schr. Wissanschaftliche Ges. Strassburg* **27**, 204–15.

Sapper, K. 1921. Die Vulkanberge New-Guineas. *Z. Vulkanol.* **6**, 1–14.

Sapper, K. 1935. *Geomorphologie der feuchten Tropen*, 150. Berlin: Teubner.

Sarmiento, R. and R. A. Kirby 1962. Recent sediments of Lake Maracaibo. *J. Sed. Petrol.* **32**, 698–724.

Sarnthein, M. 1978. Sand deserts during glacial maximum and climatic optimum. *Nature* **272**, 43–5.

Sarnthein, M., F. Tetzlaff, B. Koopman, K. Wolter and U. Pflaumann 1981. Glacial and interglacial wind regimes over the eastern suptropical Atlantic and North-West Africa. *Nature* **293**, 193–6.

Scheidegger, A. E. 1979. The principle of antagonism in the Earth's evolution. *Tectonophysics* **55**, T7–10.

Schnütgen, A. 1981. Analysen zur Verwitterung und Bodenbildung in den Tropen an Proben von Sri Lanka. *Relief, Boden, Palaoklima* **1**, 239–75.

Schnütgen, A. and H. Späth 1978. Der Vergleich von Altschottern der nördlichen Eifel mit tropischen Schottern aus dem nordwestlichen Sri Lanka. *Kölner Geogr. Abh.* **36**, 155–85.

Schubert, C. 1980. Contribution to the paleolimnology of Lake Valencia, Venezuela: seismic stratigraphy. *Catena* **7**, 275–92.

Schulz, J. P. 1960. Ecological studies on rain forest in Northern Suriname. *Proc. K. Ned. Acad. Wet., Sect. C* **53**, 1–267.

Schumm, S. A. 1956. The evolution of drainage systems and slopes in badlands at Perth Amboy, New Jersey. *Bull. Geol Soc. Am.* **67**, 597–646

Schumm, S. A. 1965. Quaternary palaeohydrology. In *Quaternary of the United States*, H. E. Wright and D. G. Frey (eds), 783–94. Princeton: Princeton University Press.

Schumm, S. A. 1968. Speculations concerning paleohydrologic controls of terrestrial sedimentation. *Bull. Geol Soc. Am.* **79**, 1573–88.

Schumm, S. A. 1969. River metamorphosis. *J. Hydraul. Div. Am. Soc. Civil Engrs. Proc.* **95**, 255–73.

Schumm, S. A. 1977. *The Fluvial System*. New York: Wiley.

Schumm, S. A. and R. S. Parker 1973. Implications of complex response of drainage systems for quaternary alluvial stratigraphy. *Nature Phys. Sci.* **243**, 99–100.

Schwan, W. 1980. Geodynamic peaks in Alpinotype orogenesis and changes in ocean-floor spreading during Late Jurassic–Late Tertiary time. *Bull. Am. Assoc. Petrol. Geol.* **64**, 359–73.

Scott, G. A. J. and Street, R. M. 1976. The role of chemical weathering in the formation of Hawaiian Amphitheatre-headed valleys. *Z. Geomorph.* NF **20**, 171–89.

Scrivenor, J. B. 1931. *The Geology of Malaya*. London.

Scrutton, R. A. and R. V. Dingle 1976. Observations on the processes of sedimentary basin formation at the margins of southern Africa. *Tectonophysics* **36**, 143–56.

Segalen, P. 1969. Le remaniement des sols et le mise en place de la stoneline en Afrique. *Cah. ORSTOM. Sér. Pédol.* **7**, 113–31.

Semmel, A. and H. Rohdenburg. 1979. Untersuchungen zur Boden und Reliefentwicklung in Süd-Brasilien. *Catena* **6**, 203–17.

Servant, M. 1973. *Séquences continentales et variations climatiques: évolution du Bassin du Tchad au Cénozoique Supérieur*. Thèse, ORSTOM, Paris.

Servant, M. 1974. Les variations climatiques des régions intertropicales du continent africain depuis la fin du Pléistocène. *Soc. Hydrotech. Fr., 13e J. Hydraul. Paris* **1** (8), 1–10.

Servant, M. and S. Servant-Vildary 1972. Nouvelles données pour une interprétation paléoclimatique de series continentales du Bassin Tchadien. *Palaeoecol. Afr.* **6**, 87–92.

Servant, M. and S. Servant-Vildary 1980. L'environment Quaternaire du bassin du Tchad. In *The Sahara and the Nile*, M. A. J. Williams and H. Faure (eds). Rotterdam: Balkema.

Servant-Vildary, S. 1978. *Étude des diatomées et paléolimnologie du bassin du Tchad au Cénozoique Supérieur*. Trav. Doc. ORSTOM 84, vols I & II.

Seuffert, O. 1973. Die Laterite Sudindiens als Klimazeugen. *Z. Geomorph.* NF Suppl. **17**, 242–59.

Shackleton, N. J. and J. P. Kennett 1975. Paleotemperature history of the Cenozoic and the initiation of Antarctic glaciation: oxygen and carbon isotope analysis in DSDP sites 277, 279 and 281. In *Initial reports of the deep sea drilling project*, J. P. Kennett *et al.* (eds), no. 29, 743–55. Washington: US Govt Printing Office.

Shackleton, N. J. and N. D. Opdyke 1976. Oxygen-isotope and paleomagnetic stratigraphy of Pacific Core V28-239 Late Pliocene to Latest Pleistocene. *Geol Soc. Am. Mem.* **145**, 449–64.

Sharma, G. R., V. D. Misra, D. Mandal, B. B. Misra and J. M. Pal 1980. *Beginnings of agriculture*. Allahabad: Abinash Prakashan.

Sheppard, C. R. C. 1979. Interspecific aggression between reef corals with reference to their distribution. *Marine Ecol. Prog. Ser.* **1**, 237–47.

Shi, Y. and J. Wang 1981. The fluctuations of climate, glaciers and sea level since late Pleistocene in China. In *Sea level, ice and climatic change*, J. Allison (ed.), 281–93. Int. Assoc. Hydrol. Sci. Publ. 131.

Shinn, E. A. 1976. Coral reef recovery in Florida and the Persian Gulf. *Environ. Geol.* **1**, 241–54.

Sigurdsson, H. 1972. Partly-welded pyroclast flow deposits in Dominica, Lesser Antilles. *Bull. Volcan.* **26**(I), 148–63.

Simonett, D. A. 1967. Landslide distribution and earthquakes in the Bewani and Torricelli Mountains, New Guinea. In *Landform studies from Australia and New Guinea*, J. N. Jennings and J. A. Mabbutt (eds), 64–68. Cambridge: Cambridge University Press.

Singh, G. 1971. The Indus Valley culture seen in the context of post-glacial climatic and ecological studies in north-west India. *Arch. Phys. Anthrop. Oceania* **6**, 177–89.

Singh, G. 1981. Late Quaternary pollen records and seasonal palaeoclimates of Lake Frome, South Australia. *Hydrobiologia* **82**, 419–30.

Singh, G. 1982. Environmental upheaval: the vegetation of Australasia during the Quaternary. In *A history of Australasian vegetation*, J. M. B. Smith (ed.), 90–108. Sydney: McGraw-Hill.

Singh, G. and D. P. Agrawal 1976. Radiocarbon evidence for deglaciation in north-western Himalaya, India. *Nature* **260**, 232.

Singh, G., R. D. Joshi and A. B. Singh 1972. Stratigraphic and radiocarbon evidence for the age and development of three salt lake deposits in Rajasthan, India. *Quat. Res.* **2**, 496–505.

Singh, G., A. P. Kershaw and R. Clark 1981. Quaternary vegetation and fire history in Australia. In *Fire and the Australian biota*, 23–54. Canberra: Australian Academy of Science.

Singh, G., R. D. Joshi, S. K. Chopra and A. B. Singh 1974. Late Quaternary history of vegetation and climate of the Rajasthan Desert, India. *Phil Trans R. Soc. Lond. B* **267**, 467–501.

Singhvi, A. K., Y. P. Sharma and D. P. Agrawal 1982. Thermoluminescence dating of sand dunes in Rajasthan, India. *Nature* **295**, 313–15.

Sioli, H. 1957. Sedimentation in Amazonasgebiet. *Geol. Rundsch* **45**, 608–33.

Sioli, H. 1961. Landschaftsökologischer Beitrag aus Amazonien. *Natur Landsch. Mainz* **36**(5), 73–7.

Sioli, H. 1973. Principais biòtopos de producão primària nas àguas da Amazônia. *Bol. Geogr.* **236**, 118–27.

Sioli, H. 1975. Tropical rivers as expressions of their terrestrial environments. In *Tropical ecological systems: trends in terrestrial and aquatic research*, F. B. Golley and E. Medina (eds), 275–88. Berlin: Springer.

Sivarajasingham, S. 1968. *Soil and land use survey in the Eastern Province*. Rep. to Govt of Sierra Leone, UNDP/FAO, TA 2584.

Sloss, L. L. and R. C. Speed 1974. Relationships of cratonic and continental-margin tectonic episodes. In *Tectonics and sedimentation*, W. R. Dickinson (ed.), 98–119. Soc. Econ., Paleont., Min., Tulsa, Spec. Publn 22.

Smart, P. L. and H. Friederich 1982. An assessment of the methods used and the results obtained from water tracing experiments in the Gunung Mulu National Park, Sarawak. *Trans. Br. Cave Res. Assoc.* **9**, 100–12.

Smith, A. B. 1980a. Saharan and Sahel zone environmental conditions in the Late Pleistocene and Early Holocene. *Proc. 8th Panafr. Congr. Prehist. Quat. Studies*, Nairobi, Sepember 1977, R. E. Leakey and B. A. Ogot (eds), 139–42.

Smith, A. B. 1980b. Domesticated cattle in the Sahara and their introduction into West Africa. In *The Sahara and the Nile*, M. A. J. Williams and H. Faure (eds), 489–501. Rotterdam: Balkema.

Smith, A. G. and J. C. Briden 1977. *Mesozoic and Cenozoic palaeocontinental maps*. Cambridge: Cambridge University Press.

Smith, A. G., A. M. Hurley and J. Briden 1981. *Phanerozoic palaeocontinental world maps*. Cambridge: Cambridge University Press.

Smith, B. J. 1982. Effects of climate and land-use change on gully development: an example from Northern Nigeria. *Z. Geomorph.* NF Suppl. **44**, 33–51.

Smith, D. I. and T. C. Atkinson 1976. Process, landforms and climate in limestone regions. In *Geomorphology and climate*, E. Derbyshire (ed.), 367–409. London: Wiley Interscience.

Snowy Mountains Engineering Corporation 1974. *Serayu River Basin study – feasibility report*. Djakarta: Dept. Public Works and Elec. Power.

Snyder, C. T. and W. B. Langbein 1962. The Pleistocene lake in Spring Valley, Nevada and its climatic implications. *J. Geo. Res.* **67**, 2385–94.

Sonntag, C., U. Thorweihe, J. Rudolph, E. P. Lohnert, C. Junghans, K. O. Munnich, E. Klitsch, E. M. El Shazly and F. M. Swailem 1980. Isotopic identification of Saharian groundwaters, groundwater formation in the past. *Palaeoecol. Afr.* **12**, 159–71.

Sowunmi, M. A. 1981. Aspects of Late Quaternary vegetational changes in West Africa. *J. Biogeog.* **8**, 457–74.

Späth, H. 1977. Rezente Verwitterung und Abtragung an Schicht- und Rumpfstufen in semiariden Westaustralien. *Z. Geomorph.* NF, Suppl. **28**, 81–100.

Späth, H. 1981. Bodenbildung und Reliefentwicklung in Sri Lanka. *Relief, Boden, Palaoklima* **1**, 185–238.

Späth, H. 1982. Aussagemöglichkeiten von Böden fur die Reliefentwicklung in Sri Lanka. In *Tagungsber. und Wiss. Abh. Dtsch. Geographentag Mannheim* 1981. Wiesbaden: Steiner (in press).

Stace, H. C. *et al.* 1968. *A handbook of Australian soils*. Adelaide: Rellim Technical.

Stauffer, P. H. 1973. Late Pleistocene age indicated for volcanic ash in West Malaysia. *Geol. Soc. Malay. Newslett.* **40**, 1–4.

Stauffer, P. H., S. Nishimura and B. C. Batchelor 1980. Volcanic ash in Malaya from a catastrophic eruption of Toba, Sumatra, 30 000 years ago. In *Physical geology of Indonesian island areas*, S. Nishimura (ed.), 156–64. Kyoto University, Japan.

Steers, J. A. 1929. The Queensland coast and the Great Barrier Reefs. *Geogr. J.* **124**, 232–57, 340–67 (Discussion: 367–70).

Steers, J. A. 1940. The coral cays of Jamaica. *Geogr. J.* **95**, 30–42.

Stephenson, P. J., T. J. Griffin and F. C. Sutherland 1980. Cainozoic volcanism in north eastern Australia. In *The geology and geophysics of north eastern Australia*, R. A. Henderson and P. J. Stephenson (eds), 349–74. Brisbane: Geological Society of Australia.

Stocker, G. C. 1971. The age of charcoal from old jungle-fowl nests and vegetation change on Melville Island. *Search* **2**, 28–30.

Stocker, G. C. and J. J. Mott, 1981. Fire in the tropical forests and woodlands of northern Australia. In *Fire and the Australian biota*, A. M. Gill, R. H. Groves and I. R. Noble (eds), 425–39. Canberra: Australian Academy of Science.

Stoddart, D. R. 1963. Effects of Hurricane Hattie on the British Honduras reefs and cays, Oct. 30–31, 1961. *Atoll Res. Bull.* **95**, 142.

Stoddart, D. R. 1965. Re-survey of hurricane effects on the British Honduras reefs and cays. *Nature* **207**, 589–92.

Stoddart, D. R. 1969a. Climatic geomorphology: review and re-assessment. *Progr. Geog.* **1**, 159–222.

Stoddard, D. R. 1969b. Post-hurricane changes on the British Honduras reefs and cays: re-survey of 1965. *Atoll Res. Bull.* **131**, 1–25.

Stoddart, D. R. 1971. Coral reefs and islands and catastrophic storms. In *Applied coastal geomorphology*, J. A. Steers (ed.), 155–97. London: Macmillan.

Stoddart, D. R. 1974. Post-hurricane changes on the British Honduras reefs: re-survey of 1972. *Proc. 2nd Int. Coral Reef Symp.* Vol. 2, 473–83.

Stoddart, D. R. 1976. Climatic variability, seabirds and guano in the Central Pacific: a model for the Pleistocene. *Int. Geog.* **3**, 93–8.

Stoddart, D. R. 1977. Structure and ecology of Caribbean coral reefs. *Symp. Prog. Marine Res. Caribbean and Adjacent Regions.* FAO Fisheries Rept. 200, 427–48.

Stoddart, D. R. and T. P. Scoffin 1983. Phosphate rock on coral reef islands. In *Chemical sediments and geomorphology*, A. S. Goudie and K. Pye (eds), 369–400. London: Academic Press.

Stott, P. A. 1978. Tropical rain forest in recent ecological thought: the re-assessment of a non-renewable resource. *Progr. Phys. Geog.* **2**, 80–98.

Strahler, A. N. and Strahler, A. H. 1977. *Geography and man's environment*. New York: Wiley.

Street, F. A. 1979a. Late Quaternary precipitation estimates for the Ziway-Shala Basin, southern Ethiopia. *Palaeoecol. Afr.* **11**, 135–43.

Street, F. A. 1979b. *Late Quaternary lakes in the Ziway–Shala Basin, southern Ethiopia*. PhD Thesis, Cambridge University.

Street, F. A. 1980. The relative importance of climate and hydrogeological factors in influencing lake-level fluctuations. *Palaeoecol. Afr.* **12**, 137–58.

Street, F. A. 1981. Tropical palaeoenvironments. *Progr. Phys. Geog.* **5**, 157–85.

Street, F. A. and F. Gasse 1981. Recent developments in research into the climatic history of the Sahara. In *The Sahara: ecological change and early economic history*, J. A. Allen (ed.), 7–28. London: Menas Press.

Street, F. A. and A. T. Grove 1976. Environmental and climatic implications of Late Quaternary lake-level fluctuations in Africa. *Nature* **261**, 385–90.

Street, F. A. and A. T. Grove 1979. Global maps of lake-level fluctuations since 20 000 yr BP. *Quat. Res.* **12**, 83–118.

Street-Perrott, F. A. and N. Roberts 1983. Fluctuations in closed basin lakes as an indicator of past atmospheric circulation patterns. In *Variations in the global water budget*, F. A. Street-Perrott, M. Beran and R. A. S. Ratcliffe (eds). Dordrecht: Reidel.

Stromquist, L. 1981. Recent studies in soil erosion, sediment transport, and reservoir sedimentation in semi-arid central Tanzania. In *Tropical agricultural hydrology*, R. Lal and E. W. Russell (eds), 189–200. Chichester: Wiley.

Strong, D. R. 1977. Epiphyte loads, tree falls and perennial forest disruption: a mechanism for maintaining higher tree species richness in the tropics. *J. Biogeog.* **4**, 215–18.

Sudres, A. 1947. La dégradation des sols au Fouta Djalon. *Agron. Trop. Fr.* **2**, 226–46.

Summerfield, M. A. 1981. Macroscale geomorphology. *Area* **13**, 3–8.

Sunartadirdja, M. A. and H. Lehmann 1960. Der tropische Karst von Maros und Nord-Bone in S W Sulawesi (Celebes). *Z. Geomorph.* NF suppl. **2**, 49–65.

Swan, B. 1977. Sri Lanka. In *The encyclopedia of world regional geology*, Part II: *Eastern Hemisphere*, R. W. Fairbridge (ed.). New York: Arnold.

Swan, B. 1979. Sand dunes in the humid tropics: Sri Lanka. *Z. Geomorph*. NF **23**, 152–71.

Sykes, L. R. 1978. Intraplate seismicity, reactivation of pre-existing zones of weakness, alkaline magmatism, and other tectonism postdating continental fragmentation. *Rev. Geophys Space Phys* **16**, 621–88.

Szabo, B. J., W. C. Ward, A. E. Weide and M. J. Brady 1978. Age and magnitude of Late Pleistocene sea-level rise on the eastern Yucatan Peninsula. *Geology* **6**, 713–5.

Talbot, M. R. 1980. Environmental responses to climatic change in the West African Sahel over the past 20 000 years. In *The Sahara and the Nile*, M. A. J. Williams and H. Faure (eds), 37–62. Rotterdam: Balkema.

Talbot, M. R. 1981. Holocene changes in tropical wind intensity and rainfall: evidence from south east Ghana. *Quat. Res*. **16**, 201–20.

Talbot, M. R. 1982. Holocene chronostratigraphy of tropical Africa. *Striaie* **16**, 17–20.

Talbot, M. R. and G. Delibrias 1977. Holocene variations in the level of Lake Bosumtwi, Ghana. *Nature* **268**, 722–4.

Talbot, M. R. and F. Delibrias 1980. A new Late Pleistocene–Holocene water-level curve for Lake Bosumtwi, Ghana. *Earth Planet. Sci. Lett*. **47**, 336–44.

Talbot, M. R. and J. B. Hall 1981. Further Late Quaternary leaf fossils from Lake Bosumtwi, Ghana. *Palaeoecol. Afr*., **13**, 83–92.

Temple, P. H. and A. Sundborg 1972. The Rufiji River, Tanzania, hydrology and sediment transport. *Geogr. Ann*. **54A**, 345–68.

Thiessen, R., K. Burke and W. S. F. Kidd 1979. African hotspots and their relation to the underlying mantle. *Geology* **7**, 263–6.

Thom, B. G. and J. Chappell 1975. Holocene sea levels relative to Australia. *Search* **6**, 90–3.

Thomas, D. B., K. A. Edwards, R. G. Barber and I. G. G. Hogg 1981. Runoff, erosion and conservation in a representative catchment in Machakos District, Kenya. In *Tropical agricultural hydrology*, R. Lal and E. W. Russell (eds), 395–417. Chichester: Wiley.

Thomas, M. F. 1965. An approach to some problems of landforms analysis in tropical environments. In *Essays in geography for A. Austin Miller*, P. D. Wood and J. B. Whittow (eds), 118–44. University of Reading.

Thomas, M. F. 1974. *Tropical geomorphology*. London: Macmillan.

Thomas, M. F. 1980. Timescales of landform development on tropical shields – a study from Sierra Leone. In *Timescales in geomorphology*, R. A. Cullingford, D. A. Davidson and J. Lewin (eds), 333–54. London: Wiley.

Thomas, M. F. and M. B. Thorp 1980. Some aspects of the geomorphological interpretation of Quaternary alluvial sediments in Sierra Leone. *Z. Geomorph*. NF Suppl. **36**, 140–61.

Thompson, R. and F. Oldfield 1978. Evidence for recent palaeomagnetic variation in lake sediments from New Guinea. *Phys. Earth Planet. Int*. **17**, 300–6.

Thomson, J. 1882. Notes on the basin of the River Rovuma, East Africa. *Proc. R. Geogr. Soc. N. S.* **4**, 65–79.

Thorbecke, F. 1911. Das Manenguba Hochland. *Mitt. Dtsch. Schutzgebieten* **24**, 279–304.

Thorbecke, F. 1914. Geographische Arbeiten in Tikar und Wute. *Verh. Dtsch. Geographentag, Strassburg*, 147–65.

Thorbeche, F. 1927. Klima und Oberflachenformen: die Stellung des Problems. *Dusseldorfer geographische Vortrage und Erorterungen*, 3 Teil: *Morphologie der Klimazonen*, 1–3 Breslau: Ferdinand Hirt.

Thornbury, W. D. 1954. *Principles of geomorphology*. New York: Wiley.

Tillmanns, W. 1981. Tonmineralogische Untersuchungen von Proben aus Sri Lanka. *Relief, Boden, Paläoklima* 1, 277–80.

Timmermann, O. F. 1935. *Ceylon. Seine natürlichen Landschaftsbilder und Landschaftstypen*. Mitt. Geog. Ges. München 28.

Tomblin, J. F. 1964. *The volcanic history and petrology of the Soufriere region, St Lucia*. DPhil Thesis, Oxford University.

Tomblin, J. F. 1965. The geology of the Soufrière volcanic centre, St Lucia. *Trans 4th Caribb. Geol Conf.*, Trinidad 1965, 367–76.

Tomblin, J. F. 1971. Geological field guide to St Lucia. In *Int. Field Inst. Guidebook to the Caribbean Arc System*. Washington DC: Am. Geol. Inst.

Tomlinson, P. B. 1978. Branching and axis differentiation in tropical trees. In *Tropical trees as living systems*. P. B. Tomlinson and M. H. Zimmerman (eds), 187–207. Cambridge: Cambridge University Press.

Torgersen, T., M. F. Hutchinson, D. E. Searle and H. A. Nix 1983. General bathymetry of the Gulf of Carpentaria and the Quaternary physiography of Lake Carpentaria. *Palaeogeog. Palaeoclim. Palaeoecol.* **41**, 207–26.

Torquato, J. R. and V. G. Cordani 1981. Brazil–Africa geological links. *Earth Sci. Rev.* **17**, 155–76.

Toupet, Ch. and P. Michel 1979. Sécheresse et aridité. L'exemple de la Mauritanie et du Sénégal. *Géo. Éco. Trop.* **3**(2), 137–57.

Tremenheere, C. W. 1867. On the physical geography of the Lower Indus. *Proc. R. Geogr. Soc.* **9**, 22–31.

Tricart, J. 1954. Influence des sols salés sur la déflation éolienne en Basse-Mauritanie et dans le delta du Sénégal. *Rev. Géomorph. Dyn.* **5**(3), 124–32.

Tricart, J. 1955. Nouvelles observations sur les sebkhas de l'Aftout es Sahel mauritanien et du delta du Sénégal. *Rev. Géomorph. Dyn.* **6**(4), 177–87.

Tricart, J. 1956a. Dégradation du milieu naturel et problèmes d'aménagement au Fouta Djalon (Guinée). *Rev. Géogr. Alp.*, 7–36.

Tricart, J. 1956b. Types de fleuves et systèmes morphologiques en Afrique occidentale. *Com. Trav. Hist. Sci. Bull. Sect. Géogr.* **68**, 303–44.

Tricart, J. 1965. *Traité de géomorphologie*, Vol. 5: *Le modelé des régions chaudes (Forêts et savanes)*. Paris: SEDES.

Tricart, J. 1972. *Landforms of the humid tropics, forests and savannas* (Engl. trans.). London: Longman.

Tricart, J. 1974a. Existence de périodes sèches au Quaternaire en Amazonie et dans les régions voisines. *Rev. Géomorph. Dyn.* **23**, 145–58.

Tricart, J. 1974b. *Apports de Erts-1 à notre connaissance ecogenetique des Llanos de l'Orénoque (Colombie et Venezuela)*, 317–24.

Tricart, J. 1975. Influence des oscillations climatiques récentes sur le modelé en Amazonie Orientale (région de Santarém) d'après les images de radar latéral. *Z. Geomorph.* NF **19**, 140–63.

Tricart, J. 1976. Quelques aspects de l'utilisation des images multispectrales du satellite Landsat-1 (Erts) dans l'étude écologique des pays tropicaux (Mali, Colombie, Vénézuéla). *Trav. Doc. Ceget*, **25**, 79–119.

Tricart, J. 1977a. Types de lits fluvieux en Amazonie brésilienne. *Ann. Géog.* **473**, 1–54.

Tricart, J. 1977b. Aperçus sur le Quaternaire amazonien. *'Recherches françaises sur le Quaternaire'*. Bull. Ass. Fr. Étude. Quat. Suppl. 50, 265–71.

Tricart, J. 1977c. La région d'Obidos (Amazonie brésilienne) sur les images Landsat, comparaison avec les mosaiques-radar. *Photointerpretation* **76**(2), 22–35.

Tricart, J. 1979. Altération et dissection différentielles dans un socle cratonisé: l'exemple de la dorsale brésilienne. *Ann. Géog.* **487**, 265–314.

Tricart, J. and P. Alfonsi 1981. Action éoliennes récentes aux abords du delta de l'Orénoque. *Bull. AGG*, **476**, 75–82.

Tricart, J. and A. Cailleux 1965. *Introduction à la géomorphologie climatique*. Paris: SEDES.

Tricart, J. and M. Michel 1965. Monographie et carte géomorphologique de la région de Langunillas (Andes Vénézuéliennes). *Rev. Géomorph. Dyn.* **15**, 1–33.

Tricart, J. and A. Millies-Lacroix 1962. Les terrasses quaternaires des Andes Vénézuéliennes. *Bull. Soc. Géol. Fr.* **7**(4), 201–18.

Troll, C. and J. H. Paffen 1964. Karte der Jahreszeiten – Klimate der Erde. *Erdkunde* **18**, 5–28.

Tunnicliffe, V. 1981. Breakage and propagation of the stony coral, *Acropora cervicornis*. *Proc. Natl Acad. Sci. (USA)* **78**, 2471–31.

Turcotte, D. L. and E. R. Oxburgh 1973. Mid-plate tectonics. *Nature* **244**, 337–9.

Turcotte, D. L. and R. E. Oxburgh 1978. Intra-plate volcanism. *Phil Trans R. Soc. Lond. A* **288**, 561–78.

Turvey, N. D. 1974. *Nutrient cycling under tropical rainforest in central Papua*. Univ. Papua New Guinea Dept. Geog. Occ. Paper 10.

Twidale, C. R. 1978. An origin of Ayers Rock, central Australia. *Z. Geomorph*. NF Suppl. **31**, 177–206.

Twidale, C. R. 1982. *Granite landforms*. Amsterdam: Elsevier.

Twidale, C. R. and J. A. Bourne. 1975. Episodic exposure of inselbergs. *Bull. Geol Soc. Am*. **86**, 1473–81.

ULP-CGA (Blanck J.-P, A.-R. Cloots-Hirsch and A. Gobert) 1977. *Unité écologique expérimentale de la région de Maradi (Niger)*. DGRST-ULP Strasbourg.

Umbgrove, J. H. F. 1947. *The pulse of the Earth*. The Hague: Martinus Nijhoff.

Vageler, P. 1911. Bodenkundliche Skizzen aus Ugogo. *Der Pflanzer* **7**, 565–638.

Vageler, P. 1912. Ugogo I. *Beihefte Tropenpflanz*. **13**, 53–66.

Vail, P. R., R. M. Mitchum Jr and S. Thompson III 1977. Seismic stratigraphy and global changes of sea level, 4. Global cycles of relative sea level change. In *Seismic stratigraphy – applications to hydrocarbon exploration*, C. E. Payton (ed.), 83-39. Mem. Am. Assoc. Pet. Geol. 26.

Valentin, C. 1978a. *Divers aspects des dynamiques actuelles de quelques sols ferrallitiques de Côte d'Ivoire. Recherches méthodologiques. Résultats et interprétation agronomiques*. ORSTOM, Abidjan.

Valentin, C. 1978b. Problèmes méthodologiques de la simulation des pluies. Application à l'étude de l'érodibilité des sols. In *Coll. sur l'érosion agricole des sols en milieu tempéré non-méditerranéen*, Strasbourg-Colmar, 20-23 Sept. ULP Strasbourg et INRA.

Van Andel, T. H., G. R. Heath, T. C. Moore and D. F. R. McGeary 1967. Late Quaternary history, climate, and oceanography of the Timor Sea, northwestern Australia. *Am. J. Sci*. **265**, 737–58.

Van der Hammen, T. 1972. Changes in vegetation and climate in the Amazon basin and surrounding areas during the Pleistocene. *Geol. Mijnb*., **51**, 641–3.

Van der Hammen, T. 1974. The Pleistocene changes of vegetation and climate in tropical South America. *J. Biogeog*. **1**, 3–26.

Van der Linden, P. 1978. *Contemporary soil erosion in the Sanggreman River basin related to the Quaternary landscape development. A pedogeomorphic and hydro-geomorphicological case study in Middle-Java, Indonesia*. Publns Fys. Geogr. Bodemkundig Lab., Univ. Amsterdam, 25.

Van Devender, R. R. and W. G. Spaulding 1979. Development of the vegetation and climate of the southwestern United States. *Science* **204**, 701–10.

Van Dijk, J. W. and V. K. R. Ehrecron 1949. *The different rate of erosion in two adjacent basins in Java*. Meded. Alg. Proefstn. Land. Buitenz. 84.

Van Geel, G. and T. van der Hammen 1973. Upper Quaternary vegetational and climatic sequences of the Fuquene area (Eastern Cordillera, Colombia). *Palaeogeog. Palaeoclim. Palaeoecol*. **14**, 9–92.

Van Lengkerke, H. J. 1981. Die Batticaloa-Zyklone 1978: Eine Naturkatastrophe in Sri Lanka und ihre Folgen. *Erdkd. Wiss. Geog. Z. Biehft*. **54**, 85–115.

Van Steenis, C. G. G. J. 1972. *The mountain flora of Java*. Leiden: E. J. Brill.

Van Zinderen Bakker, E. M. 1976. The evolution of Late-Quaternary palaeoclimates of Africa. *Palaeoecol. Afr*. **9**, 160–202.

Vazquez-Yanes, C. 1974. Studies on the germination of seeds: *Ochroma lagopus* Swartz. *Turrialba* **24**, 176–9.

Veatch, A. C. 1935. *Evolution of the Congo basin*. Geol. Soc. Am. Mem. 3.
Veeh, H. H. and J. Chappell 1970. Astronomical theory of climatic change: support from New Guinea. *Science* **167**, 862–5.
Veevers, J. J. 1977. Rifted arch basins and post-breakup rim basins on passive continental margins. *Tectonophysics* **41**, T1–5.
Veevers, J. J. 1981. Morphotectonics of rifted continental margins in embryo (East Africa), youth (Africa–Arabia), and maturity (Australia). *J. Geol.* **89**, 57–82.
Veevers, J. J. and D. Cotterill 1978. Western margin of Australia: evolution of a rifted arch system. *Bull. Geol Soc. Am.* **89**, 337–55.
Veillon, J.-M. 1978. Architecture of the New Caledonian species of Araucaria. In *Tropical trees as living systems*, P. B. Tomlinson and M. H. Zimmerman (eds), 233–45. Cambridge: Cambridge University Press.
Verstappen, H. Th. 1975. On palaeoclimates and land form development in Malaysia. In *Modern Quaternary research in southeast Asia*, G. Bartstra and Casparie W. A. (eds), 3–35. Rotterdam. Balkema.
Vogt, J. 1959. Aspects de l'évolution morphologique récente de l'Ouest africain. *Ann. Géog.*, 367.
Vogt, J. 1961. *Badlands du Nord-Dahomey*. Actes du 85ème Congr. Nat. Soc. Sav., Chambéry-Annecy 1960, Sect. Géog. 227–39.
Vogt, P. R. 1972. Evidence for global synchronism in mantle plume convection, and possible significance for geology. *Nature* **240**, 338–42.
Vogt, P. R. 1979. Global magmatic episodes: new evidence and implications for the steady-state mid-oceanic ridge. *Geology* **7**, 93–8.
Vogt, P. R. 1981. On the applicability of thermal conduction models to mid-plate volcanism: Comments on a paper by Gass *et al*. *J. Geophys. Res.* **86B**, 950–60.
Von Humboldt, A. 1811. *Essai politique sur la Royaume de la Nouvelle-Espagne*. Paris.
Von Post, L. 1916. Om stogsträdpollen i sydsvenska torfmossagerfoljder. *Geol. För. Stockh. Förh.* **38**, 384–94.
Vuillaume, G. 1969. Analyse quantitative du rôle du milieu physico-climatique sur le ruissellement et l'érosion à l'issue de bassins de quelques hectares en zone sahélienne (Bassin du Kountkouzout, Niger). *Cah. ORSTOM Sér. Hydrol.* **6**(4), 87–132.
Vyas, L. N., R. K. Garg and N. L. Vyas, 1976. Litter production and nutrient release in deciduous forest of Bansi, Udaipur. *Flora* **165**, 103–11.

Wadia, D. N. 1941. The making of Ceylon. *Spolia Zeylonica* **23**, 18–20.
Wadia, D. N. 1945. *Three superposed peneplains of Ceylon – their physiography and geological structure*. Ceylon Dept Mineral. Prof. Pap. 1, 25–32.
Wadia, D. N. 1966. *Geology of India*, 3rd Edn. London: Macmillan.
Wahl, E. and R. A. Bryson 1975. Recent changes in Atlantic surface temperatures. *Nature* **254**, 45–6.
Walcott, R. I. 1970. Flexural rigidity, thickness and viscosity of the lithosphere. *J. Geophys. Res* **75**, 3941–54.
Walcott, R. I. 1972. Gravity, flexure, and the growth of sedimentary basins at a continental edge. *Bull. Geol Soc. Am.* **83**, 1845–8.
Walker, D. and J. R. Flenley 1979. Late Quaternary vegetational history of the Enga District of upland Papua New Guinea. *Phil Trans R. Soc. B* **286**, 265–344.
Walker, D. and G. Singh 1982. Vegetation history. In *Australian vegetation*, R. H. Groves (ed.), 26–43. Hong Kong: Cambridge University Press.
Walker, D. and S. R. Wilson 1978. A statistical alternative to the zoning of pollen diagrams. *J. Biogeog.* **5**, 1–21.
Walker, D. and Y. Pittelkow 1981. Some applications of the independent treatment of taxa in pollen analysis. *J. Biogeog.* **8**, 37–51.
Walker, N. D., H. H. Roberts, L. J. Rouse Jr and O. K. Huh 1982. Thermal history of reef-associated environments during a record cold-air outbreak event. *Coral Reefs* **1**, 83–7.

Wall, J. R. D. 1967. The Quaternary geomorphological history of North Sarawak with special reference to the Subis Karst, Niah. *Sarawak Mus. J.* **15**, 97–125.
Walling, D. E. 1982. Physical hydrology. *Progr. Phys. Geog.* **6**, 122–33.
Walling, D. E., M. Peart, F. Oldfield and R. Thompson 1979. Identifying suspended sediment sources by magnetic measurement on filter paper residues. *Nature* **281**, 110–13.
Walsh, R. P. D. 1977. Changes in the tracks and frequency of tropical cyclones in the Lesser Antilles from 1650 to 1975 and some geomorphological and ecological implications. *Swansea Geogr.* **15**, 4–11.
Walsh, R. P. D. 1980a. Run-off processes and models in the humid tropics. *Z. Geomorph.* NF Suppl. **36**, 176–202.
Walsh, R. P. D. 1980b. *Drainage density and hydrological processes in a humid tropical environment: the Windward Islands.* PhD thesis, University of Cambridge.
Waltham, A. C. and D. B. Brook 1980. Geomorphological observations in the limestone caves of the Gunung Mulu National Park, Sarawak. *Trans. Br. Cave Res. Assoc.* **7**, 123–39.
Wark, J. W. and F. J. Keller 1963. *Preliminary study of sediment sources and transport in the Potomac River Basin.* Interstate Commission on the Potomac River Basin.
Warren, A. 1970. Dune trends and their implications in the central Sudan. *Z. Geomorph.* NF Suppl. **10**, 154–80.
Warren, A. 1972. Observations on dunes and bi-modal sand in the Tenere Desert. *Sedimentology* **19**, 37–44.
Wasson, R. J. 1976. Holocene aeolian landforms in the Belarabon area, SW of Cobar, NSW. *J. Proc. R. Soc. NSW* **109**, 91–101.
Watts, W. A. 1975. A Late Quaternary record of vegetation from Lake Annie, south-central Florida. *Geology* **3**, 344–6
Watts, W. A. 1980. Late Quaternary vegetation history at White Pond on the inner coastal plain of South Carolina. *Quat. Res.* **13**, 187–99
Wayland, E. J. 1919. An outline of the Stone Ages of Ceylon. *Spol. Zeylon.* **11**, 85–125.
Wayland, E. J. 1921. *Some features of the drainage of Uganda.* Geol. Surv. Uganda Ann. Rep. 1920, 75–80.
Wayland, E. J. 1925. The Jurassic rocks of Tabbowa. *Ceylon J. Sci.* **13**, 195–208.
Wayland, E. J. 1934. *Peneplains and some other erosional platforms.* Geol Surv. Uganda Ann. Rep. 1934, 77–9.
Webb, B. 1982. The geology of the Melinau Limestone of the Gunung Mulu National Park. *Cave Sci.* **9**, 94–9.
Webb, L. F. 1958. Cyclones as an ecological factor in tropical lowland rainforest, north Queensland. *Austr. J. Bot.* **6**, 220–8.
Webb, T. III 1980. The reconstruction of climatic sequences from botanical data. *J. Interdiscip. Hist.* **10**, 749–72.
Webster, P. J. and N. A. Streten 1972. Aspects of late Quaternary climate in tropical Australasia. In *Bridge and barrier*, D. Walker (ed.), 39–60. Canberra: ANU Press.
Webster, P. J. and N. A. Streten 1978. Late Quaternary ice age climates of tropical Australasia: interpretations and reconstructions. *Quat. Res.* **10**, 279–309.
Webster, P. J. and N. A. Streten 1981. Comments on: 'Surface circulation of the Indian Ocean during the last glacial maximum approximately 18 000 yr BP'. *Quat. Res.* **16**, 421–3.
Wellington, G. M. 1980. Reversal of digestive interactions between Pacific reef corals: mediation by sweeper tentacles. *Oecologia (Berl.)* **47**, 340–3.
Wells, J. W. 1957. Coral reefs. In *Treatise on marine ecology*, 609–31. Mem. Geol. Soc. Am. 67.
Wells, P. V. 1976. Macrofossil analysis of wood rat (*Neotoma*) middens as a key to the Quaternary vegetation history of arid America. *Quat. Res.* **6**, 223–48.
Wendland, W. M. 1977. Tropical storm frequencies related to sea surface temperature. *J. Appl. Meteor.* **16**, 477–81.
Wendorf, F. and members of the Combined Prehistoric Expedition 1977. Late Pleistocene and recent climatic changes in the Egyptian Sahara. *Geogr. J.* **143**, 211–34.
Wendorf, F. and R. Schild 1980. *Prehistory of the Eastern Sahara.* New York: Academic Press.

Wentworth, C.K. 1927. Estimates of marine and fluvial erosion in Hawaii. *J. Geol.* **35**, 117–33.
Wentworth, C. K. 1928. Principles of stream erosion in Hawaii. *J. Geol.* **36**, 385–410.
Wentworth, C. K. 1943. Soil avalanches on Oahu, Hawaii. *Bull. Geol Soc. Am.* **54**, 53–64.
West, E. A. 1979. *The equilibrium of natural streams*. Norwich: Geobooks.
Whitehead, J. C. (now Newsome J. C.) 1984. *Late Quaternary vegetational history of the Alahan Panjang area, West Sumatra*. PhD thesis in preparation, University of Hull.
Whitehouse, F. W. 1940. *Studies in the Late Geological history of Queensland*. Univ. Qld. Geol. Pap. 2.
Whitmore, T. C. 1966. The social status of *Agathis* in a rainforest in Melanesia. *J. Ecol.* **54**, 285–301.
Whitmore, T. C. 1974. *Change with time and the role of cyclones in tropical rain forest on Kolombangara, Solomon Islands*. Commonw. Forest. Inst. Pap. 46.
Whitmore, T. C. 1975. *Tropical rain forests of the Far East*. Oxford: Clarendon Press.
Whitmore, T. C. 1978. Gaps in the forest canopy. In *Tropical trees as living systems*, P. B. Tomlinson and M. H. Zimmerman (eds), 639–55. Cambridge: Cambridge University Press.
Whitmore, T. C. 1982. On pattern and process in forests. In *The plant community as a working mechanism*, E. I. Newman (ed.), 45–59. Oxford: Blackwell Sci. Publ.
Wilford, G. E. 1961. *Geology and mineral resources of Brunei and adjacent parts of Sarawak*. Mem. Geol. Surv. Dept Br. Terr. Borneo, 10.
Wilhelmy, H. 1977. Verwitterungskleinformen als Anzeichen stabilier Grobformung. *Würzburger Geogr. Arb.* **45**, 177–98.
Williams, H. R. 1978. The Archaean geology of Sierra Leone. *Precambrian Res.* **6**, 251–68.
Williams, M. A. J. 1968. Termites and soil development near Brocks Creek, Northern Territory. *Aust. J. Sci.* **31**, 153–4.
Williams, M. A. J. 1973. The efficacy of creep and slopewash in tropical and temperate Australia. *Aust. Geogr. Stud.* **11**, 62–78.
Williams, M. A. J. 1975. Late Pleistocene tropical aridity synchronous in both hemispheres? *Nature* **253**, 617–18.
Williams, M. A. J. 1978. Termites, soils and landscape equilibrium in the Northern Territory of Australia. In *Landform evolution in Australasia*, J. L. Davies and M. A. J. Williams (eds), 128–41.
Williams, M. A. J. and D. A. Adamson 1980. Late Quaternary depositional history of the Blue and White Nile rivers in central Sudan. In *The Sahara and the Nile*, M. A. J. Williams and H. Faure (eds), 281–304. Rotterdam: Balkema.
Williams, M. A. J. and H. Faure (eds) 1980. *The Sahara and the Nile*. Rotterdam: Balkema.
Williams, M. A. J. and K. Royce 1982. Quaternary geology of the Middle Son Valley, north central India: implications for prehistoric archaeology. *Palaeogeog. Palaeoclim. Palaeoecol.* **38**, 139–62.
Williams, M. A. J., D. A. Adamson and H. H. Abdulla 1982. Landforms and soils of the Gezira: a Quaternary legacy of the Blue and White Nile rivers. In *A land between two Niles*, M. A. J. Williams and D. A. Adamson (eds), 111–42. Rotterdam: Balkema.
Williams, M. A. J., D. A. Adamson, P. M. Williams, W. H. Morton and D. E. Parry 1980. Jebel Marra Volcano: a link between the Nile Valley, the Sahara and Central Africa. In *The Sahara and the Nile*, M. A. J. Williams and H. Faure (eds), 305–37. Rotterdam: Balkema.
Williams, M. A. J., A. H. Medani, J. A. Talent and R. Mawson 1974. A note on Upper Quaternary mollusca west of Jebel Aulia. *Sudan Notes and Records* **54**, 168–72.
Williams, M. A. J., F. A. Street and F. M. Dakin 1978. Fossil periglacial deposits in the Semien highlands, Ethiopia. *Erdkunde* **32**, 40–6.
Williams, M. A. J., F. M. Williams and P. M. Bishop 1981. Late Quaternary history of Lake Besaka, Ethiopia. *Palaeoecol. Afr.* **13**, 93–104.
Willis, B. 1936. *East African Plateaus and Rift Valleys*. Stud. Compar. Seism., Carnegie Inst., Washington.
Wills, K. J. A. 1974. The geological history of southern Dominica and plutonic nodules from the Lesser Antilles. PhD thesis, University of Durham.

Windley, B. F. 1977. *The evolving continents*. Chichester: Wiley.

Wirthmann, A. 1977. Erosive Hangentwicklung in verschiedenen Klimaten. *Z. Geomorph*. NF Suppl. **28**, 42–61.

Wischmeier, W. H. and D. D. Smith 1960. A universal soil-loss estimation to guide conservation farm planning. *7th Int. Congr. of Soil Science* Vol. 1, 322–6.

Wolman, M. G. and J. P. Miller 1960. Magnitude and frequency of forces in geomorphic processes. *J. Geol*. **68**, 54–74.

Woodley, J. D. *et al*. 1981. Hurricane Allen's impact on Jamaican coral reefs. *Science* **214**, 749–55.

Woodroffe, C. D. 1980. The lowland terraces. *Geogr. J*. **146**, 21–33.

Woolnough, W. G. 1927. The duricrust of Australia. *J. Proc. R. Soc. NSW* **61**, 1–53.

Wright, L. W. 1973. Landforms of the Yaruna granite area, Viti Levu, Fiji: a morphometric study. *J. Trop. Geog*. **37**, 74–80.

Wyatt-Smith, J. 1954. Storm forest in Kelantan. *Malay. Forest* **17**, 5–11.

Wyrtki, K., E. Stroup, W. Patzert, R. Williams and W. H. Quinn 1976. Predicting and observing El Niño. *Science* **191**, 343–6.

Wyrwoll, K. H. 1979. Late Quaternary climates of Western Australia: evidence and mechanisms. *J. R. Soc. W. Aust*. **62**, 129–42.

Yamaguchi, M. 1975. Sea level fluctuations and mass mortalities of reef animals in Guam, Mariana Islands. *Micronesica*, **II**(2), 227–43.

Young, A. 1969. Nature resource survey in Malawi: some considerations of the regional method in environmental descriptions. In *Environmental and land use in Africa*, M. F. Thomas and G. Whittington (eds). London: Methuen.

Young, A. 1972. *Slopes*. Edinburgh: Oliver and Boyd.

Young, R. W. 1977. Landscape development in the Shoalhaven River catchment of southeastern New South Wales. *Z. Geomorph*. NF **21**, 262–83.

Zimmerman, M. H. 1978. Structural requirements for optimal water conduction in tree stems. In *Tropical trees as living systems*. P. B. Tomlinson and M. H. Zimmerman (eds), 517–32. Cambridge: Cambridge University Press.

Zonnefeld, J. I. S. 1968. Quaternary climatic changes in the Caribbean and northern South America. *Eiszeitalter und Gegenwart* **19**, 203–8.

Index

Numbers in italics refer to figures.

aborigines 161
Acacia nilotica 90
Acacia senegal 87
Acanthaster planci 22
accelerated erosion 75, 182
Acropora cervicornis 21, 24–5
Acropora palmata 24–5
aeolian action 76, 210–12, Table 11.2
aggradation 142–5, 147–8, 206, 233, 261
Aglaia 156
Algeria Table 11.3
allophane 97, 112–13, 121, 134, 137–8
alluvial fan 38, 141, 145, 148, 323
Amazon 57, 58–9, 64, 65, 69, 72, 197, *10.1*, 199–203, 226, 300–1, 322–3
Andes 9, 59, 65, 72, 204, 215
angiosperm 157
Angola 274, *14.1*
Aphamixis grandiflora 155
Araucaria spp. 33, 161
Ardisia lancelota 156
avicennia 155
Ayers Rock 3, 297
Azores anticyclone 95

Babinda, Queensland 52
Bahamas 22–3, 30, 323
Banco, Ivory Coast 46–50
bank erosion 67, 87, 90–1
Barbados 29, 117, 323, *5.1*, Table 5.4
basal weathering surface 264
base level 65–6, 85, 142, 144, 204, 216, 239, 292
baseflow 4, 52, 91, 324
bauxite 199, 236, 278
bedload 66–7
Belize Tables 3.2 & 3.3
Benue Rift 282, 286
Bermuda 323
bioclimatic zone *4.1*
biogeochemical cycle 51
biogeomorphic response model 182
biostasy 236, 240, 261, 266–7, 324, 326
bioturbation 308
black water rivers 51, 59, 201
blue-line network 98
Borneo 21, 123–5, 127, 146, Table D.1, 272
Bornhardt 3, 247, 264
botanic gardens 4
braided channel 67, 141, 207, 220
Brahmaputra river 223, 227–8, Table 3.6
Brazil 5, 16, 65, 197–212, 215, *13.1*, 279, 315, Tables 3.2, 3.3 & 5.5
Brunhes normal-polarity epoch 137
bulk density 113

caesium 137, 187, 194
calcrete 224, 270
Cambodia Table 3.7
Cameroons 5, 219, *13.2*, 279, *14.1*, 285, Table 3.7
campo limpo 206
campo sujo 208
carbon dioxide 8, 167, 325
carbon isotopes Table 11.1
Casuarina 33
catena 310
cationic denudation 264
cave sediments 134–8, 142
cave systems 127
Cayenne 224, 271
Cecropia 202
Celtis 156
cerrados 201
channel adjustment 122
channel metamorphosis 70
climatic aggressivity 79, *4.3*
climatic change 5, 11, 26, 117, 122, 150–1, 160, 165, 180–1, 197, 202, 206, 213, 222, 225, 281, 297, 311, 315, 321, 326
climatic classification 42
climatic geomorphology 65
cockpits 5
Colombia 19, 216, *13.1*, 279, Tables 3.2 & 3.3
complex response 264, 267
Congo river; *see also* Zaire river 72, 320
coral reefs 7, 14, 20–4, 221, Table 11.1, 323
corestone *1.1*, 250, *12.9*, 257, *12.15*, *12.16*, 261
Costa Rica 18
cratonic regime 240, 264, 266–7, 288
Crotalus durissus 201
crustal dome 285
CSIRO 272–3
cuirasse de nappe 273

Cuphea 206
curacao 22
cyclone 9, 18, 19, 23, 27–31, 39, 40–1, 51, Table 3.8, 71, 97, 117, Table 5.4, 223, 324, 326
  flora 59
  forests 51
  scrub 19
  vegetation 18, 21
cymatogen 242, 286
cymatogeny *14.2*, 287, 327
Cyrtopodium andersonii 213

dambo 310
Dana, James D. 4
Darwin, Charles 3.4
Davis, W. M. 7, 266
deep-sea core record 31, 221, *11.1*
dell 247
denudation 5, 260, 264, 266–7, *14.2*, 292–3
denudation chronology 293
denudation rate 56, 65, 68, 264
denudation system 13, 57, 85, 151, 267
Diadema antillarum 20
diamond *12.4*, *12.15–16*, 248–50
dior 86
dissolved load 58
disturbance rate 20
divergent weathering 265, 297
doldrums 9
domal uplift 282, 285–6, 327
Dominica 52–3, 57, 94–7, 99–100, 103–10, 112–19, 121, Table 5.4
double planation surfaces (double surfaces of levelling) 8, 9, *1.1*, 239, 264, 310, 324, 326
drainage density 37, 93, 97–122, *5.2–9*, Table 5.5, 124, 319–20
drainage network development 93
duplex soils 224
duricrust 7, 8, 36, 81, 199, 206, 214, 236, 241–2, 253, 264, 269–72, 292, 321
duripan phase 270, 272

earthquakes 19, 20
East African rift 282, 286, 290, 292
Easter Island 151, 153–5
ecosystem 11–14, 43, 51, 72, 151, 206, 219, 264, 323–6
Elaeocarpus 156
El Chichon 236, 321, 326
El Niño 26–7, 321, 326–7
El Verde 11, 18, 47, 49
eluviation 248, 298
emergence 22
endogenetic processes 39
epeirogenic regime 241
epeirogenic uplift 282, 292
ephemeral channels 41
equilibrium 13, 121, 133, 150, 183, 240, 267, 323
erodibility 107, 113, 181
erosion
  pin 55
  rate 151, 181, 185, 191–2, 196, 292–3, 325
  surface 7, 238, 291, 294, 327
erosivity 39, 40, 80, *4.3*, 181, 320, 328
etchplain 239–41, 260, 267, 292, 307, 327
etchplanation 8, 238–67, 326–7
etchsurface 265–6
Ethiopia 182, 183, 219, 222, 224, 226, Tables 11.2 & 3, 291–2, 320, *8.1*
Eucalyptus 19, 33, 160
Eugenia 156
evaporites 199
exfoliation 5
evapotranspiration 47
exogenetic processes 39

FAO 269–72
ferricrete 59, 236, 270, 278
Fiji 4, 23, Table 5.5
fire 19, 81, 233
Flachmuldentäler 6
Flandrian transgression 37, 201, 203, 213, 216
Florida 22–3, 31–2
fluvial processes 4
fluvial terraces 138–42, 205
forest growth cycle 16
Fouta Djalon Guinea 54–5, *4.1*
frequency–duration characteristics 57
frost 6
Funafuti Atoll 22–3

Gambia 87–8
Gamblian pluvial 222
gaps 16–18, 323
geothite 55
Gerlach trough 55
Ghana 45, 233, *12.14*, 262, Tables 3.2–4
gibbsite 55, 137, 311
glaciation 5, 6, 123, 201, 222, 232, 324
glacis 81, 245, *12.3–5*, 247–51, 257, 267
Gleichenia scrub Table 7.1
Gondwanaland 9, 39, 236, 238, 278, 282, 321
Gramineae 206
granite weathering 4, 212
graviers sous berge 88
Great Barrier Reef 7, 24, 232
Grenada 37, 94–7, 100–3, 107–8, 113, 119, Table 5.4
Grenadines 94, *5.1*
Grimaldian regression 37, 203
Guam 22–3
guano 27
Guatemala Tables 3.2 & 3.3
Guinea *4.1*, 242, 264, *14.1*

Gulf of Mexico 28, 31–2
gully erosion 324
Guyana 115, 212–14, 239, 278, Table 3.5
Guyana shield 197, 202–3, 212–15, 320
gypcrete 270

Harmattan 224
Hawaiian 4, 9, 19, 23, 27, 45, Tables 2.1, 3.1 & D.1
high spot 287
Holocene transgression 37, 72, 226
Hong Kong 4
Hortonian overland flow 53–4, 112
hot spot 287–90, 293
hurricane
  Allen 23, 25
  Betsy 18, 23
  David 23, 25
  Donna 23
  Frederic 23, 25
  Greta 25
  Hattie 23
hurricane forest 18
hurricanes 9, 11, 23, 24, 27–31
hydraulic head 130
hydrogen isotopes Table 11.1

Igapo 51
imbricate fabrics 134
Inchirian
  maximum 37
  transgression 71
Indus river 223, 228
inselberg 3, 5–7, *1.1*, 36, 81, 207, 209–11, 213, 242, 247, 264, 295–6, 300, 305–8, 327
Instituts Français de Recherches Appliquées 75
intensity, rainfall 41, 52, 54, 76, 83
intermediate disturbance hypothesis 15
International Biological Programme 38, 45
International Hydrological Decade 38
Intertropical Convergence Zone 71, 77, 95, 146
intra-plate volcanism 287
Ipomoea batatas 185
Irrawaddy river 223, Table 3.6
Ivory Coast 16, 18, 65, 81–2, 262–3, Tables 3.1, 3.2 & 3.5, 270

Jamaica 5, 16, 23–5
Java 5, 17, 36, 68–9, Table 3.2

K-cycles 261
kaolinite 55, 113, 199, 250, 308, 311
kaolinite-to-quartz ratio 32
karst 7, 124
  tropical 5, 209, Table D.1
Kenya 69, 219, 222, *14.1*, 285, Table 3.7
Kerbtäler 6
Kilimanjaro 286

Lake
  Chad and Chad Basin *8.1*, *8.2*, Tables 11.2 & 11.3, 233, *12.14*, 263, 270, *14.1*, 315
  Egari 186–92, *7.2*, *9.1*, *9.2*
  Eyre 279
  Frome 233
  Inim 155, *7.2*, *7.3*
  Ipea 186, 189–96, *7.2*
  Mobutu Sese Soko (Albert) 171, *14.1*, Tables 11.2 & 11.3
  Pipiak 186, 189, *7.2*
  levels 166–83, 225
  status 167
  Turkana 181
  Victoria 182, *14.1*, Tables 11.2 & 11.3
LANDSAT 150, 200, 209, 215
landslides 19, 20, 67–8, 124, 146, 199, 324
La Selva rainforest 18
laterite 4, 6, 199, 224, Table D.1, 266, 311, 315
lava flows 95, 113
lead 187, 191, 194, 210
lessivage 241
Liberia 242
Libya 181, Table 11.3
limestone
  hills 126
  tablet erosion 144
liquid limit 81
Lithocarpus 156
lithospheric domes 285
litter production 45, 47
litterfall 46–7
Little Ice Age 31
llanos 72, 213, 215, 271, 274
loess 221 Table 11.1, 228
long profile 65
lunette 91, 232
Lynch's Crater, Queensland *2.5*, 69–70, 160, 161, 220, *11.1*, 232

Madagascar 65, 271, 278, 286
magnetic foliation 137
magnetic susceptibility *9.3*, *9.4*, *9.5*
magnitude and frequency 16, 95, 107, 112, 264
Malaysia 16, 18, 21, 40–1, 43–5, 50, 52, 55, 68–9, 112, 119, 328, *3.4*, Tables 3.7, 5.5 & D.1
Mali 271
mangroves 18, 47, 123, 155
Maurita 206
meandering channel 67, 141, 204, 220
Mekong river Table 3.6
membrane tectonics 290–1
METEOSAT 150

Mexico 177–9
microclimate 153
monogenetic regime 265
monsoon 9, 39, 224, 227–8, 303, 325
montane forest 97, 156, 219
Montastrea annularis 14
montmorillonite 55, 97, 112, 308
morphogenetic change 85
Mount Kenya 5, 6, 286
Mozambique 274

nebkha 85, 91
New Britain 5
New Guinea 4, 5, 9, 18–21, 65–7, 146, 151, 155–8, 160, 185–96, 221, 232, 323, Tables 2.1, 3.2, 3.7 & D.1, *6.1*
New Ireland 5
New Zealand 4, 68, 232
Nicaragua 5, 6
Niger delta 72, 182, 262, 291, 322
Niger river 71, 182, Tables 11.2 & 11.3, 270–1, 320, 322
Nigeria 239–40, 282, 299–300, 315, 320
Nile Tables 11.1, 11.2, 11.3, 183, 220, 223, 231, 270, 278
nitric acid 5
Nouakchottian 37, 90–1
nutrient cycling 11, 21, 45–51, 324

Office de la Recherche Scientifique et Technique Outre-Mer (ORSTOM) 75–6, 87, 272, 279
Ogolian 37, 87, 253, 261–3
Olduvai event 221
Olgaboli tephra 187
opal phytolith 223, Table 11.1
Orinoco 69, 72, 213, 215, 278
overgrazing 91
oxygen isotope record 31, 119, 221, *11.1*, Table 11.1, 227, 233

palaeo-environmental reconstruction 153
palynology 153–64
Panama 19, 21, Tables 2.1 & 3.3
panplanation 250
Paraguay 208–9
Pasoh Reserve, Malaysia 11, 47, Tables 3.2 & 3.5, 52, 55–7
pebble lithology 134, Table 6.2
pedi planation 264–6, 292
pediment 207–11, 214, 216, 265, 267
peneplain 239
peritropical zone 236
permeability 52, 57, 113
Persian Gulf 22
petric phase 270
petroferric phase 270
Philippines 7, 21, 27, 41, Tables 3.5 & 3.7
phosphate rock 27
photosynthesis 21, 49
phreatic caves, cave passages 129–36, 143, 147
pipeflow 54, 216
placer deposits 123
planation surface 236, *12.2*, 265, 292, 298, 305–7, 310, 313–15, 326
plastic limit 81
plate tectonics *1.2*, 9–11, 39, 236, 269, 278, 281, 319, 321
plinthite 270
Pocillopra damicornis 25
Podocarpus imbricatus 157, 159–60
podsols 202, 224
pollen
  analysis 226, 232
  diagram *7.1*, *7.3*, 159, 182
  influx 162–4, 196
  rain 150, 153, 155
  record 32–3, 69, 182
  spectra 155, 221
polygenetic development 265, 314–15
population biology 162
porewater pressure 19
Post-Inchirian regression 37
Post-Nouakchottian 90
potassium–argon dating 282
principal components analysis 162
Puerto Rico 7, 18, 22, 66, Table 3.2
pyroclast deposits 95, 100, 113, 121

Quaternary 5, 11, 123, 150, 153, 165, 203, 206–16, 219, 261, 281, 299, 306, 321
Quercus 156

radar, sideways looking airborne 11, 150, 200
radiocarbon dates 147, 166–7, 187, 199, 225, 253, 256–7, *12.14*, 261, 313
rainforest 5, *1.2*, 14–16, 21, 43, 45–51, 72, 97–8, 124, 150, 155, 160, 197, 201, 207, 219, 232, 245, 261–2, 265, 271, 312, 315, 323–4
Rajasthan 222, 224, 227–8, 233
recovery time 20, 265
Red Sea 22, 222–4, *14.1*, 286, 291
refuge theory 14, 182, 202, 324
regolith 19, 240, 247, 261, 297–9, 324
relief
  generation 295, 297, 303–15
  zones 9
remanent magnetism 137
response time 93
Rhaphidophora oophaga 136
rhexistasy 151, 236, 240, 261, 264, 324, 326
Rhizophora 155
rift valley 7, 199, 271, 278
Rio Negro 51, 59, 197, 202, 204
rock bar 88, 91

Rovuma valley 3
ruggedness index 99
runoff
  concentrated 84
  diffuse 84
  embryonic 84
  incipient 83
  plot 55
  unconcentrated 201, 215

St Lucia 94–7, 100–4, 107–8, 113, 115, 119, 121, Table 5.4
St Vincent 37, 94–7, 100, 103, 106, 110–11, 113, 115, 119, 121, Table 5.4
Sabah 123, 146, *6.1*
sabkha 91, 208
Saguinus 202
Sahara desert 176–7
Saharan dust 223, 224, Tables 11.1 & 11.2, 262
sahel 78, 85–6, 90, 321, *4.1*
sahelian drought 26, 36, 86, 328
Sahul shelf 232–3, 323
salt crystallisation 91
salt lakes 166
saprolite 239, *12.4*, *12.19*, 247, 253, 265–7, 326
Sarawak 18, 37, 50–1, 124–48, 323, *6.1*, Tables 3.2 & 3.3
satellite imagery 11, 87
saturated overland flow 52–3, 57, 112–13
Saudi Arabia 171
savanna 41, 76, 86, 155, 206–8, 262, 270–1
sclerophyll forest 19
scanning electron microscope 72
sea-floor spreading *14.2*, 293
sea-level change 11, 119, 124, 145, 155, 200, 206, 221, 292, 312–13, 319, 322–3, 325
sea surface temperature 29, 156, 221, 223
sediment budget 186, Table 9.3
sediment discharge 44
sediment flows 22
sediment transport 65, 141, 300
sediment yield 36, Table 3.7, 69, 71, 181–2, 203–4, 263
seed bank 17
seed rain 17
seismic activity 21
selva 66
Senegal 38, 81, 86–7, 224, *12.14*, Table 3.5
Senegal river 36, 71, 87–8, 90, Tables 11.2 & 11.3, 270
sericite 250
Shorea spp. 50
sideways-looking airborne radar (SLAR) 11, 150, 200
Sierra Leone 38, Table 11.3, 233, 242, *12.1–14*, 245, 247–50, 253, 256, 315, 325, *12.15–19* 260–7
silcrete 270
silicification 306
Singapore 4, 41
slopewash 67
smectoid clays (*see also* montmorillonite) 97, 112
snowline 156
sohlenkerbtal 250
soil creep 76
soil erosion 4, 81–2, 320
Solomon Islands 18
solonchak 308
solonetz 308
Sophora 155
Southern Oscillation 26–7, 327
speleothem
  dating 124, 142, 148
  deposition 134, 323
splash erosion 201
Sri Lanka 7, 41, 44, 119, 238, *13.3*, 279, Tables 5.5 & Table D.1, 272, 293–301, 303–15, 320, 322
stability 13, 14, 236, 295
stemflow 47, 49, 52
stoneline 214–15, 247, *12.7*, *12.19*, 267, 298, 308, *16.3*
Streckhange 308
stretch slope 298–9
submontane forest 156, Table 7.1
subsurface flow 52, 320
Sudan 226, *13.2*, 297, *14.1*, Tables 11.2 & 11.3
Sulawesi 8, 272, *6.1*
Sumatra 7, 21, 156–7, 159–60, 272, *6.1*, Table 7.1
Sunda shelf 123, *6.1*
surface wash 52, 55, 310, 324
Surinam 18, 65, 155, 213–14, 271
suspended sediment 44, 65, 66, 204
Symingtonia populnea 157

tabuleiros 201–2
Tahiti 4
Tanzania Tables 3.2 & 3.7
Tasmania 232
tephra 187
terra rossa 306, *16.2*
thalossocratic regime 241
thermal optimum 167
thermoluminescence 226
threshold 151, 171
throughfall 47, 49, 51
throughflow 54, 57, 113, 248
timescale 38, 221, 262, 264–5, 267, 320, 326
time-series analysis 162
tor 242, 247, 308
tower karst 129
trade winds 9, 26, 91, 95, 215, 223, 226

traditional development (Traditionale Weiterbildung) 310, 315, 327
tree fall 20
Trinidad 16, 29, Table 3.2
triple junction 282, *14.1*, 285–6
Triumfetta 155
tropical ridge relief 310
tropics, *Immerfeuchten* 6
tropics, *Wechselfeuchten* 6
Tropicurus torquatus hispidus Spix 213
typhoons 9

Uganda 7, 219, 249, *14.1*, Table 5.5
underground drainage 131
UNESCO 269–72
Upper Volta 82, 242, *13.2*, 279, Tables 3.5 & 4.1
uranium-series dating 124, 131

vadose caves, cave passages 129–38, 147
variability, rainfall 40, 78
variable-source area concept 36
Varzea 51
Venezuela 31, 213–15, 217, *5.1*, *13.1*, 279, Table 3.2
volcano 5, 99, 153, 236

wash depression 307–8, *16.3*, 310, 312
wash divides 307–10, *16.3*
water balance 165–6
weathering 4–6, 20, 51, 55, 66, 80, 113, 134, 146, 153, 181, 212, 239, 242, 257, 262, 265–7, 297–8, 306
Weinmannia blumei 157
Western Ghats, India Tables 3.2 & 3.3
whaleback 250, 295, 297, 299
wind action 85
wind erosion 86–7, 91
Windward Islands 93–122
Wischmeier equation 75, 79

X-radiography 136–7

Yucatan 178, 323
Yunnan, China Tables 3.2 & 3.3, 272

Zaire 219, 226, *13.2*, 279, *14.1*, 309, Tables 3.2 & 3.3
Zaire river (*see also* Congo river) 71, 270, 275, 278, 320, Table 3.6
Zambezi river 271, 275, 282, *14.1*, Table 3.6
Zimbabwe *14.1*
zooanthellae 21

BIBL. UNIV. LAUR. UNIV. LIB. DUPL
3 0007 00185760 5